Juergen Hahn (Ed.)

Modeling and Analysis of Signal Transduction Networks

MDPI

This book is a reprint of the Special Issue that appeared in the online, open access journal, *Processes* (ISSN 2227-9717) from 2014–2015 (available at: http://www.mdpi.com/journal/processes/special_issues/signal_transduction_networks).

Guest Editor
Juergen Hahn
Rensselaer Polytechnic Institute
USA

Editorial Office
MDPI AG
Klybeckstrasse 64
Basel, Switzerland

Publisher
Shu-Kun Lin

Managing Editor
Jennifer Li

1. Edition 2016

MDPI • Basel • Beijing • Wuhan

ISBN 978-3-03842-141-2 (Hbk)
ISBN 978-3-03842-142-9 (PDF)

Table of Contents

List of Contributors

Robert C. Alaniz: Department of Microbial Pathogenesis and Immunology, Texas A & M University System Health Science Center, College Station, TX 77843, USA.

Ipsita Banerjee: Department of Chemical and Petroleum Engineering, University of Pittsburgh, Pittsburgh, PA 15261, USA; Department of Bioengineering, University of Pittsburgh, Pittsburgh, PA 15219, USA; McGowan Institute for Regenerative Medicine, University of Pittsburgh, Pittsburgh, PA 15219, USA.

Gregery T. Buzzard: Department of Mathematics, Purdue University, 150 N. University St., West Lafayette, IN 47907, USA.

Christina Byrne-Hoffman: Department of Basic Pharmaceutical Sciences, West Virginia University, Morgantown, WV 26506, USA.

Wei Dai: Center for Biotechnology and Interdisciplinary Studies, Rensselaer Polytechnic Institute, 110 8th Street, Troy, NY 12180, USA; Department of Biomedical Engineering and Department of Chemical & Biological Engineering, Rensselaer Polytechnic Institute, 110 8th Street, Troy, NY 12180, USA.

Robert L. Geahlen: Department of Medicinal Chemistry and Molecular Pharmacology, Purdue University, 201 S. University Street, West Lafayette, IN 47907, USA.

Juergen Hahn: Guest Editor of "Modeling and Analysis of Signal Transduction Networks", Department of Biomedical Engineering and Department of Chemical & Biological Engineering, Rensselaer Polytechnic Institute, 110 8th Street, Troy, NY 12180, USA.

Marietta L. Harrison: Department of Medicinal Chemistry and Molecular Pharmacology, Purdue University, 575 Stadium Mall Dr., West Lafayette, IN 47906, USA.

Zuyi Huang: Department of Chemical Engineering, Villanova University, Villanova, PA 19085, USA; The Center for Nonlinear Dynamics& Control (CENDAC), Villanova University, Villanova, PA 19085, USA; The Villanova Center for the Advancement of Sustainability in Engineering (VCASE), Villanova University, Villanova, PA 19085, USA.

David J. Klinke II: Department of Chemical Engineering and Mary Babb Randolph Cancer Center, West Virginia University, Morgantown, WV 26506, USA; Department of Microbiology, Immunology, and Cellular Biology, West Virginia University Morgantown, WV 26506, USA.

Arul Jayaraman: Artie McFerrin Department of Chemical Engineering, Texas A & M University, College Station, TX 77843, USA.

Jens O. M. Karlsson: Department of Mechanical Engineering, Villanova University, Villanova, PA 19085, USA.

Mariya O. Krisenko: Department of Medicinal Chemistry and Molecular Pharmacology, Purdue University, 201 S. University Street, West Lafayette, IN 47907, USA.

Lakshmi Kuttippurathu: Daniel Baugh Institute for Functional Genomics and Computational Biology, Department of Pathology, Anatomy and Cell Biology, Sidney Kimmel Medical College, Thomas Jefferson University, Philadelphia, PA 19107, USA.

Ji Liu: Key Laboratory of Advanced Control and Optimization for Chemical Processes of Ministry of Education, East China University of Science and Technology, Shanghai 200237, China; Center for Biotechnology and Interdisciplinary Studies, Rensselaer Polytechnic Institute, 110 8th Street, Troy, NY 12180, USA.

Shreya Maiti: Artie McFerrin Department of Chemical Engineering, Texas A & M University, College Station, TX 77843, USA.

Shibin Mathew: Department of Chemical and Petroleum Engineering, University of Pittsburgh, Pittsburgh, PA 15261, USA.

Reginald L. McGee: Department of Mathematics, Purdue University, 150 N. University St., West Lafayette, IN 47907, USA.

Judith Mikolajczak: Department of Medicinal Chemistry and Molecular Pharmacology, Purdue University, 575 Stadium Mall Dr., West Lafayette, IN 47906, USA.

Austin Parrish: Daniel Baugh Institute for Functional Genomics and Computational Biology, Department of Pathology, Anatomy and Cell Biology, Sidney Kimmel Medical College, Thomas Jefferson University, Philadelphia, PA 19107, USA.

Jeffrey P. Perley: Weldon School of Biomedical Engineering, Purdue University, 206 S. Martin Jischke Dr., West Lafayette, IN 47907, USA.

Ann E. Rundell: Weldon School of Biomedical Engineering, Purdue University, 206 S. Martin Jischke Dr., West Lafayette, IN 47907, USA.

Adithya Sagar: School of Chemical and Biomolecular Engineering, 244 Olin Hall, Cornell University, Ithaca, NY 14853, USA.

Sankaramanivel Sundararaj: Department of Chemical and Petroleum Engineering, University of Pittsburgh, Pittsburgh, PA 15261, USA.

Rajanikanth Vadigepalli: Daniel Baugh Institute for Functional Genomics and Computational Biology, Department of Pathology, Anatomy and Cell Biology, Sidney Kimmel Medical College, Thomas Jefferson University, Philadelphia, PA 19107, USA.

Jeffrey D. Varner: School of Chemical and Biomolecular Engineering, 244 Olin Hall, Cornell University, Ithaca, NY 14853, USA.

Zhaobin Xu: Department of Chemical Engineering, Villanova University, Villanova, PA 19085, USA.

About the Guest Editor

Juergen Hahn is the department head of the Department of Biomedical Engineering at Rensselaer Polytechnic Institute in addition to holding an appointment in the Department of Chemical & Biological Engineering. He received his Diploma degree in engineering from RWTH Aachen, Germany, in 1997, and his MS and Ph.D. degrees in chemical engineering from the University of Texas, Austin, in 1998 and 2002, respectively. He was a post-doctoral researcher at the Chair for Process Systems Engineering at RWTH Aachen, Germany, before joining the Department of Chemical Engineering at Texas A&M University, College Station, in 2003 and moving to the Rensselaer Polytechnic Institute in 2012. His research interests include systems biology and process modeling and analysis with over 80 articles and book chapters in print. Dr. Hahn is a recipient of a Fulbright scholarship (1995/96), received the Best Referee Award for 2004 from the Journal of Process Control, the CPC 7 Outstanding Contributed Paper Award in 2006, was named Outstanding Reviewer by the journal Automatica in 2005, 2006, 2007, and 2010 CAST Outstanding Young Researcher. He is a fellow of AIMBE since 2013. He is currently serving as editor of the Journal of Process Control and as associate editor for the journal Control Engineering Practice and is on the editorial board of the journal Processes.

Preface

Biological pathways, such as signaling networks, are a key component of biological systems of each living cell. In fact, malfunctions of signaling pathways are linked to a number of diseases, and components of signaling pathways are used as potential drug targets. Elucidating the dynamic behavior of the components of pathways, and their interactions, is one of the key research areas of systems biology.

Biological signaling networks are characterized by a large number of components and an even larger number of parameters describing the network. Furthermore, investigations of signaling networks are characterized by large uncertainties of the network as well as limited availability of data due to expensive and time-consuming experiments. As such, techniques derived from systems analysis, e.g., sensitivity analysis, experimental design, and parameter estimation, are important tools for elucidating the mechanisms involved in signaling networks. This Special Issue contains papers that investigate a variety of different signaling networks via established, as well as newly developed modeling and analysis techniques.

Dr. Juergen Hahn
Guest Editor

Integrated Computational Model of Intracellular Signaling and microRNA Regulation Predicts the Network Balances and Timing Constraints Critical to the Hepatic Stellate Cell Activation Process

Lakshmi Kuttippurathu, Austin Parrish and Rajanikanth Vadigepalli

Abstract: Activation and deactivation of hepatic stellate cells (HSCs) is an important mechanism contributing to both healthy liver function and development of liver diseases, which relies on the interplay between numerous signaling pathways. There is accumulating evidence for the regulatory role of microRNAs that are downstream from these pathways in HSC activation. However, the relative contribution of these pathways and interacting microRNA regulators to the activation process is unknown. We pursued a computational modeling approach to explore the timing and regulatory balances that are critical to HSC activation and quiescence. We developed an integrated model incorporating three signaling pathways with crosstalk (NF-κB, STAT3 and TGF-β) and two microRNAs (miR-146a, miR-21) that are differentially regulated by these pathways. Simulations demonstrated that TGF-β-mediated regulation of microRNAs is critical to drive the HSC phenotypic switch from quiescence (miR-146a^{high} miR-21^{low}) to an activated state (miR-146a^{low} miR-21^{high}). We found that the relative timing between peak NF-κB and STAT3 activation plays a key role driving the initial dynamics of miR-146a. We observed re-quiescence from the activated HSC state upon termination of cytokine stimuli. Our integrated model of signaling and microRNA regulation provides a new computational platform for investigating the mechanisms driving HSC molecular state phenotypes in normal and pathological liver physiology.

Reprinted from *Processes*. Cite as: Kuttippurathu, L.; Parrish, A.; Vadigepalli, R. Integrated Computational Model of Intracellular Signaling and microRNA Regulation Predicts the Network Balances and Timing Constraints Critical to the Hepatic Stellate Cell Activation Process. *Processes* **2014**, *2*, 773-794.

1. Introduction

Hepatic stellate cells (HSC), although important for many aspects of liver function in healthy tissue, are primary drivers of liver diseases such as fibrosis and cirrhosis [1,2]. In healthy livers, the main role of HSCs appears to be storage and transport of retinoids as well as modulating the innate immune response [3–5]. Activation of HSCs occurs in response to numerous stimuli, including induction of activating factors or loss of repressive signals due to genetic or post-transcriptional regulation. During liver regeneration induced by chemical injury (e.g., CCl₄ exposure) or partial hepatectomy, HSCs activate and secrete several pro-regenerative proteins, including hepatocyte growth factor (HGF) and epidermal growth factor (EGF) [6–8]. During diseases associated with chronic inflammation, HSCs alter their gene and protein expression profiles, change their morphology, and deposit fibrous extracellular matrix (ECM), causing scarring and leading to fibrosis and cirrhosis [1].

The process of HSC activation during chronic inflammation is governed by autocrine and paracrine signaling factors from other non-parenchymal and parenchymal cells and is likely highly dependent on the local microenvironment [2,9]. It is known that the initial changes in HSCs in response to injury could be a result of paracrine stimulation from other non-parenchymal cells. Several transcription factors play a role in HSC activation. To initiate HSC activation, inflammatory molecules such as IL-1, IL-6, and TGF-β bind to receptors on the stellate cell membrane [10]. The presence of constitutive NF-κB activating inflammatory pathways has been reported in HSCs [11]. A recent study found IL-1 driven NF-κB inducing multiple MMPs provoking HSC activation [12]. IL-6 stimulation of STAT3 increases collagen production in HSCs leading to acceleration of fibrosis. STAT3 directly up-regulates TGF-β expression in HSCs [10]. TGF-β initiates a signaling cascade that results in the phosphorylation of SMAD2 (pSMAD2) and further upregulates SMAD7 expression. SMAD7 acts as an anti-fibrogenic factor through an auto-inhibitory feedback loop by inhibiting TGF-β mediated phosphorylation of SMAD2. This feedback loop leads to a restriction of the intensity and duration of TGF-β signaling [13]. pSMAD2 forms a complex with SMAD4 and the complex translocates to the nucleus, after which they interact with other factors to regulate gene expression of collagen and other fibrogenic factors [10,14]. Interaction between SMAD, NF-κB and STAT3 pathways providing positive and negative feedback modulates HSC activation [4,10–12,14,15]. Increased TGF-β expression, lead to increased ECM production by stellate cells and altered ECM composition, with higher fraction of fibrous collagens and lower fractions of basement membrane collagens Activation of HSCs also results in increased levels of growth factors (PDGF, EGF, FGF-2, and others), which induce HSC proliferation, and increased levels of chemotactic proteins (PDGF, MCP-1, CXCR3), which attract HSCs to an activation site [16]. Increased proliferation, chemotaxis, and production of activating signals by active HSCs act as positive feedbacks to amplify HSC activation once it has begun to occur in a subset of cells [2,8].

MicroRNAs are small, non-coding RNAs approximately 21 nucleotides in length which are involved in the regulation of multiple cellular processes [17]. Recently, the role of microRNAs governing the activation and function of HSCs has begun to be appreciated. Knockdown of miR-27a and miR-27b in activated HSCs was shown to allow reversion to a quiescent phenotype [18]. miR-15b and miR-16 are known to be downregulated during the activation of HSCs [19]. Recent studies have revealed the role of miR-29b in HSC activation [20]. Another microRNA, miR-146a, has emerged as a potential modulator of HSC activation. miR-146a expression is shown to be downregulated during rat HSC activation [21]. A recent *in vitro* study reported miR-146a expression to be decreased by almost 17-fold in activated HSCs [22]. In contrast, when miR-146a was overexpressed in activated HSCs *in vitro*, the cells appear to take on an anti-fibrotic phenotype, expressing low mRNA levels of IL-6, high mRNA levels of basement collagen (Col1a1), and lower levels of NF-κB binding activity (a mediator of inflammatory response). miR-21 is one of the well-studied microRNAs that modulates cell cycle progression and proliferation during liver regeneration. The role of miR-21 in hepatic stellate cell activation has also been studied [23]. Higher expression of miR-21 and the existence of an auto-feedback loop connecting miR-21 to TGF-β pathway activation in activated HSCs have been reported [24]. In contrast to miR-146a, miR-21 has been associated with increased cardiac fibrosis *in vivo* and increased ECM deposition *in vitro* [25].

In HSCs activated *in vitro*, miR-21 expression increased by over five-fold indicating that miR-21 may play a role modulating fibrosis in HSCs [22]. Several other papers have reported such a differential activation of microRNAs miR-21 and miR-146a in the context of liver and other diseases [26–28]. Alterations in the relative balance between this pair of microRNAs, here termed "microRNA switch", can therefore act a marker for HSC activation state. Activated HSCs have a relatively high expression of miR-21 to miR-146a, whereas quiescent HSCs have a low miR-21 to miR-146a ratio of expression. In summary, the opposing expression and regulation of these microRNAs and their interaction with the cytokine induced signaling pathways are associated with HSC activation. Therefore, we used these two microRNAs as representative markers of HSC phenotype in our model and formulated our analysis based on their differential regulation.

Although the microRNAs miR-146a and miR-21 appear to be useful as biomarkers for HSC activation state, their role in HSC activation remains unclear. The dynamics of miR-21 and miR-146a during HSC activation have not been fully characterized. Several feed forward and feedback interactions between these microRNAs and signaling pathways have been demonstrated in multiple studies [23,25,29,30]. The known interactions between signaling pathways and these microRNAs lead to multiple possibilities for how these microRNA biomarkers likely vary during the early stages of HSC activation. We pursued a computational model based approach to explore these possibilities. Based on an array of *in vitro* and *in vivo* literature, we developed a novel computational model of HSC activation through IL-1β, IL-6, and TGF-β signaling with feedback from miR-21 and miR-146a. We utilized this model to identify the network balances and timing constraints that may be critical to altering the dynamic switch between miR-21 and miR-146a during HSC activation.

Activation of HSCs is an inherently dynamic multi-scale process due to cell–cell interactions and the tissue microenvironment regulating paracrine signaling and activating intracellular pathways. For the present purpose, we approximated the multi-scale nature of the regulation by considering microRNA based molecular markers as indicators of the HSC cellular state at physiological scales. This approximation constrains our model formulation to focus on key pathway components regulating the switch between microRNAs. By creating an integrated model based on observed molecular and phenotypic states, we are able to study how changes at the molecular level likely translate to changes at the cellular level.

2. Methods

The integrated signaling pathways and microRNA regulation model were developed as a Petri net system [31]. Petri nets are weighted directed graphs that make use of graph theoretical methods, which can efficiently simulate biological signal transduction, metabolic and gene networks. In Petri nets, molecular components are represented as place nodes and biochemical reactions are considered transition nodes. To better represent the complexity of our biological system, we modeled the network as a stochastic Petri net, where stochasticity was introduced as an exponentially distributed firing rate for each of the components (*i.e.*, the likelihood of the waiting times of the nodes to activate had an exponential distribution). The Petri net model was implemented using the software Snoopy (version 1.13, Brandenburg University of Technology, Cottbus, Germany, 2014), which provided a graphical platform for model construction, computational modeling and simulation of the network [32].

Transition nodes were modeled to operate by either mass action or Michaelis-Menten kinetics. The kinetic parameters were manually tuned through an iterative process. Initial value ranges were selected from several sources—TGF-β-receptor binding values were estimated from [33], activation of STAT3 and inhibition by SOCS3 from [34], and downstream SMAD/SMURF signaling in the TGF-β pathway from [35]. A total of 250 Monte Carlo simulations were run using the Gillespie algorithm for each analysis scenario. In the case of transient signaling, cytokine inputs were modeled as pulse functions using pre-scheduled transitions. For persistent signaling, the initial step input signals were held throughout the simulation.

3. Results and Discussion

3.1. microRNAs as Markers of Hepatic Stellate Cell Molecular Phenotypes

Activated stellate cells play a major role in the development of liver fibrosis by contributing to the pro-fibrogenic pathways and other signaling pathways leading to ECM production. It has been shown that autoregulatory feedback loops between these pathways and microRNAs influence the fate of the activated stellate cells. Altered microRNA expression levels have been previously reported in activated HSCs [22,23]. In particular, miR-21 has been shown to be upregulated in both activated HSCs and regenerating livers, and is known to be over expressed in certain cancer types [22,36]. In contrast, miR-146a was shown to be significantly down regulated during HSC activation (~17-fold decrease in expression levels) [22]. We analyzed the intensity values obtained from the microarray data for quiescent and activated hepatic stellate cells to generate relative microRNA levels [22]. The expression level of these microRNAs varied in opposite fashion during HSC activation (Figure 1). Based on these findings, we postulated that anti-correlation between these two microRNAs and their relative expression levels may be playing a role in regulating the central signaling pathways in activated HSCs and may be used as representative markers of the HSC activation state. Our approach is the first step towards addressing the testable hypothesis that the microRNA behavior based kinetics may contribute to the cellular phenotypic changes leading to HSC activation.

Figure 1. Changes in microRNA expression levels between quiescent and activated hepatic stellate cells. Data are derived from published microarray data on microRNA expression in primary hepatic stellate cells collected from Wistar rats (Gene Expression Omnibus accession GSE19462) [22].

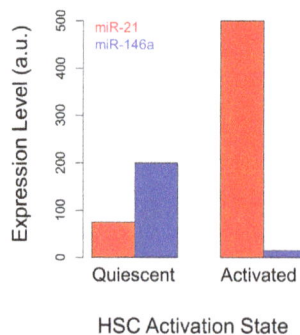

3.2. Integrated Model of Signaling and microRNA Regulation

An overview of the modeled signaling pathways and components is shown in Figure 2. Activation or inhibition arrows ending on species nodes represent influences on production or degradation, respectively. The arrows ending on other arrows represent modification of the corresponding reaction/interaction. For example, PEL1/TRAF6 levels positively influence the transformation of NF-κB ^{inactive} into NF-κB ^{active} form. NF-κB^{active} species lead to higher production of miR-146a and miR-21. Similarly, miR-146a levels negatively influence the activation of PEL1/TRAF6 complex, and also reduce levels of SMAD4. In order to explore the dynamics of these microRNAs leading to an activated HSC phenotype, we developed a stochastic Petri net model incorporating several well-characterized pathways that have been shown to play a role in the activation of HSCs and stimulation of liver fibrogenesis.

A Petri net implementation of the integrated network is shown in Figure 3. Our model includes IL-1/NF-κB, IL-6/STAT3, and TGF-β/SMAD signaling pathways interacting with miR-146a and miR-21. We explored the contributions of individual pathways to the differential regulation of microRNAs to identify necessary and sufficient regulatory relationships to account for the microRNA switch during HSC activation. We started with the IL-1/NF-κB pathway alone and continued to sequentially expand our model to include IL-6/STAT3 and TGF-β/SMAD pathways.

Figure 2. Schematic view of the integrated network model of signaling pathways and microRNA regulatory interactions. Black lines represent activating interactions, while red lines represent inhibitory interactions. Hypothesized interactions are represented by dashed lines.

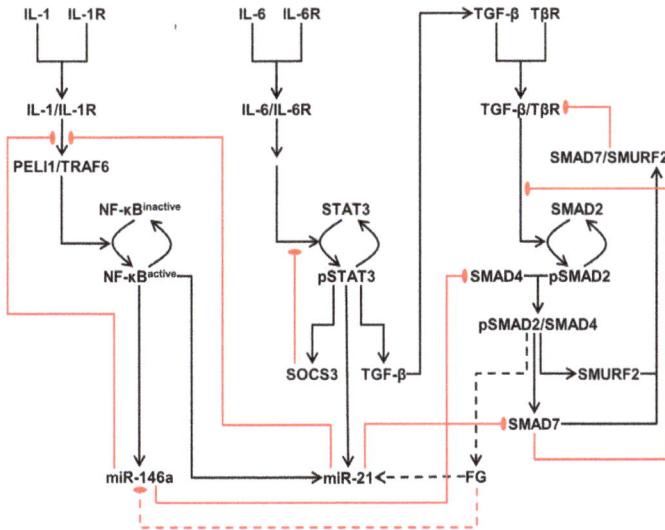

Figure 3. Graphical representation of the Petri net model. Light gray nodes represent "logical" elements that allow for crosstalk; dark gray nodes represent input stimuli. Edges with arrows correspond to mass flux. Edges with round ends ("read edges") correspond to influences without affecting the levels of the source nodes. Dashed edges ("modifier edges") correspond to inhibitory influences.

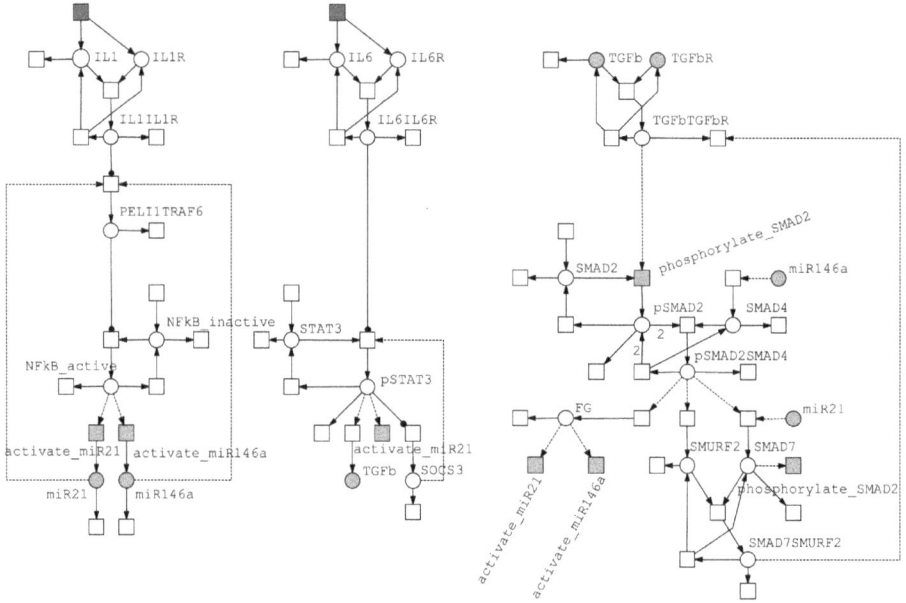

3.2.1. Model of the IL-1/NF-κB Pathway

Constitutive activity of the transcription factor nuclear factor-κB (NF-κB) has been strongly implicated in the promotion of liver fibrogenesis and HSC activation [37,38]. In our model, the proinflammatory cytokine IL-1 activates the NF-κB pathway (Figure A1). The production of IL-1 and IL-1R leads to the formation of an IL-1/IL-1R complex in a concentration-dependent manner. This complex can either dissociate into its individual components or proceed to the activation of the next component in the pathway via signal transduction [39]. During persistent signal transduction IL-1/IL-1R activates the PELI1/TRAF6 (Pellino 1/tumor necrosis factor receptor-associated factor 6) complex. NF-κB is represented in the model as distributed across two likely states: an inactive species that is activated by PELI1/TRAF6 complex, and an active species that is capable of upregulating its target genes. The balance between the production and degradation maintains the steady state levels of the inactive species. In the presence of active PELI1/TRAF6 complex, the transition from inactive to active form of NF-κB is modeled using mass action kinetics. The activation of NF-κB is modeled as a reversible reaction approximately accounting for shuttling between cytoplasm and nucleus that is not explicitly represented in the model. In biological systems, the PELI1/TRAF6 complex is the result of interactions between several other proteins, such as IL-1 receptor-associated kinases (IRAK1/IRAK4), and causes activation of NF-κB by mediating its release from inhibitory protein IκB [39]. Our model approximates these intermediary reactions as a single reaction with apparent kinetic parameter values yielding an NF-κB activation profile similar to experimental observations.

Once activated, NF-κB increases expression of miR-21 and miR-146a [29,30]. In the model, microRNA production by activated NF-κB is governed by mass action kinetics. Previous experimental studies have shown that miR-21 targets Peli1, thereby creating a potential negative feedback targeting NF-κB signaling [29]. miR-146a has been shown to modulate the IL-1 signaling pathway by targeting IRAK1 and TRAF6 to induce their down regulation [29,30]. Based on these observations, miR-21 and miR-146a was modeled as inhibiting the PELI1/TRAF6 complex. This results in a negative feedback loop between the microRNAs and NF-κB activation.

We tested whether activation of NF-κB pathway is sufficient to lead to the experimentally observed switch between miR-21 and miR-146a levels. Simulations showed that both miR-21 and miR-146a levels closely followed the profile of active NF-κB (Figure 4A). Both microRNAs exhibited an initial increase in their levels, followed by a drop-off to steady state levels higher than baseline. Despite the presence of inhibitory feedback, there was no phenotypic switch from a miR-21^{low} miR-146a^{high} state to a miR-21^{high} miR-146a^{low} state. Based on the network structure in which NF-κB serves as an activator of both microRNAs, the switch from a miR-21^{low} miR-146a^{high} state to a miR-21^{high} miR-146a^{low} state can occur only with differential rates of activation of the two microRNAs. The model was expanded to include additional signaling pathways in order to accomplish such differential microRNA regulation.

It is known that NF-κB acts primarily as a mediator of immune and inflammatory responses in liver cells [37]. However, NF-κB in has been shown to function as an anti-apoptotic factor and promote the expression of numerous growth factors and cytokines, including IL-6, in HSCs [2,40]. IL-6 plays a key role in activating the transcription factor STAT3, which has recently been found to be a necessary factor in liver fibrosis [10,40]. In addition, STAT3 has been shown to induce miR-21 expression in a cancer context to inhibit tumor suppressor pathways [41]. We tested whether the addition of the IL-6/STAT3 pathway to our model was sufficient to yield a microRNA switch.

3.2.2. Model of the IL-6/STAT3 Pathway

STAT3 activation in liver non-parenchymal cells, such as HSCs, occurs in response to IL-6 signaling, among other cytokines [42]. IL-6 binds to its receptor, IL-6R, to form the IL-6/IL-6R complex (Figure A2). Binding of ligand-receptor complex IL-6/IL-6R and its dissociation are modeled using mass action kinetics. The IL-6/IL-6R complex leads to activation of STAT3. Our model approximates an intermediate step whereby IL-6/IL-6R activates the protein Janus kinase (JAK), causing STAT3 phosphorylation [42]. Once activated, phosphorylated STAT3 (pSTAT3) is able to upregulate the expression of its target genes. Suppressor of cytokine signaling 3 (SOCS3), activated by pSTAT3, targets the signaling intermediary gp130 protein to inhibit STAT3 activation [42]. The kinetics of this negative feedback were modeled using Michaelis-Menten formalism with SOCS3 acting as a competitive inhibitor that reduces STAT3 activation. STAT3 activation in HSCs increases miR-21 expression significantly [43]. Our model approximates the upregulation of miR-21 by pSTAT3 using mass action kinetics.

Figure 4. Dynamics of microRNA and transcription factor expression levels during HSC activation. (**A**) Expression levels considering the IL-1/NF-κB signaling pathway alone; (**B**) Expression levels considering IL-1/NF-κB and IL-6/STAT3 signaling pathways regulating the microRNAs. Solid arrowheads mark the start of input cytokine stimuli.

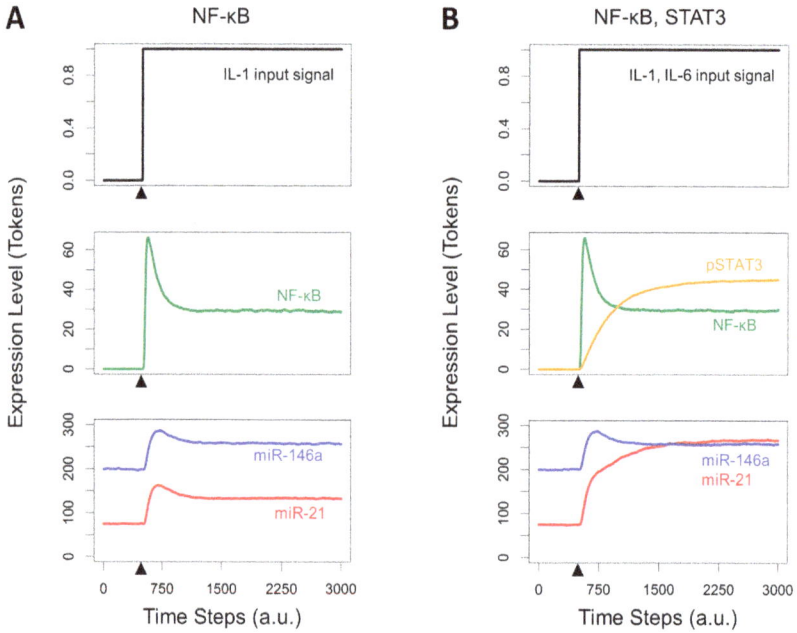

Simultaneous activation of NF-κB and STAT3 pathways led to higher levels of miR-21 and a modest increase in steady state miR-146a levels (Figure 4B). However, miR-146a levels remained high, contrary to experimental observations [8], indicating that the activation of these pathways is insufficient to induce the microRNA switch. These results indicated the requirement for negative regulation of one or both of the microRNAs to account for the switch during HSC activation.

Previous experimental studies have shown that STAT3 promotes production of TGF-β, a key factor in driving HSC-mediated fibrogenesis [10]. Based on these results, the STAT3 pathway was modified as leading to production of TGF-β. Additionally, crosstalk between TGF-β pathway and microRNAs in liver has been well studied [44,45]. Based on these findings we expanded the model to include a TGF-β-mediated pathway leading to activation of genes driving fibrogenesis and likely regulate miR-21 and miR-146a expression.

3.2.3. Model of the TGF-β/SMAD Pathway

Active TGF-β binds to TGF-β receptor I (TβRI), forming TGF-β:TβRI complex (Figure A3). The ligand receptor complex is then able to dissociate, or promote the activation of SMAD2. This activation step accounts for the ligand-receptor complex-mediated phosphorylation of SMAD2 [33,46]. Inactive SMAD2 is present at a steady state level balanced by its constitutive production and degradation. The active, phosphorylated SMAD2 species (pSMAD2) is able to bind to another protein, SMAD4, with a 2:1 stoichiometry of SMAD2 *vs.* SMAD4. This complex, termed pSMAD2/SMAD4 in the model, leads to increase in expression of SMURF2 and SMAD7, which act as negative regulators of TGF-β signaling. The translocation of pSMAD2/SMAD4 into the nucleus has been approximated in the activation rates specified in the model. SMAD7 exerts its influence by blocking SMAD2 phosphorylation and is also able to form a complex with SMURF2 to promote degradation of TβRI [47,48]. miR-146a is involved in the inhibition of TGF-β-induced HSC proliferation and activation by decreasing SMAD4 protein expression [45]. miR-21 is known to directly target and inhibit SMAD7, thereby creating as a positive feedback to the pro-fibrogenic TGF-β pathway [44,45].

Simulations of the expanded model did not impact the dynamic profiles of microRNAs due to the lack of regulation by the TGF-β pathway. This led us to hypothesize that the shift in balance of the microRNAs is driven by the pro-fibrotic target genes of the TGF-β pathway. Previous studies identified the activation of fibrotic genes by the TGF-β pathway [10]. Experimental studies in tissues other than liver have demonstrated an opposing influence of the TGF-β pathway on miR-21 and miR-146a expression [45,49]. However, downstream effectors of the TGF-β/SMAD pathway on these microRNAs in HSCs are presently unknown. Based on this phenomenological relationship, we introduced an empirical pro-fibrotic gene (FG) as a downstream target of the TGF-β/SMAD pathway. In our model, activation of FG modulates microRNA expression by upregulating miR-21 and downregulating miR-146a via mass action kinetics.

3.3. Crosstalk between Signaling Pathways and Effect on microRNA Levels

Simulations of the integrated network with crosstalk between STAT3 and TGF-β pathway as well as the microRNAs were able to reproduce the expected phenotypic switch of microRNA levels (Figure 5). Our results were consistent with the experimentally observed dynamics of microRNA levels. miR-21 levels showed robust increase upon cytokine stimuli. Following the peak expression, miR-21 levels approached a steady state and were maintained well above baseline expression. However, the initial upregulation of miR-146a due to NF-κB activation was followed by a decline due to the negative regulation mediated by increase in FG levels. Over the course of the simulation, miR-146a expression increased steadily to below baseline levels, closely mimicking the *in vitro* experimental data on HSC activation [22]. These results emphasized the role of crosstalk between the TGF-β pathway and microRNAs to lead to a pro-fibrogenic activated HSC phenotype.

Our results highlight the relative dynamic contribution of exogenous signaling molecules such as cytokines IL-1 and IL-6 produced by other liver cells, as well as endogenous molecules such as TGF-β produced by HSCs, to the microRNA switch underlying HSC activation.

Figure 5. Dynamics of microRNA and transcription factor expression levels of the integrated model with all three signaling pathways.

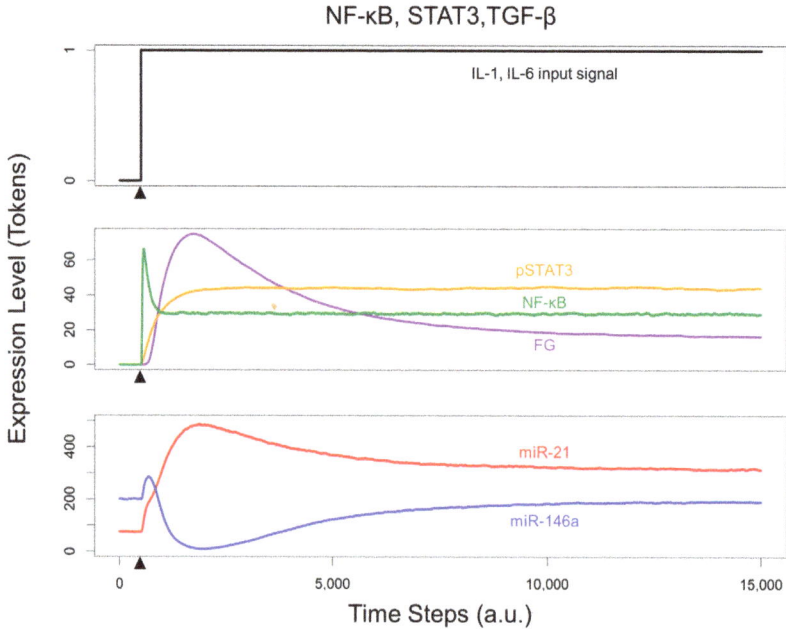

NF-κB, STAT3,TGF-β

3.4. Effects of Timing Differences in NF-κB and STAT3 Activation

We examined the effect of differential timing in activation of NF-κB and STAT3 pathways on the dynamic regulation of microRNAs. Previous experimental results indicated that NF-κB activation occurs earlier, with activity spiking rapidly before decreasing to a lower steady state, whereas peak STAT3 activation occurs later followed by a steady decline [50]. However, we considered an alternative scenario in which peak NF-κB activation occurred after that of STAT3, which is likely to happen in the case of altered cell–cell signaling interactions, to evaluate the effects on downstream microRNA expression. Our simulation results revealed that earlier NF-κB peak activity than that of STAT3 resulted in an initial transient increase of both miR-146a and miR-21 expression along with a delayed microRNA switch (Figure 6A). Upon STAT3 activation, miR-21 expression levels further increased and miR-146a levels decreased before both reached a steady state. The alternative case of later NF-κB peak activity led to an initial monotonic decrease in miR-146a levels to levels below the baseline expression, and a persistent increase in miR-21 levels (Figure 6B). Both microRNAs showed a slight increase in expression levels upon NF-κB activation, eventually reaching the same steady state levels as the first scenario. These results suggest that the timing of cytokine stimuli is responsible for controlling the initial dynamics of microRNA expression, but does not significantly alter the overall expression profile or steady state levels. Experimental validation of initial microRNA dynamics could provide insight into the physiological relevance of differential cytokine signaling.

Figure 6. Dynamics of microRNA expression levels in response to differential timing of cytokine stimuli. (**A**) Expression levels with synchronized cytokine production; (**B**) Expression levels with staggered cytokine production (IL-6 produced before IL-1). Solid arrowheads denote the introduction of input cytokine signals.

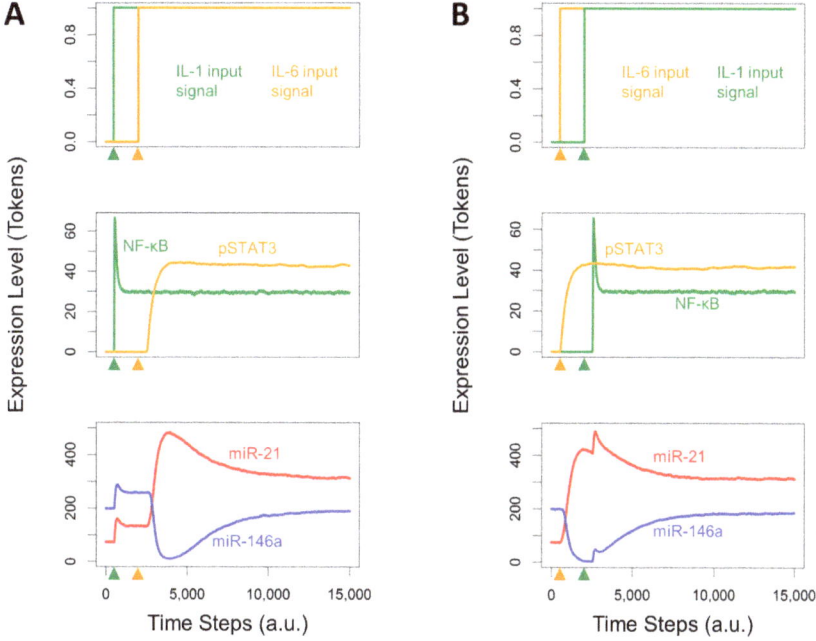

3.5. Re-Quiescence from the Activated HSC State

Finally, we explored the dynamics of microRNAs as the system returned from an activated to a quiescent phenotype. The simulation was allowed to reach an activated state phenotype followed by the termination of cytokine stimuli. During re-quiescence, both miR-146a and miR-21 levels decreased initially due to the rapid decline of NF-κB and STAT3 activity (Figure 7). Unlike miR-21, which steadily decreased to baseline levels as STAT3 and FG activity declines, miR-146a recovered from the drop in expression and eventually returned to baseline levels. Although the miR-146a expression level patterns were similar to those previously observed *in vitro* by Maubach *et al.*, the return of miR-146a to baseline levels is insufficient to completely restore a quiescent HSC phenotype [22]. This indicates the involvement of additional factors participating in the complex biological process of re-quiescence. Therefore, accounting for additional microRNAs and other pathways regulating the gene expression might be necessary to capture the multi-scale regulation of HSC phenotypes leading to re-quiescence.

Figure 7. Dynamics of microRNA and transcription factor expression levels during re-quiescence. Solid arrowheads denote the introduction of input cytokine signals. Hollow arrowheads denote the cessation of cytokine signals.

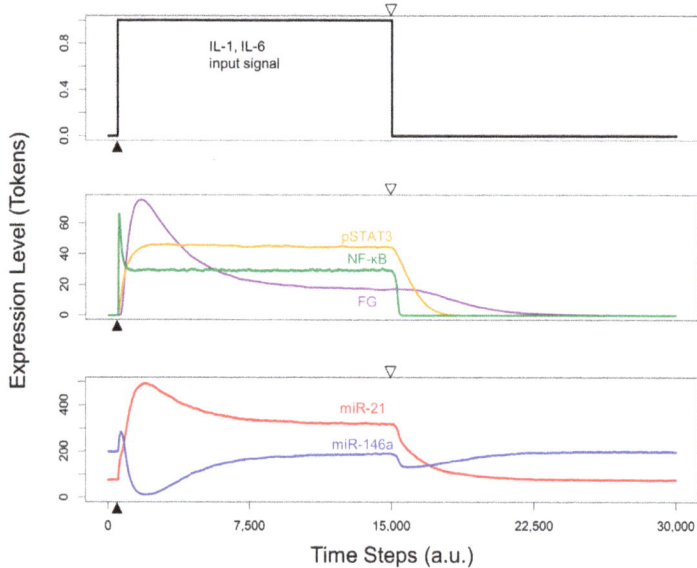

4. Conclusions

Activation of HSCs, which relies on a variety of interconnected signaling pathways, plays a critical role in the development of liver fibrosis. There is increasing evidence for the role of microRNAs in regulating the transformation of quiescent into activated HSCs [23]. Although many of the fibrogenic pathways have been previously studied, further investigation is required to better understand the role of crosstalk between pathways and their regulation mediated by the microRNAs. To address this gap in knowledge, we developed a Petri net system-based model to study regulatory interactions between the microRNAs miR-21 and miR-146a that may occur through as yet unknown direct or indirect mechanisms [22,36].

We performed model simulations to test whether known interactions between signaling and microRNAs can explain the differential regulation of miR-21 and miR-146a. Our results indicate that NF-κB and STAT3 signaling pathways are not sufficient to account for the anti-parallel regulation of these microRNAs. Activation of the TGF-β/SMAD pathway by STAT3 also appears to be insufficient to mimic the measured phenotype. Therefore, we introduced a phenomenological relationship between downstream fibrotic genes (FG) and microRNAs into our model as an additional component that upregulated miR-21 expression and downregulated miR-146a expression. The introduction of TGF-β/SMAD pathway-mediated differential microRNA regulation was necessary to induce a switch between microRNA levels. Upon termination of the cytokine stimuli, the system returned to its quiescent state. This ability of HSCs to switch between activated and quiescent states is critical for maintaining proper liver function, and dysregulation of this process can play a role in the development of liver fibrosis.

Timing of the activation of cytokines plays a significant role in the activation of inflammatory markers during liver regeneration. It is known that timing of cytokine signaling in HSCs is dependent on the type of damage. NF-κB and STAT3 pathways are likely to be activated simultaneously in the case of damage due to infection or alcohol intake. In contrast, in the case of drug-induced liver injury and liver regeneration, IL-1 stimulus is known to precede the IL-6-dependent activation of STAT3 pathway [51]. Cytokine signaling in HSCs is the result of paracrine factors from other cells, and the timing of cell communication can determine HSC response to injury. In order to determine how differential activation of cytokine stimuli influences the dynamics of microRNA expression, we analyzed two possible scenarios. These are primarily applicable to the whole liver tissue, however testing these possibilities for HSCs using our model can lead to predictions reconciling the potential difference in temporal relationships between the activation of NF-κB and STAT3 pathways in *in vivo vs. in vitro* conditions. Our results show that the relative timing between cytokine stimuli in HSCs alters the initial dynamics of microRNA expression. Earlier activation of NF-κB resulted in an initial transient increase in expression of both microRNAs, whereas later activation of NF-κB *versus* STAT3 resulted in an initial monotonic decrease in miR-146a while miR-21 gradually increased and reach steady state levels.

There is a paucity of evidence in the literature regarding the kinetics of HSC activation. We have used empirical kinetic approximations in order to bridge this gap in knowledge. For parameters, we relied on a range of values obtained from previous modeling studies. However, most of our kinetic parameters were derived through an iterative process. This approach provided us with responses that qualitatively matched the expected kinetics. However, a formal global sensitivity analysis will enable us to fine-tune the model parameters to develop more accurate HSC specific predictions [52]. We believe that such a model based on the HSC specific biology can lead to testable hypothesis on HSC activation that can be validated by experiments. For example, the use of exogenous cytokines for *in vitro* studies has been previously reported for HSCs, and could be easily adapted to investigate the possible role of cell–cell communication in microRNA expression dynamics using HSC cultures [53]. Briefly, the cells would be pre-treated with IL-1 or IL-6, followed by addition of the alternate cytokine at a later time point, and microRNA levels would be quantified over this time series. Upregulation of microRNAs can be accomplished by addition of a precursor for the microRNA of interest, and could help determine, for example, the effect of miR-146a overexpression on re-quiescence [43]. In addition to overexpression, knockout studies can be utilized both *in vitro* and *in vivo* to investigate the role of key components in HSC activation. For example, transfection of cultured HSCs with anti-miR sequences or siRNA against a gene can be used to investigate the relative importance of specific microRNAs and genes in HSC activation, as well as the consequences of their removal. This could be used to determine whether or not, for instance, TGF-β or miR-21 is necessary for HSC activation. These studies can be expanded to include transgenic knockout models, in the case of gene targets, or with the introduction of anti-miR pharmacological compounds, e.g., Vivo-Morpholino [43]. In order to study the effects of these changes in activated HSCs, it is important to ensure that the cells achieve total differentiation from the quiescent state. To this end, it has been reported that HSCs require a stiff culture environment in order to undergo activation [54].

However, it is not entirely clear how the biomechanical aspects specifically regulate signaling pathways and microRNAs considered in the present model.

Our model accounts for a representative marker of HSC activation representing the switch in miR-21 and miR-146a levels, and can be expanded to consider additional mechanisms of microRNA regulation of target genes, e.g., transcriptional degradation *versus* translational repression. Further avenues for model expansion include incorporation of co-regulated microRNAs as well as signaling and metabolic pathways involved in HSC activation and re-quiescence [2,22]. Our model provides a new computational platform for studying HSC activation and the mechanisms where microRNAs act as fine-tuning markers modulating the system behavior, thereby contributing to potential therapeutic approaches.

Acknowledgments

Financial support for this study was provided by National Institutes of Health grants from National Institute on Alcohol Abuse and Alcoholism R01 AA018873, R21 AA022417, T32 AA007463. L.K. and A.P. acknowledge Daniel Cook for help in formulating the Introduction and Hiren Makadia for helpful discussion on model development.

Author Contributions

Lakshmi Kuttippurathu constructed the model and contributed to model implementation, analysis, and discussion; Austin Parrish implemented the model, and contributed to the analysis and discussion; Rajanikanth Vadigepalli designed the study and initial model, contributed to manuscript writing, and oversaw all aspects of the research.

Appendix

A detailed list of the equations used in our Petri net model, as well as graphical views of the simulated pathways, are presented here.

Table A1. Table of all reactions modeled in the Petri net system using either mass action or Michaelis-Menten kinetics, along with their respective parameters.

Reaction	Equation	Parameters
IL-1/IL-1R association	$k_1[IL1][IL1R]$	$k_1 = 0.001$
IL-1/IL-1R dissociation	$k_1^*[IL1/IL1R]$	$k_1^* = 1$
IL-1 degradation	$k_2[IL1]$	$k_2 = 0.001$
IL-1/IL-1R degradation	$k_3[IL1/IL1R]$	$k_3 = 0.01$
PELI1/TRAF6 activation	$k_4[IL1/IL1R] - k_5[miR21] - k_6[miR146a]$	$k_4 = 0.5$ $k_5 = 0.01$ $k_6 = 0.01$
PELI1/TRAF6 degradation	$k_7[PELI1/TRAF6]$	$k_7 = 0.1$
NF-κB^{inactive} production	k_8	$k_8 = 0.3$
NF-κB activation	$\dfrac{v_1[PELI1/TRAF6][NF\kappa B^{inactive}]}{[NF\kappa B^{inactive}] + k_9}$	$v_1 = 0.5$ $k_9 = 100$

Table A1. *Cont.*

Reaction	Equation	Parameters
NF-κB^{inactive} degradation	$k_{10}[NF\kappa B^{inactive}]$	$k_{10} = 0.003$
NF-κB^{active} inactivation	$k_{11}[NF\kappa B^{active}]$	$k_{11} = 0.1$
NF-κB^{active} degradation	$k_{12}[NF\kappa B^{active}]$	$k_{12} = 0.01$
IL-6/IL-6R association	$k_{13}[IL6][IL6R]$	$k_{13} = 0.001$
IL-6/IL-6R dissociation	$k_{13}^*[IL6/IL6R]$	$k_{13}^* = 1$
IL-6 degradation	$k_{14}[IL6]$	$k_{14} = 0.001$
IL-6/IL-6R degradation	$k_{15}[IL6/IL6R]$	$k_{15} = 0.01$
STAT3 phosphorylation	$\dfrac{v_2[IL6/IL6R][STAT3]}{STAT3 + k_{16}\left(1 + \dfrac{SOCS3}{k_{17}}\right)}$	$v_2 = 0.008$ $k_{16} = 50$ $k_{17} = 3$
STAT3 production	k_{18}	$k_{18} = 0.08$
STAT3 degradation	$k_{19}[STAT3]$	$k_{19} = 0.0008$
pSTAT3 degradation	$\dfrac{v_3[pSTAT3]}{[pSTAT3] + k_{20}}$	$v_3 = 0.025$ $k_{20} = 0.00007$
SOCS3 production	$\dfrac{v_4[pSTAT3]}{[pSTAT3] + k_{21}}$	$v_4 = 0.1$ $k_{21} = 0.00007$
SOCS3 degradation	$k_{22}[SOCS3]$	$k_{22} = 0.001$
TGF-β production	$k_{23}[pSTAT3]$	$k_{23} = 0.04$
TGF-β degradation	$k_{24}[TGF\beta]$	$k_{24} = 0.001$
TGF-β/TβR1 association	$k_{25}[TGF\beta][T\beta RI]$	$k_{25} = 0.001$
TGF-β/TβR1 dissociation	$k_{25}^*[TGF\beta/T\beta RI]$	$k_{25}^* = 1$
TGF-β/TβR1 degradation	$k_{26}[TGF\beta/T\beta RI] + k_{27}[SMAD7/SMURF2]$	$k_{26} = 0.01$ $k_{27} = 0.1$
SMAD2 production	k_{28}	$k_{28} = 0.01$
SMAD2 degradation	$k_{29}[SMAD2]$	$k_{29} = 0.0001$
SMAD2 phosphorylation	$\dfrac{v_5[TGF\beta/T\beta RI][SMAD2]}{SMAD2 + k_{30}\left(1 + \dfrac{SMAD7}{k_{31}}\right)}$	$v_5 = 10$ $k_{30} = 5000$ $k_{31} = 0.01$
SMAD2 dephosphorylation	$\dfrac{v_6[pSMAD2]}{[pSMAD2] + k_{32}}$	$v_6 = 0.001$ $k_{32} = 5000$
pSMAD2 degradation	$k_{33}[pSMAD2]$	$k_{33} = 0.0005$
SMAD4 production	$k_{34} - k_{35}[miR146a]$	$k_{34} = 0.024$ $k_{35} = 0.00008$
SMAD4 degradation	$k_{36}[SMAD4]$	$k_{36} = 0.0001$
pSMAD2/SMAD4 association	$k_{37}[pSMAD2]^2[SMAD4]$	$k_{37} = 0.0005$
pSMAD2/SMAD4 dissociation	$k_{37}^*[pSMAD2/SMAD4]$	$k_{37}^* = 10$
pSMAD2/SMAD4 degradation	$k_{38}[pSMAD2/SMAD4]$	$k_{38} = 0.01$
SMAD7 production	$k_{39} + k_{40}[pSMAD2/SMAD4] - k_{41}[miR21]$	$k_{39} = 0.057$ $k_{40} = 0.003$ $k_{41} = 0.0001$
SMAD7 degradation	$k_{42}[SMAD7]$	$k_{42} = 0.001$
SMURF2 production	$k_{43} + k_{44}[pSMAD2/SMAD4]$	$k_{43} = 0.05$ $k_{44} = 0.003$

Table A1. *Cont.*

Reaction	Equation	Parameters
SMURF2 degradation	$k_{45}[SMURF2]$	$k_{45} = 0.001$
SMAD7/SMURF2 association	$k_{46}[SMAD7][SMURF2]$	$k_{46} = 2.9$
SMAD7/SMURF2 dissociation	$k_{46}^{*}[SMAD7/SMURF2]$	$k_{46}^{*} = 0.2$
SMAD7/SMURF2 degradation	$k_{47}[SMAD7/SMURF2]$	$k_{47} = 0.008$
FG production	$k_{48}[pSMAD2/SMAD4]$	$k_{48} = 0.03$
FG degradation	$k_{49}[FG]$	$k_{49} = 0.0005$
miR-21 activation	$k_{50} + k_{51}[NF\kappa B^{active}] + k_{52}[pSTAT3] + k_{53}[FG]$	$k_{50} = 0.75$ $k_{51} = 0.02$ $k_{52} = 0.03$ $k_{53} = 0.03$
miR-146a activation	$k_{54} + k_{55}[NF\kappa B^{active}] - k_{56}[FG]$	$k_{54} = 2$ $k_{55} = 0.02$ $k_{56} = 0.04$
miR-21 degradation	$k_{57}[miR21]$	$k_{57} = 0.01$
miR-146a degradation	$k_{58}[miR146a]$	$k_{58} = 0.01$

Figure A1. Schematic (**A**) and Petri net (**B**) representations of the IL-1/NF-κB network simulated in Figure 4A.

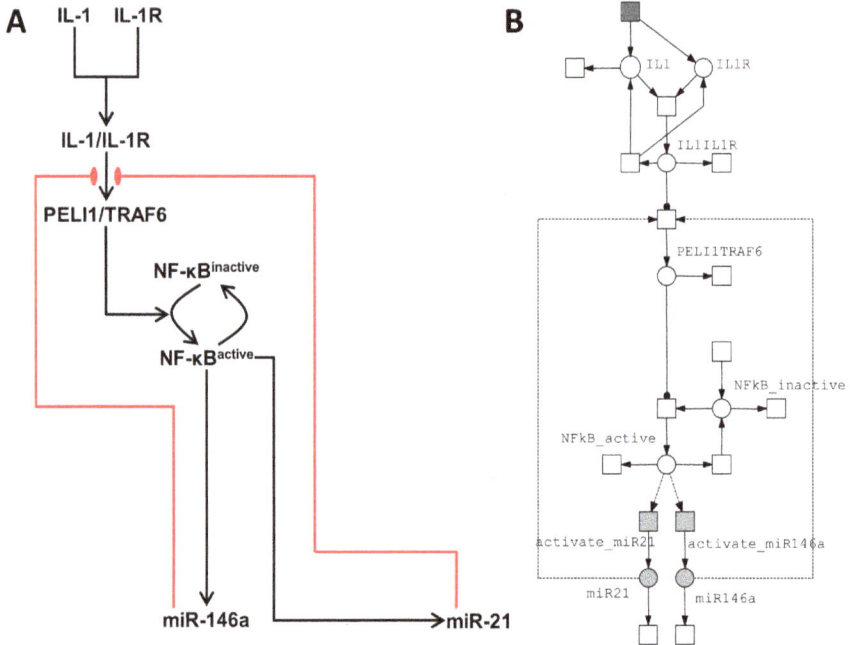

Figure A2. Schematic (**A**) and Petri net (**B**) representations of the IL-1/NF-κB and IL-6/STAT3 network simulated in Figure 4B.

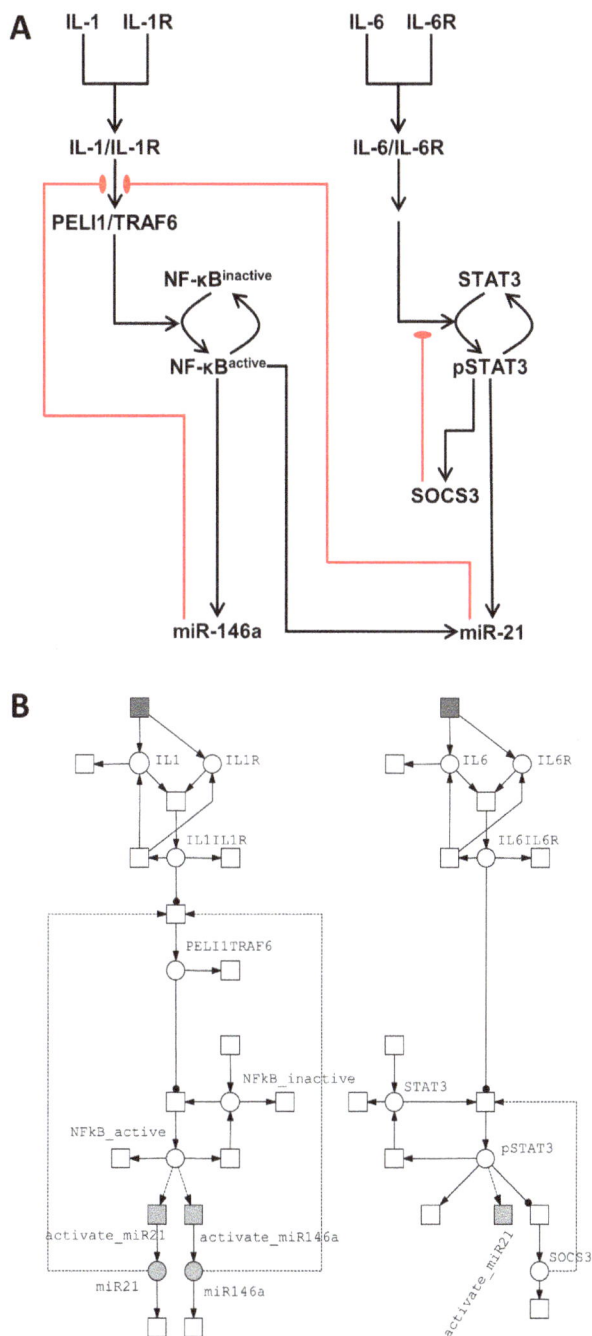

Figure A3. Schematic (**A**) and Petri net (**B**) representations of the model simulated in Figure 5, excluding the fibrotic gene (FG) node.

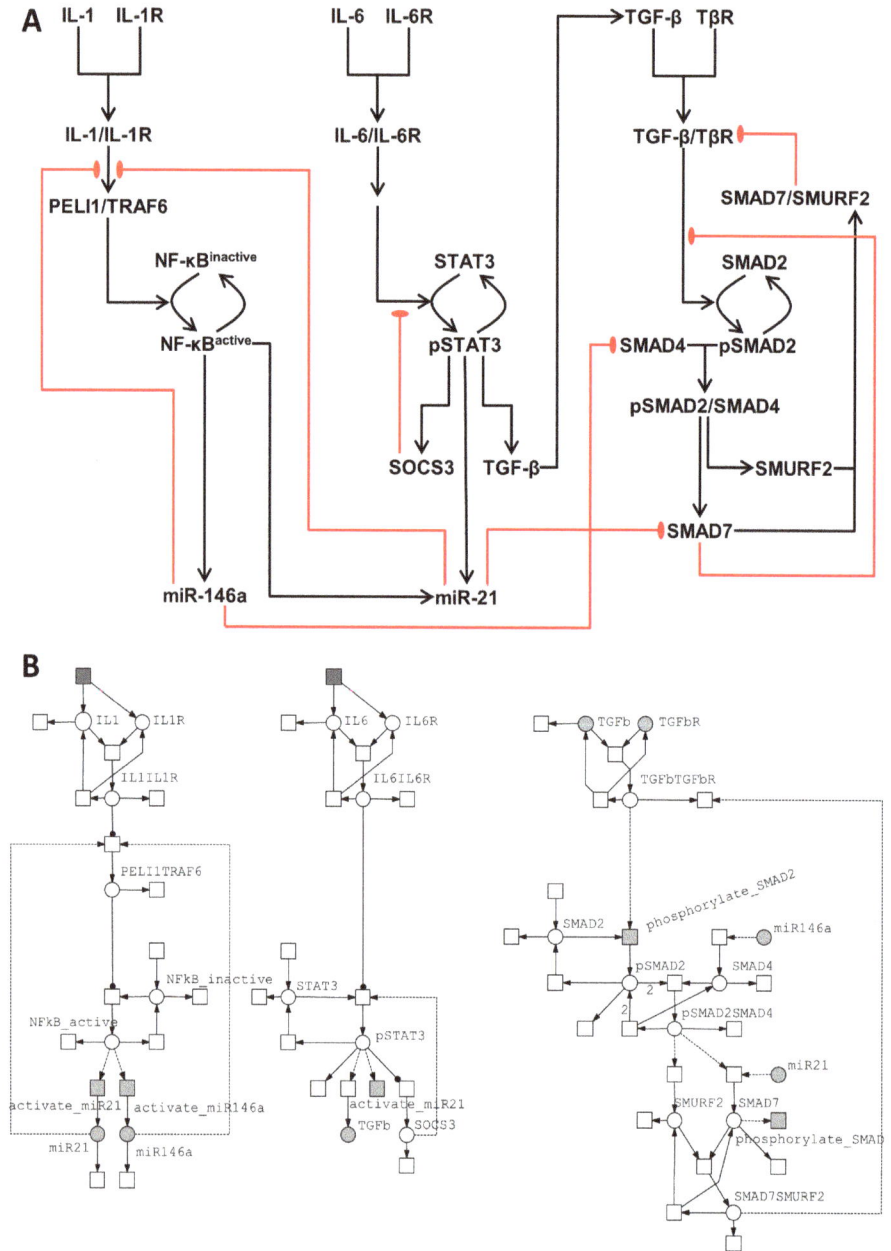

Conflicts of Interest

The authors declare no conflict of interest.

References

1. Bataller, R.; Brenner, D. Liver fibrosis. *J. Clin. Investig.* **2005**, *115*, 209–218.

2. Friedman, S. Hepatic stellate cells: Protean, multifunctional, and enigmatic cells of the liver. *Physiol. Rev.* **2008**, *88*, 125–172.

3. Hendriks, H.; Verhoofstad, W. Perisinusoidal fat-storing cells are the main vitamin A storage sites in rat liver. *Exp. Cell Res.* **1985**, *160*, 138–149.

4. Paik, Y.-H.; Lee, K.S.; Lee, H.J.; Yang, K.M.; Lee, S.J.; Lee, D.K.; Han, K.-H.; Chon, C.Y.; Lee, S.I.; Moon, Y.M.; Brenner, D.A. Hepatic stellate cells primed with cytokines upregulate inflammation in response to peptidoglycan or lipoteichoic acid. *Lab. Invest.* **2006**, *86*, 676–686.

5. Paik, Y.-H.; Schwabe, R.F.; Bataller, R.; Russo, M.P.; Jobin, C.; Brenner, D.A. Toll-like receptor 4 mediates inflammatory signaling by bacterial lipopolysaccharide in human hepatic stellate cells. *Hepatology* **2003**, *37*, 1043–1055.

6. Maher, J.J. Cell-specific expression of hepatocyte growth factor in liver. *J. Clin. Invest.* **1993**, *91*, 2244–2252.

7. Malik, R.; Selden, C.; Hodgson, H. The role of non-parenchymal cells in liver growth. *Semin. Cell Dev. Biol.* **2002**, *13*, 425–431.

8. Mullhaupt, B.; Feren, A.; Fodor, E.; Jones, A. Liver expression of epidermal growth factor RNA. Rapid increases in immediate-early phase of liver regeneration. *J. Biol. Chem.* **1994**, *2883*, 19667–19670.

9. Yin, C.; Evason, K.J.; Asahina, K.; Stainier, D.Y.R. Hepatic stellate cells in liver development, regeneration, and cancer. *J. Clin. Invest.* **2013**, *123*, 1902–1910.

10. Xu, M.-Y.; Hu, J.-J.; Shen, J.; Wang, M.-L.; Zhang, Q.-Q.; Qu, Y.; Lu, L.-G. Stat3 signaling activation crosslinking of TGF-β1 in hepatic stellate cell exacerbates liver injury and fibrosis. *Biochim. Biophys. Acta* **2014**, *1842*, 2237–2245.

11. Hellerbrand, C.; Jobin, C.; Licato, L.L.; Sartor, R.B.; Brenner, D.A. Cytokines induce NF-κB in activated but not in quiescent rat hepatic stellate cells. *Am. J. Physiol.* **1998**, *275*, G269–G278.

12. Gieling, R.G.; Wallace, K.; Han, Y.-P. Interleukin-1 participates in the progression from liver injury to fibrosis. *Am. J. Physiol. Gastrointest. Liver Physiol.* **2009**, *296*, G1324–G1331.

13. Yoshida, K.; Matsuzaki, K. Differential regulation of TGF-β/Smad signaling in hepatic stellate cells between acute and chronic liver injuries. *Front. Physiol.* **2012**, *3*, 1–7.

14. Tahashi, Y.; Matsuzaki, K.; Date, M.; Yoshida, K.; Furukawa, F.; Sugano, Y.; Matsushita, M.; Himeno, Y.; Inagaki, Y.; Inoue, K. Differential regulation of TGF-beta signal in hepatic stellate cells between acute and chronic rat liver injury. *Hepatology* **2002**, *35*, 49–61.

15. Shigekawa, M.; Takehara, T.; Kodama, T.; Hikita, H.; Shimizu, S.; Li, W.; Miyagi, T.; Hosui, A.; Tatsumi, T.; Ishida, H.; *et al.* Involvement of STAT3-regulated hepatic soluble factors in attenuation of stellate cell activity and liver fibrogenesis in mice. *Biochem. Biophys. Res. Commun.* **2011**, *406*, 614–620.

16. Tsukamoto, H. Cytokine regulation of hepatic stellate cells in liver fibrosis. *Alcohol. Clin. Exp. Res.* **1999**, *23*, 911–916.

17. Kerr, T.A.; Korenblat, K.M.; Davidson, N.O. MicroRNAs and liver disease. *Transl. Res.* **2011**, *157*, 241–252.

18. Ji, J.; Zhang, J.; Huang, G.; Qian, J.; Wang, X.; Mei, S. Over-expressed microRNA-27a and 27b influence fat accumulation and cell proliferation during rat hepatic stellate cell activation. *FEBS Lett.* **2009**, *583*, 759–766.

19. Guo, C.-J.; Pan, Q.; Li, D.-G.; Sun, H.; Liu, B.-W. miR-15b and miR-16 are implicated in activation of the rat hepatic stellate cell: An essential role for apoptosis. *J. Hepatol.* **2009**, *50*, 766–778.

20. Sekiya, Y.; Ogawa, T.; Yoshizato, K.; Ikeda, K.; Kawada, N. Suppression of hepatic stellate cell activation by microRNA-29b. *Biochem. Biophys. Res. Commun.* **2011**, *412*, 74–79.

21. Guo, C.J.; Pan, Q.; Cheng, T.; Jiang, B.; Chen, G.Y.; Li, D.G. Changes in microRNAs associated with hepatic stellate cell activation status identify signaling pathways. *FEBS J.* **2009**, *276*, 5163–5176.

22. Maubach, G.; Chin, M.; Lim, C.; Chen, J.; Yang, H.; Zhuo, L. miRNA studies in *in vitro* and *in vivo* activated hepatic stellate cells. *World J. Gastroenterol.* **2011**, *17*, 2748–2773.

23. Huang, J.; Yu, X.; Fries, J.W.U.; Zhang, L.; Odenthal, M. MicroRNA function in the profibrogenic interplay upon chronic liver disease. *Int. J. Mol. Sci.* **2014**, *15*, 9360–9371.

24. Zhang, Z.; Zha, Y.; Hu, W.; Huang, Z.; Gao, Z.; Zang, Y.; Chen, J.; Dong, L.; Zhang, J. The autoregulatory feedback loop of MicroRNA-21/programmed cell death protein 4/Activation protein-1 (MiR-21/PDCD4/AP-1) as a driving force for hepatic fibrosis development. *J. Biol. Chem.* **2013**, *288*, 37082–37093.

25. Jiang, X.; Tsitsiou, E.; Herrick, S.E.; Lindsay, M.A. MicroRNAs and the regulation of fibrosis. *FEBS J.* **2010**, *277*, 2015–2021.

26. Liang, G.; Li, G.; Wang, Y.; Lei, W.; Xiao, Z. Aberrant miRNA expression response to UV irradiation in human liver cancer cells. *Mol. Med. Rep.* **2014**, *9*, 904–910.

27. Karakatsanis, A.; Papaconstantinou, I.; Gazouli, M.; Lyberopoulou, A.; Polymeneas, G.; Voros, D. Expression of microRNAs, miR-21, miR-31, miR-122, miR-145, miR-146a, miR-200c, miR-221, miR-222, and miR-223 in patients with hepatocellular carcinoma or intrahepatic cholangiocarcinoma and its prognostic significance. *Mol. Carcinog.* **2013**, *52*, 297–303.

28. Rosato, P.; Anastasiadou, E.; Garg, N.; Lenze, D.; Boccellato, F.; Vincenti, S.; Severa, M.; Coccia, E.M.; Bigi, R.; Cirone, M.; *et al.* Differential regulation of miR-21 and miR-146a by Epstein-Barr virus-encoded EBNA2. *Leukemia* **2012**, *26*, 2343–2352.

29. Marquez, R.T.; Wendlandt, E.; Galle, C.S.; Keck, K.; Mccaffrey, A.P. MicroRNA-21 is upregulated during the proliferative phase of liver regeneration, targets Pellino-1, and inhibits NF-κB signaling. *Am. J. Physiol. Gastrointest. Liver Physiol.* **2010**, *298*, 535–541.

30. Taganov, K.; Boldin, M. NF-κB-dependent induction of microRNA miR-146, an inhibitor targeted to signaling proteins of innate immune responses. *Proc. Natl. Acad. Sci. USA* **2006**, *103*, 12481–12486.

31. Petri, C. *Kommunikation mit Automaten*; Technische Universität Darmstadt: Darmstadt, Germany, 1962; p. 128.

32. Rohr, C.; Marwan, W.; Heiner, M. Snoopy—A unifying Petri net framework to investigate biomolecular networks. *Bioinformatics* **2010**, *26*, 974–975.

33. Chung, S.-W.; Miles, F.L.; Sikes, R.A.; Cooper, C.R.; Farach-Carson, M.C.; Ogunnaike, B.A. Quantitative modeling and analysis of the transforming growth factor beta signaling pathway. *Biophys. J.* **2009**, *96*, 1733–1750.

34. Furchtgott, L.A.; Chow, C.C.; Periwal, V. A model of liver regeneration. *Biophys. J.* **2009**, *96*, 3926–3935.

35. Wegner, K.; Bachmann, A.; Schad, J.-U.; Lucarelli, P.; Sahle, S.; Nickel, P.; Meyer, C.; Klingmüller, U.; Dooley, S.; Kummer, U. Dynamics and feedback loops in the transforming growth factor β signaling pathway. *Biophys. Chem.* **2012**, *162*, 22–34.

36. Dippold, R.P.; Vadigepalli, R.; Gonye, G.E.; Hoek, J.B. Chronic ethanol feeding enhances miR-21 induction during liver regeneration while inhibiting proliferation in rats. *Am. J. Physiol. Gastrointest. Liver Physiol.* **2012**, *303*, G733–G743.

37. Elsharkawy, A.M.; Mann, D.A. Nuclear factor-κB and the hepatic inflammation-fibrosis-cancer axis. *Hepatology* **2007**, *46*, 590–597.

38. Lee, K.S.; Buck, M.; Houglum, K.; Chojkier, M. Activation of Hepatic Stellate Cells by TGFa and Collagen Type I Is Mediated by Oxidative Stress Through c-myb Expression. *J. Clin. Invest.* **1995**, *96*, 2461–2468.

39. Jiang, Z.; Johnson, H.J.; Nie, H.; Qin, J.; Bird, T.A.; Li, X. Pellino 1 is required for interleukin-1 (IL-1)-mediated signaling through its interaction with the IL-1 receptor-associated kinase 4 (IRAK4)-IRAK-tumor necrosis factor receptor-associated factor 6 (TRAF6) complex. *J. Biol. Chem.* **2003**, *278*, 10952–10956.

40. He, G.; Karin, M. NF-κB and STAT3—Key players in liver inflammation and cancer. *Cell Res.* **2011**, *21*, 159–168.

41. Iliopoulos, D.; Jaeger, S.A.; Hirsch, H.A.; Bulyk, M.L.; Struhl, K. STAT3 activation of miR-21 and miR-181b-1 via PTEN and CYLD are part of the epigenetic switch linking inflammation to cancer. *Mol. Cell* **2010**, *39*, 493–506.

42. Wang, H.; Lafdil, F.; Kong, X.; Gao, B. Signal transducer and activator of transcription 3 in liver diseases: A novel therapeutic target. *Int. J. Biol. Sci.* **2011**, *7*, 536–550.

43. Francis, H.; McDaniel, K.; Han, Y.; Liu, X.; Kennedy, L.; Yang, F.; McCarra, J.; Zhou, T.; Glaser, S.; Venter, J.; *et al.* Regulation of the extrinsic apoptotic pathway by microRNA-21 in alcoholic liver injury. *J. Biol. Chem.* **2014**, doi:10.1074/jbc.M114.602383.

44. Marquez, R.T.; Bandyopadhyay, S.; Wendlandt, E.B.; Keck, K.; Hoffer, B.A.; Icardi, M.S.; Christensen, R.N.; Schmidt, W.N.; McCaffrey, A.P. Correlation between microRNA expression levels and clinical parameters associated with chronic hepatitis C viral infection in humans. *Lab. Invest.* **2010**, *90*, 1727–1736.

45. He, Y.; Huang, C.; Sun, X.; Long, X.; Lv, X.; Li, J. MicroRNA-146a modulates TGF-beta1-induced hepatic stellate cell proliferation by targeting SMAD4. *Cell. Signal.* **2012**, *24*, 1923–1930.

46. Yoshida, K.; Murata, M.; Yamaguchi, T.; Matsuzaki, K. TGF-β/Smad signaling during hepatic fibro-carcinogenesis (Review). *Int. J. Oncol.* **2014**, *45*, 1363–1371.

47. Hayashi, H.; Abdollah, S.; Qiu, Y.; Cai, J.; Xu, Y.Y.; Grinnell, B.W.; Richardson, M.A.; Topper, J.N.; Gimbrone, M.A.; Wrana, J.L.; *et al.* The MAD-related protein Smad7 associates with the TGFbeta receptor and functions as an antagonist of TGFbeta signaling. *Cell* **1997**, *89*, 1165–1173.

48. Yan, X.; Liu, Z.; Chen, Y. Regulation of TGF- b signaling by Smad7 Overview of TGF-b Signaling Pathways. *Acta Biochim. Biophys. Sin. (Shanghai)* **2009**, *41*, 263–272.

49. Davis, B.N.; Hilyard, A.C.; Lagna, G.; Hata, A. SMAD proteins control DROSHA-mediated microRNA maturation. *Nature* **2008**, *454*, 56–61.

50. Salazar-Montes, A.; Ruiz-Corro, L.; Sandoval-Rodriguez, A.; Lopez-Reyes, A.; Armendariz-Borunda, J. Increased DNA binding activity of NF-kappaB, STAT-3, SMAD3 and AP-1 in acutely damaged liver. *World J. Gastroenterol.* **2006**, *12*, 5995–6001.

51. Kurinna, S.; Barton, M.C. Cascades of transcription regulation during liver regeneration. *Int. J. Biochem. Cell Biol.* **2011**, *43*, 189–197.

52. Miller, G.M.; Ogunnaike, B.A.; Schwaber, J.S.; Vadigepalli, R. Robust dynamic balance of AP-1 transcription factors in a neuronal gene regulatory network. *BMC Syst. Biol.* **2010**, *4*, 171.

53. Kong, X.; Feng, D.; Wang, H.; Hong, F.; Bertola, A.; Wang, F.-S.; Gao, B. Interleukin-22 induces hepatic stellate cell senescence and restricts liver fibrosis in mice. *Hepatology* **2012**, *56*, 1150–1159.

54. Olsen, A.L.; Bloomer, S.A.; Chan, E.P.; Gaça, M.D.A.; Georges, P.C.; Sackey, B.; Uemura, M.; Janmey, P.A.; Wells, R.G. Hepatic stellate cells require a stiff environment for myofibroblastic differentiation. *Am. J. Physiol. Gastrointest. Liver Physiol.* **2011**, *301*, G110–G118.

Mathematical Modeling and Analysis of Crosstalk between MAPK Pathway and Smad-Dependent TGF-β Signal Transduction

Ji Liu, Wei Dai and Juergen Hahn

Abstract: Broad evidence exists for cross talk between the Mitogen-activated protein kinases (MAPK) pathway and Smad-dependent TGF-β signal transduction. A variety of studies, oftentimes involving different cell types, have identified several potential mechanisms for the crosstalk. However, there is no clear consensus on the actual mechanism(s) responsible for the crosstalk. This work develops a model of the pathway, including several hypothesized crosstalk mechanisms, and discusses which of the potential mechanisms can appropriately describe observed behaviors. Simulation results show a good agreement of the findings with results reported in the literature.

Reprinted from *Processes*. Cite as: Liu, J.; Dai, W.; Hahn, J. Mathematical Modeling and Analysis of Crosstalk between MAPK Pathway and Smad-Dependent TGF-β Signal Transduction. *Processes* **2014**, *2*, 570-595.

1. Introduction

The Transforming Growth Factor β (TGF-β) family of proteins is involved in regulating a variety of cellular processes such as cell proliferation, apoptosis, differentiation, immune functions and tumor invasion/metastasis [1–5]. TGF-β signals through a receptor complex of serine/threonine kinase type I (TbRI) and type II (TbRII) receptors. Upon ligand-binding, this receptor complex activation leads to phosphorylation of cytoplasmic mediated Smad2 and/or Smad3 (R-Smad). This phosphorylation at the *C*-terminal SSXS motif of Smad2/3 allows them to form a heteromeric complex with the common mediator Smad (C-Smad), Smad4. These complexes translocate into the nucleus and act as transcription factors to regulate target gene expression [1]. The nuclear Smad complexes are then dephosphorylated by Smad phosphatases [6]. It should be noted that while TGF-β activates Smad signaling, Smad plays an important role in the epithelial to mesenchymal transition, which is important for tumor metastasis [7].

While the Smad pathway is considered one of the main pathways for TGF-β signaling, it nevertheless has to be considered in a broader context as other pathways directly influence Smad signaling activity [8,9]. The fact that some of this crosstalk is also activated by TGF-β makes the problem of investigating Smad signaling even more challenging [10].

Mitogen activated protein kinase (MAPK), including Erk1/2, JNK1/2/3, and p38/MAPKs, is a family of kinases. Multiple extracellular stimuli, such as cytokines, Epidermal Growth Factor (EGF), hepatocyte growth factor (HGF), as well as TGF-β, can initiate a cascade of serial phosphorylation activation from MAP kinase kinase kinases (MAPKKKs) to MAP kinase kinases (MAPKKs) and finally to MAPKs [8]. Broad evidence exists for crosstalk of Smad signaling with the MAPK pathway, which has a significant effect on TGF-β induced signal transduction. The crosstalk is highly cell-type dependent and either inhibits or enhances the TGF-β/Smad pathway [8].

Erk activation by TGF-β has been observed in mesenchymal cells. Linker region phosphorylation of Smad2 and 3 by activated Erk resulted in increased half-life of *C*-terminal phosphorylated Smad2 and 3 and increased duration of Smad target gene transcription [1]. MAPKs play an important role in TGF-β mediated extracellular matrix production in mesangial cells [9]. Hayashida *et al.* [11] showed that blocking Erk inhibited TGF-β stimulation of Smad2 *C*-terminal phosphorylation and Smad2 association with Smad4 in human mesangial cells. The interaction is through Erk-induced phosphorylation of Smad2 in the linker region. Funaba *et al.* [12] reported that constitutively active MEK induced by EGF increased the half-life of Smad2 and enhanced the Smad2-dependent transcription activity in mink lung epithelial cells and COS7 cells. It has been shown that TGF-β induced receptor activation stimulated a parallel p42/p44 MAPK pathway that targeted Smad2 for an increased nuclear translocation and enhanced gene activation [13]. A high level of phosphorylated Smad2 in the linker region (PSmadL) was detected in a fraction of TGF-β treated vascular smooth-muscle cells [13]. In addition, the inhibitors of Erk and p38 attenuated the effect of TGF-β on increasing PSmadL.

Conversely, extensive observations have been reported that Erk negatively regulates the TGF-β induced gene transcription in some hyperactive Ras signaling cells [14–17]. Kretzschmar *et al.* [14,15] suggested that oncogenic Ras signaling can directly interfere with Smad-dependent responses by attenuating the agonist-induced nuclear accumulation of Smad1, Smad2, and Smad3 in mammary and lung epithelial cells. This effect is mediated by the phosphorylation of Smad1, Smad2, and Smad3 in the linker region by Ras-activated Erk1 and Erk2 protein kinases. A possible explanation for the contradictory observations is that responses, which depend on a certain level of nuclear Smad activity, may require the counterbalancing effect of Ras signaling to achieve a suitable level of nuclear Smad activity [18]. In a majority of human cancers carrying oncogenic Ras mutations, the activity rather than the levels of Ras expression is elevated. Thus, the levels of Ras activity are affected by the levels of MAPK activity, ultimately shaping the TGF-β response via crosstalk between Smad and the Ras-MAPK pathway [19].

Although crosstalk between the MAPK pathway and TGF-β/Smad signal transduction has been widely described, most of the studies only investigated this relationship for one specific cell type. As it is possible, and even quite likely, that different mechanisms are dominant in different cell types, this paper seeks to model and analyze the different proposed mechanisms. A model of the pathways, including different potential crosstalk mechanisms, is developed and thoroughly analyzed with respect to the effect that these mechanisms have on the observed response. Specific emphasis has been put on uncertainty in the model, as modeling potential mechanisms implies that the model structure is not well known. The goal of the presented work is to contribute to the understanding of crosstalk between the MAPK pathway and TGF-β/Smad signaling.

2. Materials and Methods

2.1. Model of the TGF-β/Smad Pathway

The part of the model representing the TGF-β/Smad pathway used in this work was developed by Zi *et al.* [6]. This model takes into account the receptor trafficking and endocytosis, Smads nucleocytoplasmic shuttling, Smad complex formation and dissociation, and the ligand induced negative feedback. There are 16 states and 20 parameters in the model. The nominal values of the model parameters were derived from experimental analysis in epithelial cells. The equations and the nominal parameter values for this model can be obtained from the literature [6]. Simulation results which agree well with experimental data include: The level of nuclear Smad complex, *i.e.*, the transcription factor, peaked at approximately 45 min after TGF-β addition and then declined to saturate above the basal level after about 5–6 h; the type I receptor kinase inhibitor SB431542 added at 30 min will cause rapid decrease of the nuclear Smad complex level to the basal level; also, TGF-β caused a change in the overall Smad2 distribution from predominantly in the cytoplasm to predominantly in the nucleus, which suggests TGF-β induces a change in the intercellular location of the majority of Smad2.

Similar to the simulation by Zi *et al.* two different types of stimulations are implemented in this work. The first stimulation is exposure to a constant level of 2 ng/mL of TGF-β. The second stimulation adds 2 ng/mL of TGF-β for 60 min after which the signal is terminated, which can experimentally be achieved via the addition of the type I receptor kinase inhibitor SB431542.

2.2. Model of the MAPK Pathway

The stimulus-response behavior of the MAPK pathway exhibits ultra-sensitivity and a Sigmoid-like shape [20–22]. When the ligand concentration is above a specific value, the variation of the output signal is negligible. Because most of experiments on the crosstalk between the two pathways focused on the low-stimulus response, this work will focus on low stimulus levels where the stimulus-response curves are less steep than those at high stimulation.

Multiple extracellular stimuli (cytokines, EGF, HGF), as well as TGF-β itself, can initiate a cascade of serial phosphorylation activation from MKKKs to MKKs and finally MAPKs. MAPKs such as Erk, JNK or p38 are the kinases of the cascade. The kinases of the first level, MKKKs, are activated by several mechanisms involving Ras (in the case of Erk) activation or other MAP kinase kinase kinases (MKKKKs) activation [22]. Schoeberl *et al.* established a detailed model of the EGF-stimulated MAPK pathway, which consists of 94 states and 95 parameters using mass action kinetics [23]. However, this model is quite complex and includes a significant amount of detail, which is not necessary for this study, and can possibly even be detrimental due to uncertainties in the interactions of the pathways. In contrast to this, Kholodenko's model [22], consisting of 8 states and 22 parameters, is used in this work for the three-level cascade from Ras/MKKKKs to MAPKs. The input stimulus of this MAPK cascade is the concentration of activated receptors (Ras/MKKKKs). However, the receptor activation mechanism in response to extracellular stimuli is not well understood for the pathway investigated in this work. In order to address this point, the input stimulus of the MAPK cascade, namely the concentration of activated receptors, can be written as

$$V_1 = V_0 \exp\left(-\gamma\left(t - t_0\right)\right) \tag{1}$$

where, $1/\gamma$ is the characteristic time of activated receptors (Ras/MKKKKs). The smaller the value of γ, the longer the receptor will be in the activated state. In the limiting case of $\gamma \rightarrow 0$, the pathway is permanently activated [20]. Here, t_0 is introduced to represent the delayed response to extracellular stimuli and V_0 is the concentration of activated receptors at $t = 0$.

Evidence has indicated that the time, level and duration of MAPKs activation are not correlated with a specific cellular function, but it appears to be regulated in a cell-type and stimulus-specific manner [24]. For example, EGF results in a transient activation of MAPK pathway in PC12 pheochromocytoma cells, whereas the nerve growth factor (NGF) is associated with a sustained activation of the MAPK pathway in PC12 cells. Although the exact mechanisms of Erk, JNK, or p38 MAPK activation by TGF-β are poorly understood, it has been observed that TGF-β-induced MAPKs activation also varies with cell type. In some cell lines, delayed activation suggests an indirect response requiring protein translation, whereas in others, activation is rapid and comparable to signaling by EGF [25]. Rapid and transient p38 MAPK activation by TGF-β has been described in certain cell types, including human neutrophils, HEK293, and C2C12 cells, whereas the prolonged and sustained p38 MAPK activation was observed in pancreatic carcinoma cells, hepatocytes and osteoblasts [26]. Hough *et al.* reported that Erk phosphorylation in mesenchymal cells became significant between 60 and 90 min and increased for at least 3 h after TGF-β addition [1]. In mesenchymal AKR-2B cells, Erk phosphorylation became detectable between 20 and 40 min of TGF-β treatment and remained elevated for at least 120 min [23]. Blanchette *et al.* [27] observed that TGF-β stimulation of HepG2 cells resulted in a relatively delayed (detectable at the 30 min time point), but sustained, p42/p44 MAPK phosphorylation with a maximum at the 2 h time point.

It is clear that the parameters γ, t_0 and V_0 are dependent on the cell types and extracellular stimuli used. V_0 is dependent on the concentration of the extracellular stimulus. The protein dynamics in the MAPK pathway has a similar nature over the entire range of values for V_0. As such, V_0 is set to be proportional to the initial concentration of extracellular stimuli. In this case the combination of Equation (1) and the three-level cascade model presented by Kholodenho [22] shows good agreement with the model described by Schoeberl *et al.* [23]. The stimulus-response behaviors of the two models for proteins of interest to this work have been found to be similar in simulations (data not shown).

Therefore, the combination of Equation (1) and the three-level cascade model presented by Kholodenko [22] is used to describe the MAPK pathway in this work. The nominal values of other parameters of this model can be found in the literature [22].

2.3. Interaction Mechanisms Reported in the Literature

As mentioned above, there is significant evidence of interactions between the MAPK and the TGF-β/Smad pathway. A summary of interaction mechanisms reported in the literature is provided below.

A number of studies mention that the linker region phosphorylation of Smad proteins plays an important role on how MAPKs, *i.e.*, Erk, p38, or JNK, affect the TGF-β/Smad pathway [1,2,9–15]. Reported mechanisms include activated MAPKs phosphorylating R-Smad proteins, regulating nuclear translocation or DNA binding of Smads [9], all of which either enhance or inhibit the TGF-β induced gene expression. Although the exact mechanisms whereby MAPKs regulate Smads nuclear translocation and DNA binding are still unknown, it is clear that multiple interactions between MAPK and Smad pathways might occur depending on the cell types and possibly the extent of MAPK activation. The reported mechanisms from the literature fall into the following categories:

I.   Activated Erk, p38 or JNK, phosphorylates R-Smad in linker region. This process attenuates the R-Smad concentration and inhibits R-Smad association with C-Smad in the cytoplasm [2,25].

II. Activated Erk, p38 or JNK regulates the translocation of the Smad complex from the cytoplasm to the nucleus [10].

III. Activated Erk, p38 or JNK phosphorylates R-Smad in the linker region. This phosphorylated Smad (PSmadL) translocates into the nucleus and then inhibits the dephosphorylation, which is linked to the dissociation of the nuclear Smad complex [12].

IV. The nuclear transcription factors of the MAPK pathway, c-Fos and c-Jun, bind to the nuclear Smad complex to regulate its transcription activity [9].

The first three interaction mechanisms explain the regulation of Smads nuclear translocation via proteins involved in the MAPK pathway, and the last one describes the regulation of DNA binding of Smads. Interaction IV, the Smad complex association with other transcription factors, was thought to be a possible crosstalk point by many researchers [2,8–11]. While c-Jun interacts with the Smad transcriptional co-repressor TG-interacting factor (TGIF) to suppress Smad transcriptional activity [28], inhibition of TGIF by Ras is not sufficient to significantly affect the response to TGF-β stimulation [29]. This implies that interactions other than interaction IV must also contribute to the inhibition. Because Smads translocate into the nucleus in response to TGF-β, the interactions between Smads and MAPKs may be regulated largely by the alteration of Smad intercellular localization [30]. Due to this, the focus of this work is on the regulation of Smads nuclear translocation by MAPKs, which is given by the first three interaction mechanisms listed above.

A simplified model of the crosstalk including the four potential interactions is depicted in Figure 1, followed by a survey table (Table 1) of literature studies of the possible interaction mechanisms. The column "assumed mechanisms" in Table 1 refers to the first three interactions affecting Smads nuclear translocation. Some of literature [4–6,9,23,27] mentioned evident changes in Smads nuclear translocation due to MAPKs activation, but no detailed kinetic mechanism is provided. This is represented by "*" in Table 1.

2.4. Global Sensitivity Analysis

Parameter sensitivity analysis provides a powerful tool to analyze mathematical models of biological signaling networks. The analysis can improve the understanding of the signaling networks by identifying the contribution of individual factors to the signaling response. Furthermore it can be used to recognize the identifiability of key parameters through their sensitivity vectors, which is essential for estimation of these parameters [32,33].

Local sensitivity analysis addresses model behavior near a particular point in the parameter space—a nominal point, whereas global sensitivity analysis does so over a wide range of parameter values. The Morris method is an example of a global sensitivity analysis method [34] which is more suitable for analysis signal transduction pathway models than a local method. The method consists of a repetition of local evaluations whereby the output derivative d_i (t) is calculated for each parameter p_i by adding a small change Δ_i to the parameter:

$$d_i(t) = \frac{y(t, p_1, p_2, \cdots, p_i + \Delta_i, \cdots, p_k) - y(t, p_1, p_2, \cdots, p_i, \cdots, p_k)}{\Delta_i}, i = 1, \cdots, k \qquad (2)$$

where, d_i (t) is called the elementary effect of the i-th parameter at time t, p_1, p_2, ..., p_k are the model parameters analyzed, and y is a model output.

Figure 1. Crosstalk between the MAPK pathway and TGF-β stimulated Smad pathway. I, II, III, IV represent four potential mechanisms for crosstalk.

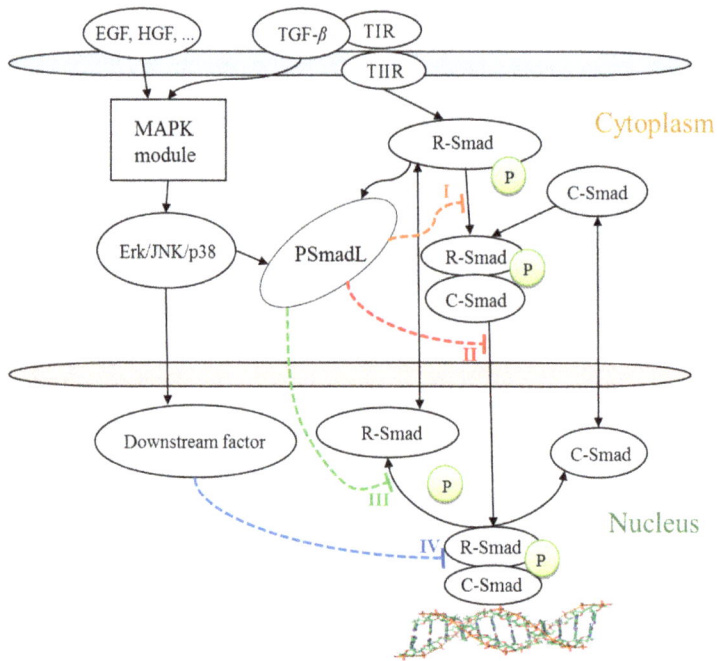

Table 1. Overview of potential interaction mechanisms reported in the literature.

Literatures	I	II	III	IV	Details	Assumed Mechanism
Hough et al. [1]			√		Experimental paper, PSmad2/3 level and gene expression were enhanced in mesenchymal cells	PSmadL stabilizes R-Smad phosphorylation
Derynck et al. [2]	√		√	√	Review paper, Transcription activity was inhibited in hyperactive Ras cells and was enhanced in normal cells	MAPK pathway changes R-Smad phosphorylation, association with Smad4 and nuclear translocation
Javelaud et al. [8]	*	*	*	√	Review paper, Smad signaling was enhanced or inhibited	PSmadL inhibits or enhances nuclear translocation
Inoki et al. [9]	*	*	*	√	Experimental paper, gene expression was enhanced in mesangial cells	MAPKs regulate the nuclear translocation
Guo et al. [10]	*	*	*	√	Review paper: Smad signaling was enhanced or inhibited	MAPKs phosphorylate R-Smad to control their intracellular redistribution

Table 1. *Cont.*

Literatures	I	II	III	IV	Details	Assumed Mechanism
Hayashida *et al.* [11]	√			√	Experimental paper, PSmad2/3 level was enhanced in human mesangial cells	MAPKs increase the R-Smad phosphorylation and its association with Smad4
Funaba *et al.* [12]		√			Experimental paper, transcription activity was enhanced in mink lung epithelial cells and COS7 cells	MAPKs increase the half-life of Smad2 and the amount of Smad complex
Burch *et al.* [13]			*		Experimental paper, transcription activity was enhanced in vascular smooth-muscle cells	PSmadL accumulates in nucleus and upregulates transcription
Kretzschmar *et al.* [14]	√				Experimental paper, nuclear PSmad2/3 level and transcription activity was inhibited in oncogenic Ras mammary and lung epithelial cells	PSmadL attenuates R-Smad nuclear accumulation
Chapnick *et al.* [19]	√			√	Review paper, Smad signaling was inhibited in cancer cells	PSmadL attenuates R-Smad phosphorylation
Zhang [25]	√			√	Review paper, Smad signaling was inhibited or enhanced	PSmadL attenuates R-Smad to inhibit its association with Smad4
Blanchette *et al.* [27]	*	*	*	√	Experimental paper, nuclear Smad2 level and gene expression were enhanced in mammalian fur cell	MAPKs increase nuclear translocation
Massagué [31]	√	*	*	√	Review paper, Smad signaling was inhibited by Erk in I, IV and enhanced by p38 and JNK in IV	MAPKs attenuates Smad accumulation in nucleus

√: Detailed kinetic mechanism is provided; * Evident changes in Smads nuclear translocation due to MAPKs activation is mentioned, but no detailed kinetic mechanism is provided.

The Morris method computes the average of the elementary effects over a number of points in parameter space and will therefore reflect an average of the sensitivity over a region of the parameter space. The mean of the elementary effects is defined as the sensitivity measure

$$S_i = \frac{1}{h}\sum_{j=1}^{h}\left|d_{ij}(t)\right|$$
(3)

where, S_i is the sensitivity measure of output y with respect to the i-th parameter. $d_{ij}(t)$ is the elementary effect of the i-th parameter at the j-th sampling point, h is the number of sampling points. The sensitivity is a function of time and sensitivities at different time points are usually concatenated into a sensitivity vector. The sensitivity can be normalized to ensure that the use of different units does not affect the sensitivity analysis results.

2.5. Measures for Evaluating Output Signals

Three key characteristics are used to evaluate output signals, *i.e.*, the transcription factor concentrations, of the pathways. These three characteristic parameters are the signal time τ, signal duration ϑ and signal amplitude I, and are calculated numerically from the output signal $y(t)$ [20].

$$\tau = \frac{\int_0^\infty ty(t)}{\int_0^\infty y(t)}$$
(4)

$$\vartheta = \sqrt{\frac{\int_0^\infty t^2 y(t)}{\int_0^\infty y(t)}} - \tau \qquad (5)$$

$$I = \frac{\sqrt{\int_0^\infty y(t)}}{2\vartheta} \qquad (6)$$

τ and ϑ represent the average time to activate the output element and the average time during which this output component is activated, respectively [20]. I is defined as the ratio of total amount of the output signal and duration ϑ of the signal, providing a measurement of the average concentration of the output element [20]. These parameters are difficult to interpret if the output signals do not return to zero, as τ and ϑ will grow over time in this case. In order to deal with this situation, another measure, $\tau_{0.9}$, is defined which evaluates how fast a system responds. $\tau_{0.9}$ is defined as the time at which 90% of maximal output signal is reached [21]. Figure 2 shows the geometric interpretation of the three measures used [20].

Figure 2. Geometric representation of signal time, signal duration and signal amplitude.

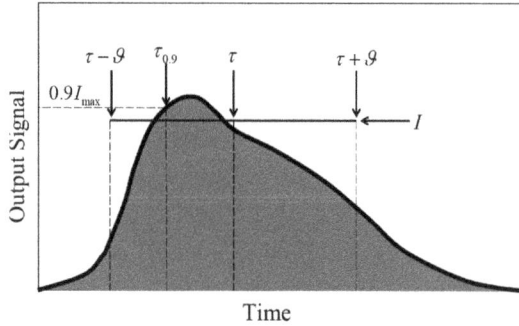

3. Model Development

As mentioned above, the individual models of the TGF-β/Smad pathway and the MAPK pathway have been described in detail in the literature [6,22]. In addition, several potential interactions have been proposed which can be responsible for crosstalk between these two pathways. However, the crosstalk has not been modeled in published studies. Furthermore, most of the studies only investigated one specific type of crosstalk, usually the one dominant in a particular cell type, and the underlying interaction kinetics was not determined. This paper seeks to address these points by developing a model of the pathways, including different potential crosstalk mechanisms, and thoroughly analyzing the effect that these different mechanisms have on the observed response.

3.1. Potential Crosstalk Mechanisms

Three potential interaction mechanisms have been reported in the literature for the regulation of Smads nuclear translocation via proteins involved in the MAPK pathway. As shown in Table 1, the first interaction (point I in Figure 1) is supported by most of studies, whereas there are fewer references mentioning the second potential interaction (point II in Figure 1). Some references discussing the second interaction implied that MAPKs regulation of the import of cytoplasmic Smad complex to the nucleus functions by attenuating R-Smad, which results in the same kinetics as the first interaction. Similarly, there are fewer studies investigating the interaction at point III. A few experimental studies [1,2] have concluded that activated MAPKs increased stability of C-terminal phosphorylated R-Smad and the amount of the nuclear Smad complex. However, the underlying interaction kinetics have not been disclosed.

Data suggest that the linker region phosphorylation of Smad2 (PSmadL) by the MEK1-Erk pathway does not directly affect Smad2 affinity for Smad4 [12]. It is assumed that PSmadL attenuates R-Smad to inhibit its association with C-Smad which causes the negative regulation of TGF-β/Smad signaling by the MAPK pathway [19]. However, this argument cannot explain the opposite effect of the MAPK pathway on TGF-β/Smad signaling reported in several sources in the literature [1,9,11–13,27]. As previously mentioned, some studies suggest that the levels of Ras activity could dictate the outcome of the cross talk [18,19]. Responses to TGF-β may require the counterbalancing effect of Ras signaling to achieve a suitable level of nuclear Smad activity. Based on this, a possible kinetics for the positive regulation of TGF-β/Smad signaling by MAPK pathway is proposed in this paper: nuclear PSmadL inhibits the dephosphorylation (or dissociation) of the nuclear Smad complex via attenuating the nuclear phosphatase, which enhances the Smad-mediated transcription activity.

It is suggested that the increase of the nuclear Smad complex, which upregulates the gene expression, is closely associated with nuclear PSmadL. Experiments by Hough *et al.* [1] and Funaba *et al.* [12] showed that most of PSmadL was found in the nucleus of cell fibroblasts. These data do not preclude the possibility of cytoplasmic phosphorylation of Smad2 in the linker region, followed by a rapid nuclear translocation [13,35]. Burch *et al.* [13] also discussed that nuclear PSmadL upregulates the gene activity in vascular smooth-muscle cells. On the other hand, it is well known that phosphatase inhibition is the most effective way that one signaling pathway can influence another [20]. Hough *et al.* [1] suggested that the negative control of the concentration of the nuclear Smad complex is due to the dephosphorylation of the nuclear Smad complex, which implies that the nuclear phosphatase has a strong negative control on the nuclear Smad complex level. Thus, phosphorylation of Smad in the linker region directly or indirectly attenuates the nuclear phosphatase and offsets its negative control on the nuclear Smad complex level. The attenuated phosphatase decreases the dephosphorylation rate of the nuclear Smad complex and thereby affects the its concentration of the nuclear Smad complex [12].

The above discussion supports the hypothesis that nuclear PSmadL, initiated by activated MAPKs, inhibits the dephosphorylation of the nuclear Smad complex, which enhances the Smad-mediated transcription activity. Therefore, the third interaction (point III) may explain the positive regulation while the first one (point I) accounts for the negative regulation. Points I and III are assumed as the possible crosstalk points of the MAPK and TGF-β/Smad pathways.

3.2. Modeling Potential Crosstalk Mechanisms

It is generally acknowledged that PSmadL is a key component for integrating MAPK signals with the TGF-β/Smad pathway [1,2,9–15]. However, the exact kinetics of the Smad linker region phosphorylation by MAPKs is poorly understood. The time course of TGF-β-mediated Smad2 linker region phosphorylation, as well as Erk phosphorylation, have been reported in Western blots in mesenchymal cells [1]. These data indicated that the time course of Smad2 linker region phosphorylation has a similar nature as the one for Erk phosphorylation. A linear relationship is often used to link PSmadL and MAPKs [21] from Figure 1. A similar assumption was made for the relationship of activated receptor and proteins involved in the interaction when modeling the crosstalk between the MAPK pathway and other pathways [20].

The potential crosstalk discussed in Section 3.1 is modeled through incorporating PSmadL into the original model of TGF-β/Smad pathway. At point I, shown in Figure 1, PSmadL can attenuate R-Smad and consequently inhibit the R-Smad association with C-Smad in the cytoplasm, which consequently decreases the concentration of the nuclear Smad complex. This interaction function can be incorporated by rewriting the reaction rate describing the formation of the Smad complex:

$$r_{\text{form}} = \frac{k_{\text{Smad-complex}}}{1 + \left(\dfrac{[PSmadL]}{K_I}\right)^m} [Smad2]_{\text{cyt}} [Smad4]_{\text{cyt}} [LRC_{\text{EE}}] \tag{7}$$

where, $[PSmadL]$ represents the concentration of interaction component PSmadL and is directly proportional to the concentration of MAPKs. K_I describes the interaction strength, *i.e.*, the smaller the value of K_I, the stronger the interaction will be. m is an adjustable parameter to control inhibition strength, and is set to $m = 1$ in this work. $k_{smads\text{-}complex}$ is the original reaction rate constant, equal to 6.85×10^{-5} nM^2·min^{-1} [6].

At point III shown in Figure 1, PSmadL may translocate into the nucleus and inhibit the dissociation of the nuclear Smad complex, which increases the concentration of the nuclear Smad complex. This behavior can be modeled by revising the reaction rate of the dephosphorylation of the nuclear Smad complex:

$$r_{\text{diss}} = \frac{k_{\text{diss}}^{\text{Smad-complex}}}{1 + \left(\dfrac{[PSmadL]}{K_{III}}\right)^m} [Smad\_complex]_{\text{nuc}} \tag{8}$$

where K_{III} is defined in the same way as K_I, and $m = 1$. $k_{\text{diss}}^{\text{Smad-complex}}$ is the original dephosphorylation rate, equal to 0.1174 min^{-1} [6].

3.3. Integrating the Individual Components of the Model

As described as Section 3.2, the interactions at point I and III account for two opposite effect of MAPKs on the TGF-β/Smad pathway. This section integrated these two competing effects into the original models which results in a model representing the MAPK and TGF-β/Smad pathways as well as their interactions. A schematic of the model is shown in Figure 3. "R", "MAPK cascade" and "r" in the blue-shaded frames represent the modules of Ras/MKKKKs receptor activation by extracellular stimuli, three-level serial phosphorylation of MAPK cascade and the linker region phosphorylation of R-Smad by activated MAPKs, respectively. The two interactions of the MAPK and the TGF-β/Smad signaling pathways, *i.e.*, inhibition and enhancement, are also shown. It can be hypothesized that the two opposite effects define specific cellular responses to TGF-β. A detailed description of the equations of this model can be found in Appendix I.

Figure 3. Representation of crosstalk between MAPK and Smad signaling pathways.

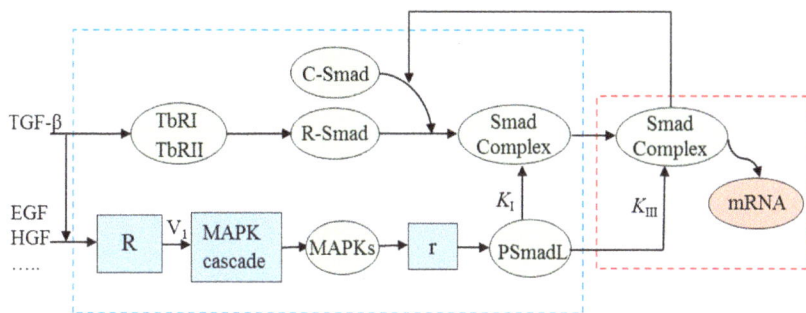

3.4. Simulation Results

3.4.1. Comparisons between Crosstalk Model and Original Model

The developed model is used in simulations in this section and the results are compared with the simulated data presented by Zi *et al.* [6]. A continuous TGF-β stimulation (2 ng/mL) is used for the simulations. MAPKs activation may be induced by any of the extracellular factors such as TGF-β, EGF, HGF and others as the model does not explicitly account for these extracellular factors. V_0 is set to 0.18 which suggests a low level of MAPKs activation. t_0 is set to zero which indicates MAPKs activation without time delay and γ is set to 0.001 which represents a transient MAPKs activation. The nuclear and cytoplasmic PSmad2 levels, calculated from the crosstalk model, are compared with the simulated data from the original TGF-β/Smad model [6]. Figure 4 shows the simulated data for the crosstalk model and original model presented by Zi *et al.* for the case where only inhibition occurs ($K_I = 1$, $K_{III} \rightarrow \infty$). Similarly Figure 5 shows the simulated data for the two models for the case where only enhancement is taking place ($K_I \rightarrow \infty$, $K_{III} = 1$). As described in [6], the shape of the profiles of nuclear and cytoplasmic PSmad2 level are very similar. For the case where only inhibition is considered, PSmanL is assumed to attenuate the Smad2 level and further inhibit the formation of the Smad complex in the cytoplasm. It will cause a decrease of the PSmad2 level in the cytoplasm as well as in the nucleus, which is consistent with the simulated data shown in Figure 4a,b. Conversely, for the case where only upregulation is considered, PSmadL is assumed to inhibit the dissociation of the nuclear Smad complex. Thus, the nuclear PSmad2 level will increase, which agrees well with the data shown in Figure 5a. This effect will cause a distinct decrease of Smad2 level in the cytoplasm as well as the nucleus (data not shown) and consequently leads to a slight decrease of the cytoplasmic PSmad2 level which can be seen in Figure 5b. Simulated data from the two models are consistent with the theoretical analysis expected from the interactions. Furthermore, simulated data of the Smad2 intercellular locations for the crosstalk model also show that TGF-β stimulation induced a change in the overall Smad2 distribution from predominantly in the cytoplasm to being mainly in the nucleus as shown in Figure 4c and Figure 5c, which is in agreement with the literature [6]. Simulation results using a pulse stimulation of TGF-β (2 ng/mL) for the first 30 min are also consistent with theoretical analysis and literature reports (data not shown).

Figure 4. Comparison of simulated data obtained from the crosstalk model and from the original model presented by Zi *et al.* [3] when only the inhibition is present for (**a**) nuclear PSmad2 level; (**b**) cytoplasmic PSmad2 level; and (**c**) Smad2 intercellular location. Note that the PSmad2 level is a relative value. Its unit is nmole when the extracellular medium volume is set to 1 Liter [6].

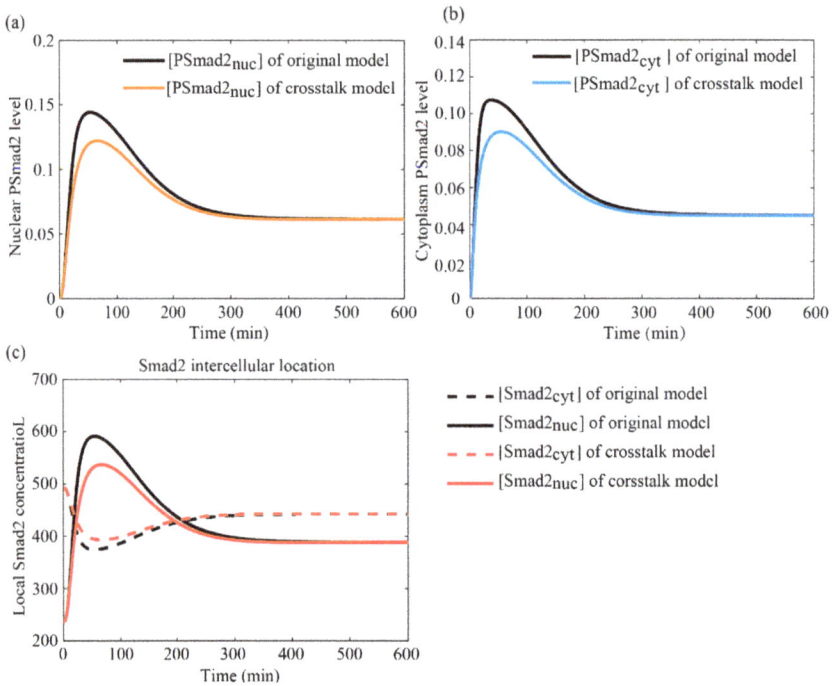

Figure 5. Comparison of simulated data obtained from the crosstalk model and from the original model presented by Zi *et al.* [3] for the case where only enhancement is present: (**a**) nuclear PSmad2 level; (**b**) cytoplasmic PSmad2 level; and (**c**) Smad2 intercellular location. Note that the PSmad2 level uses the same unit as described in the caption of Figure 4.

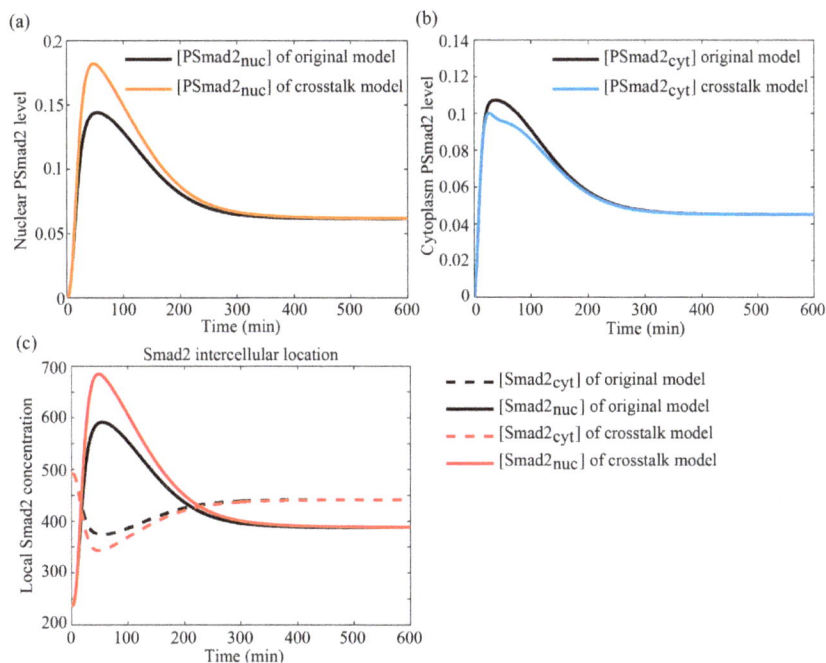

3.4.2. Comparison with Literature Data

As shown in Table 1, many studies suggest that the crosstalk between the MAPK and the TGF-β/Smad pathways either increase or decrease the nuclear PSmad2 protein level. However, the exact amounts of the increase or decrease are usually not mentioned as the data are often qualitative or semi-quantitative rather than quantitative in nature. Two sets of experimental data are used for comparison purposes here, specifically data presented by Hough *et al.* [1] and Blanchette *et al.* [27]. The decay rate of C-terminal PSmad2 in mesenchymal cells were determined experimentally by Hough *et al.* [1]. The time to decrease the signal intensity by 50% differed between treatment groups: 116 min for TGF-β + U0126 *vs.* 135 min for TGF-β alone. The intercellular location of Smad2 was observed experimentally in the human liver cell line (HepG2) [27]. With TGF-β alone 49.6% ± 3.1% of Smad2 exhibited a predominantly nuclear Smad2-specific staining compared with 10.1% ± 1.4% for unstimulated cells and 19.0% ± 0.9% for TGF-β + PD98059 (PD98059 is a MEK1-specific inhibitor). While these two sets of data cannot be used to estimate model parameters, some qualitative comparison with the simulation data can be performed.

In the experiments by Hough *et al.* [1], the cell lysates were pulsed for 10 min with TGF-β (2 ng/mL) with or without U0126 (inhibitor of Erk phosphorylation). Relative signal intensity of the *C*-terminal PSmad2 with TGF-β alone and TFG-β + U0126 was shown in Figure 5d of [1] by setting the 60 min. value at 100%. The time to decrease the signal intensity by 50% was given (116 min for TGF-β + U0126 *vs.* 135 min for TGF-β alone). Here, the decay rate of *C*-terminal PSmad2 is calculated based on the crosstalk model and assumed parameter values. Due to the time delay of Erk activation described by this experiment, t_0 is set to 30 min. The concentration of PSmadL is set to zero to mimic addition of U0126 from the experiment, and at the 10 min. time point the concentration of TGF-β is changed to zero in the simulation. For the case of [*PSmadL*] = 0, the crosstalk does not occur and the scenario reduced to the one for TGF-β induced Smad signaling without crosstalk, whereas, in the case of stimulation with TGF-β alone, the crosstalk is induced by TGF-β itself. Figure 6a shows the concentration profiles of *C*-terminal PSmad2 with crosstalk and without crosstalk. The curves indicate that the TGF-β-induced Erk activation (in case of TGF-β alone) does stabilize the PSmad2 level. The decay profiles of the signal intensity are shown in Figure 6b, whose general trend is consistent with those shown in Figure 5d of [1]. The decay time increased by 16% when no U0126 was added for the simulation while the increase in time was also 16% for the experimental data reported by Hough *et al.* [1].

Figure 6. Comparison of simulation results with and without crosstalk from the crosstalk model for (**a**) nuclear PSmad2 level; and (**b**) relative signal intensity of nuclear PSmad2.

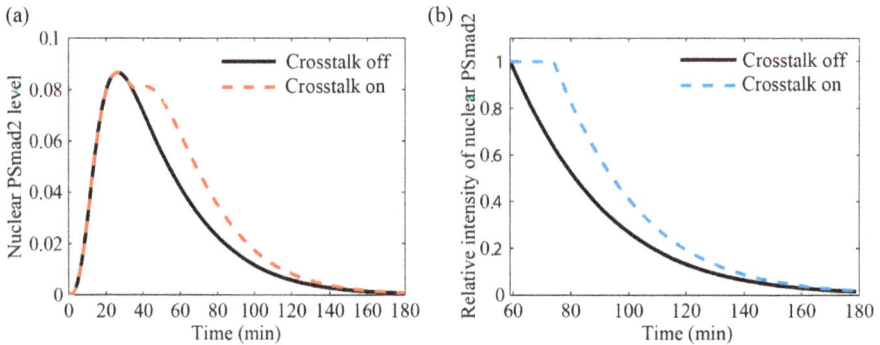

Next, the intercellular location of Smad2 is calculated based on the presented crosstalk model and compared with the experimental results presented by Blanchette *et al.* [27]. The experimental data [27] indicated that phosphorylation of p42/p44 MAPK by TGF-β was prolonged at least 30 min and sustained for at least 2 h. Thus, t_0 is set to 30 min and $\gamma = 0.0005$. Zi's experimental data showed initially 14% of Smad2 was in the nucleus in the unstimulated mesenchymal cells. When cells are stimulated by TGF-β (10 ng/mL) and PSmadL concentration is set to zero, which mimics the addition of PD98059 in an experiment, the ratio of nuclear Smad2 to total amount of Smad2 becomes 35% at the 1 h time point. When cells are treated by TGF-β (10 ng/mL) alone, the amount of nuclear PSmad2 increased 2.5-fold in simulations as compared to the case where stimulation with TGF-*β* occurred and [*PSmadL*] was set to zero. This value is consistent with the experimental values observed by Blanchette *et al.* [27] for experiments where cells were stimulated with TGF-β and, separately, with TGF-β and PD98059.

4. Results and Discussion

4.1. Sensitivity Analysis of Key Parameters

The presented model integrates the three-level cascade model of the MAPK pathway with the original model of TGF-β/Smad pathway and also includes two crosstalk mechanisms. The final output, the transcription factor, will mediate the TGF-β-induced gene expression.

The parameter set of this proposed model consists of three parts: the parameters of the TGF-β/Smad pathway model, the parameters of the three-level cascade model describing the MAPK pathway and the parameters involved in the crosstalk. The nominal parameter values found in the literature for the TGF-β/Smad pathway and the MAPK cascade models have been used in this work. The normalized sensitivity values of parameters over a period of time are calculated from the crosstalk model by the Morris method. For the parameters of the TGF-β/Smad pathway model, the uncertainty ranges are set to 50%–150% of the nominal values. The uncertainty of the parameters K_I, K_{III}, V_0 and γ involved in the crosstalk is assumed to be much larger as no values can be found in the literature. Their nominal values are assumed to be 5, 5, 0.18, 0.004, and their uncertainty ranges are defined as [0.1, 10], [0.1, 10], [0.14, 0.22], [0.001, 0.01], respectively. The value ranges of V_0 and γ were chosen to represent a transient, low level activation of the MAPK pathway. The reason for choosing the specific parameters mentioned here for the sensitivity analysis is that the interaction parameters, K_I and K_{III}, are the key parameters to determine whether the crosstalk inhibits or enhances the Smad transcription activity. Additionally, parameters describing receptor activation of the MAPK pathway by extracellular factors, e.g., the parameters V_0, γ and t_0, may affect the crosstalk. The delay time t_0 usually can be observed directly through experiments. However, there are four unknown key parameters whose effects on the output need to be investigated: K_I, K_{III}, V_0 and γ.

Here, the time-dependent sensitivity dynamics of the transcription factor concentration with respect to the parameters K_I, K_{III}, V_0 and γ are calculated and analyzed using the Morris method. A constant exposure of 2 ng/mL of TGF-β is chosen for the analysis as well as parameters values corresponding to a transient and low level activation of MAPKs.

The normalized sensitivity profiles with respect to the four parameters are depicted in Figure 7. V_0 has the biggest amplitude of the sensitivity but its sensitivity declines fast, whereas γ has a more significant long-term effect than V_0. When K_{III} is perturbed, it has a considerable impact on the amplitude and the duration of the output signal. The effect of K_I is mainly observed over the initial time interval. According to the hypothesized interaction mechanisms, MAPKs, once activated, may quickly attenuate R-Smad to inhibit the formation of the Smad complex in cytoplasm at the initial stage of interaction; this interaction strength is controlled by K_I. On the other hand, most of PSmadL remains in the nucleus and inhibits the dissociation of the Smad complex over a longer time; this interaction strength is controlled by K_{III}. The simulation results are in good agreement with the theoretical analysis.

Figure 7. Sensitivity profiles of the transcription factor concentration, *i.e.*, the nuclear PSmad2 level, with respect to K_I, K_{III}, V_0, and γ.

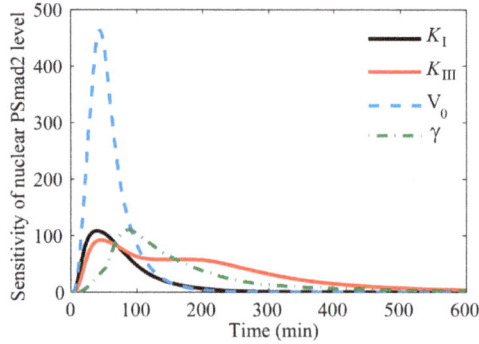

4.2. Balance between Two Competing Effects Defines the Specific Cell Responses

Based on the previous discussion, the level of the nuclear transcription factor is not only Smad-dependent, but also influenced by activated MAPKs. The effects of activated MAPKs on the Smad pathway show two opposite responses: inhibition or enhancement of the nuclear transcription factor concentration. Furthermore, it is hypothesized that the two opposite effects define specific cellular responses to TGF-β, that is to say, the balance between the two opposing effects defines specific cellular responses to TGF-β regulated by activity of the MAPK pathway.

In the literature, inhibition was observed mostly in hyperactive Ras signaling, oncogenic Ras signaling or mutant tumor cells [14–17], whereas enhancement was observed mostly in normal cells or in case of relatively low level of extracellular stimuli [1,9,11–13,27]. Massagué *et al.* hypothesized that responses to TGF-β may require balancing effects of Ras signaling to achieve a suitable level of nuclear Smad activity [18]. The previously proposed hypothesis that the balance between two opposite effects defines specific cellular responses not only agrees well with this viewpoint, but also highlights the underlying kinetics: rapid activation of Erk, in the case of Erk, induced by EGF or oncogenic Ras signaling, coupled with a high level of Ras activation, may quickly attenuate R-Smad to inhibit the formation of the Smad complex in the cytoplasm before R-Smad activation and nuclear translocation, which predominantly inhibits the Smad transcription activity. Conversely, the delayed activation of Erk induced by TGF-β or other extracellular factors in some cells, coupled with a relatively low level of Ras activation, will be negligible for the R-Smad activation and nuclear translocation. Most of PSmadL will transfer into the nucleus and inhibit the dissociation of the nuclear Smad complex, which predominantly enhances the Smad transcription activity. This hypothesized kinetics is in good agreement with the literature reports. Additionally, the simulated data from the crosstalk model, as shown in Figures 4–7, seems to align with this hypothesis.

4.3. Effects of Interaction Parameters on Cell Responses

The values of the interaction parameters K_I and K_{III} are responsible for either inhibition or enhancement of MAPKs on TGF-β/Smad signaling. This section discusses whether specific interaction outcome, either inhibition or enhancement, can be distinguished using experimental data.

4.3.1. Interaction Parameters Are Identifiable and have Distinct Effects on the Outputs

As far as a group of parameters to estimate are concerned, it is very possible that they are pairwise indistinguishable. If the effects of two parameters on the output are highly correlated then their sensitivity profiles have similar shapes. If this is the case then the parameters cannot be simultaneously estimated. The sensitivity profiles of the nuclear PSmad2 level with respect to K_I and K_{III} are shown in Figure 8a. In addition, the relative levels of the transcription factor at different values of each parameter (while the other parameters are held constant) are shown in Figure 8b,c. It is clear from Figure 8a that the sensitivity profile for K_I is distinctly different from the sensitivity profile for K_{III}. It can be concluded that K_I and K_{III} are estimable and each one has a distinct effect on the output.

Figure 8. Sensitivity analysis of nuclear PSmad2 level. (**a**) Sensitivity profiles with respect to K_I and K_{III}; (**b**) the effect of K_I on nuclear PSmad2 level; (**c**) The effect of K_{III} on nuclear PSmad2 level.

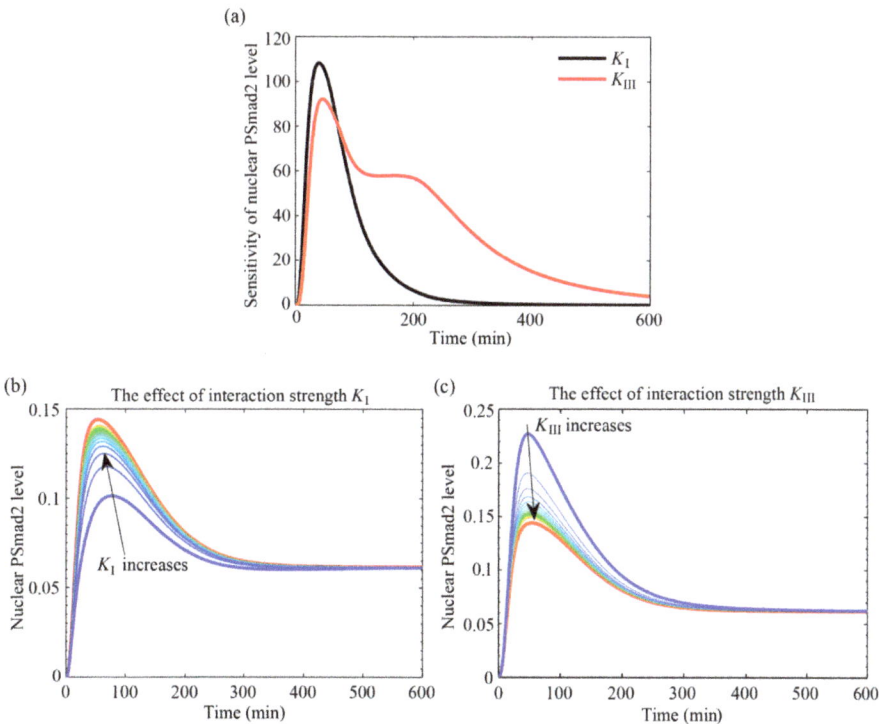

4.3.2. MAPK Pathway Extends the TGF-β/Smad Signal Transduction Duration, Which Is More Strongly Affected by Values of K_{III} than K_I

To provide further insight into the effects of the MAPK pathway on Smad signaling, signal time, duration and amplitude at different values for each interaction parameter K_I or K_{III} are calculated to quantify the effect of the crosstalk on the output signal. As previously described, signal time, duration, and amplitude will grow over time to infinity if the output signals do not return to zero. Therefore, a pulse stimulation of TGF-β (2 ng/mL) for the first 60 min is used for the following analysis. Furthermore, a transient and low-level activation of MAPKs is assumed (V_0 = 0.18, γ = 0.001) without time delay. Then the input signal, $i.e.$, the TGF-β concentration, is set to zero after 60 min.

Figure 9 shows the signal time profiles (Figure 9a), signal duration profiles (Figure 9b) for different values of the parameters K_I and K_{III}. The signal time and duration decrease as K_I or K_{III} increase and converge to basal values equal to the signal time and duration of the original TGF-β/Smad pathway model. The signal time for K_{III} is significantly larger than that for K_I as shown in Figure 9a. According to the proposed hypothesis, the inhibition effect controlled by K_I arises from PSmadL activation in the cytoplasm, whereas the enhancement effect controlled by K_{III} only becomes significant after PSmadL transfers from the cytoplasm to the nucleus. This implies that it takes longer for the enhancement interaction involving K_{III} to have a noticeable effect than for the inhibition interaction involving K_I. The results shown in Figure 9a are in good agreement with this hypothesis. On the other hand, the smaller the values of K_I and K_{III}, the larger the signal time and signal duration. That is to say, the stronger the crosstalk is, the longer it takes for the transcription factor to be activated but the activation also lasts longer. It can be concluded that the crosstalk between the pathways extends the duration of the TGF-β/Smad signal transduction, which is consistent with the literature reports [1,12]. Additionally, K_{III} has a more profound effect on the signal time and signal duration than K_I as can be seen in Figure 9a,b. A significant body of the literature has reported that positive regulation of TGF-β/Smad signaling by the MAPK pathway enhances the duration of the increased levels of nuclear phosphorylated Smad2 [1,12], whereas there are few reports mentioning down regulation. It is possibly that this is due to the fact that the signal duration of PSmad2 level is predominantly controlled by K_I and changes little under negative regulation.

4.3.3. Specific Outcomes Can Be Described through Appropriate Choices for K_I and K_{III}

The specific outcome, if signaling is enhanced or inhibited, depends upon the values of K_I and K_{III}. As such, it is important to investigate what combinations of values of these parameters have an enhancing/inhibiting effect. The signal amplitude, representing the average concentrations of the nuclear Smad complex, can be used as a measure to determine the magnitude of the effect. This signal amplitude is computed for a number of values for combinations of K_I and K_{III}. Small values of K_I, $i.e.$, stronger interactions, result in low average concentrations of the transcription factor, which illustrates the inhibition of TGF-β/Smad signaling. Conversely, small values of K_{III}, $i.e.$, strong interactions, result in large average concentrations of transcription factor, illustrating enhancement of TGF-β/Smad signaling. The variation amount of the signal amplitude I around its basal value is defined to quantify the effect when both inhibition, via K_I, and enhancement, via K_{III}, take place:

$$Var_I = I(basal) - I_I(K_I) \tag{10}$$

$$Var_{III} = I_{III}(K_{III}) - I(basal) \tag{11}$$

Var_I and Var_{III} are the variation amounts of the signal amplitudes I_I (K_I) and I_{III} (K_{III}) around their basal value I (basal). If K_I and K_{III} tend to infinity, the output signal tends to the basal value which represents the original pathway without crosstalk. I (basal) is equal to the output signal amplitude of the original TGF-β/Smad pathway.

Figure 9. Signal time (**a**) and duration; (**b**) for different values of K_I and K_{III}.

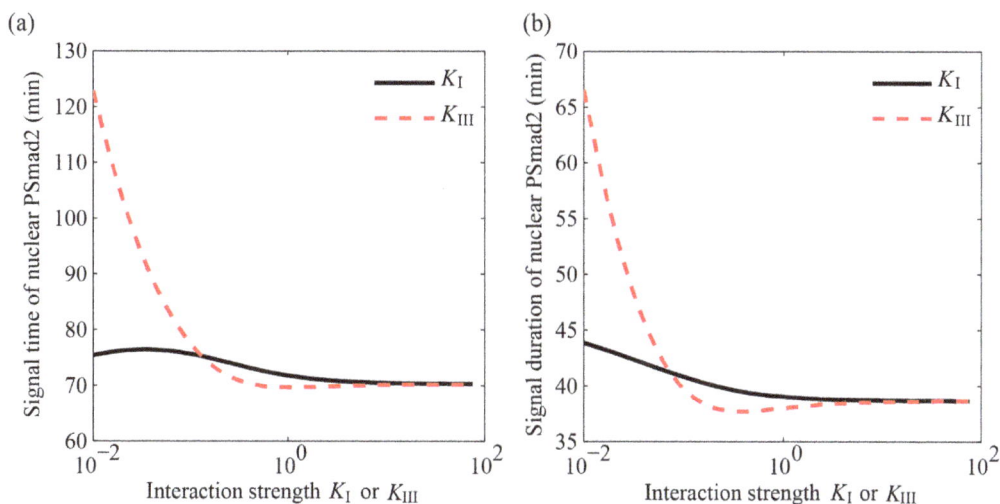

The interaction outcome is characterized by the values of Var_I and Var_{III}: (1) If $Var_I = Var_{III}$, then the two interaction will cancel each other; (2) If $Var_I > Var_{III}$, then inhibition is stronger than enhancement; (3) If $Var_I < Var_{III}$, then enhancement dominates inhibition. A plot of the surface of the difference between Var_I and Var_{III} as a function of K_I and K_{III} is shown in Figure 10b. The grey plane is added to divide the surface into two areas: overall inhibition and enhancement. The part above the plane indicates that enhancement is stronger than inhibition while the area below the plane represents the opposite effect.

Figure 10. Signal amplitude analysis for different values of K_I and K_{III}: (**a**) signal amplitude profiles; (**b**) surface plot of the difference between Var_I and Var_{III} as a function of K_I and K_{III}.

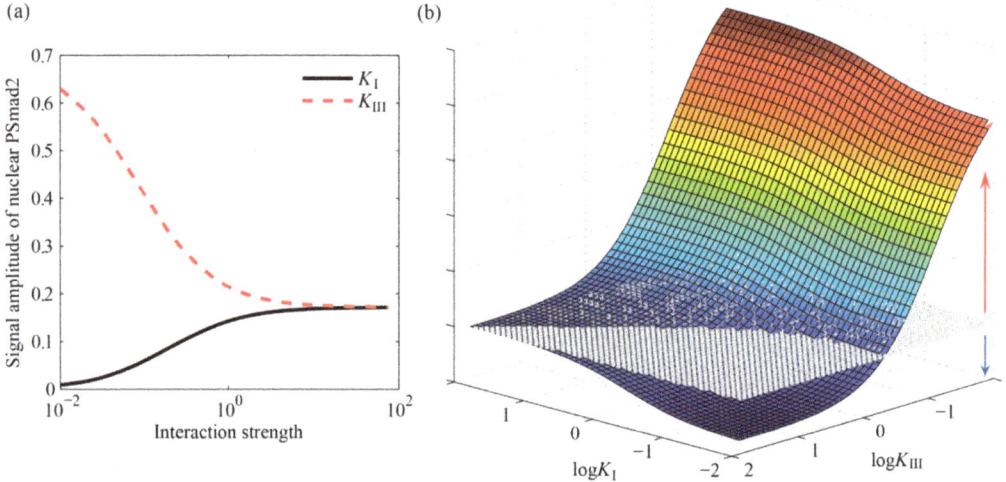

5. Conclusions

Crosstalk plays a key role for TGF-β signal transduction. Broad evidence exists for the crosstalk of Smad signaling with MAPK pathways. A variety of studies have identified different potential mechanisms for the crosstalk, however, there is no clear consensus on the actual mechanism(s) responsible for the crosstalk.

A model of the pathways, including different potential crosstalk mechanisms, is developed through integrating a model of MAPK pathway [22] with the TGF-β/Smad pathway model [6]. The effects that different interaction mechanisms have on the observed response are thoroughly analyzed and discussed. It is hypothesized that two potential interaction mechanisms account for the crosstalk between MAPK and GF-β/Smad signaling pathways: (1) PSmadL attenuates R-Smad to inhibit its association with C-Smad in the cytoplasm; (2) Nuclear PSmadL inhibits the dephosphorylation and thereby the dissociation of the nuclear Smad complex. The two mechanisms explain the inhibition and enhancement effects of MAPKs on Smad signaling. It is hypothesized that the balance between these two competing effects defines specific cellular responses to TGF-β, depending on the cell types and the extent of MAPKs activation. This hypothesis agrees with results from the literature reports [1,9,11–18,27].

The sensitivity of the transcription factor with respect to interaction parameters is calculated for the presented model. It can be concluded that interaction strengths K_I and K_{III} are estimable and each one has a distinct effect on the output. Signal time, duration and amplitude at different combinations of values of the interaction parameters K_I or K_{III} are calculated to quantify the effect of the crosstalk on the output signal. The results show that the MAPK pathway extends the TGF-β/Smad signal transduction duration, which is more strongly affected by the values of K_{III} than K_I. This conclusion agrees with the experimental reports from the literature [1,12]. Once the interaction strengths K_I and K_{III} have been estimated, the specific interaction outcome, inhibition or enhancement, can be extracted from an analysis of the signal amplitudes.

Acknowledgments

The contribution is partially supported by Natural Science Foundation of China (grant No. 61304071) through a fellowship for Ji Liu.

Author Contributions

Ji Liu performed the literature review, developed the model, and contributed to the analysis; Wei Dai focused on parameter sensitivity analysis and the analysis and discussion; Juergen Hahn oversaw all aspects of the research.

Appendix

The detailed equations of the developed model are listed here. The first eight equations explain the three-level MAPK cascade with the input $V_0 \exp(-\gamma (t - t_0))$. The remaining sixteen equations account for TGF-β/Smad signaling mediated by MAPKs (i.e., $MPAK_{pp}$).

$$\frac{d[MKKK]}{dt} = \frac{V_2[MKKK_P]}{K_2 + [MKKK_P]} - \frac{V_0 \exp(-\gamma(t-t_0))[MKKK]}{\left(K_1 + [MKKK]\right)\left(1 + \left(\frac{[MAPK_{PP}]}{K_I}\right)^n\right)} \tag{A1}$$

$$\frac{d[MKKK_P]}{dt} = \frac{V_0 \exp(-\gamma(t-t_0))[MKKK]}{\left(K_1 + [MKKK]\right)\left(1 + \left(\frac{[MAPK_{PP}]}{K_I}\right)^n\right)} - \frac{V_2[MKKK_P]}{K_2 + [MKKK_P]} \tag{A2}$$

$$\frac{d[MKK]}{dt} = \frac{V_6[MKK_P]}{K_6 + [MKK_P]} - \frac{k_3[MKKK_P][MKK]}{K_3 + [MKK]} \tag{A3}$$

$$\frac{d[MKK_P]}{dt} = \frac{k_3[MKKK_P][MKK]}{K_3 + [MKK]} + \frac{V_5[MKK_{PP}]}{K_5 + [MKK_{PP}]} - \frac{k_4[MKKK_P][MKK_P]}{K_4 + [MKK_P]} - \frac{V_6[MKK_P]}{K_6 + [MKK_P]} \tag{A4}$$

$$\frac{d[MKK_{PP}]}{dt} = \frac{k_4[MKKK_P][MKK_P]}{K_4 + [MKK_P]} - \frac{V_5[MKK_{PP}]}{K_5 + [MKK_{PP}]} \tag{A5}$$

$$\frac{d[MAPK]}{dt} = \frac{V_0 \exp(-\gamma(t-t_0))[MAPK_P]}{K_{10}+[MAPK_P]} - \frac{k_7[MKK_{PP}][MAPK]}{K_7+[MAPK]} \tag{A6}$$

$$\frac{d[MAPK_P]}{dt} = \frac{k_7[MKK_{PP}][MAPK]}{K_7+[MAPK]} + \frac{V_9[MAPK_{PP}]}{K_9+[MAPK_{PP}]} -$$
$$\frac{k_8[MKK_{PP}][MAPK_P]}{K_8+[MAPK_P]} - \frac{V_0 \exp(-\gamma(t-t_0))[MAPK_P]}{K_{10}+[MAPK_P]} \tag{A7}$$

$$\frac{d[MAPK_{PP}]}{dt} = \frac{k_8[MKK_{PP}][MAPK_P]}{K_8+[MAPK_P]} - \frac{V_9[MAPK_{PP}]}{K_9+[MAPK_{PP}]} \tag{A8}$$

$$[PSmadL] = r[MAPK_{PP}] \tag{A9}$$

$$\frac{d[T1R_{surf}]}{dt} = v_{T1R} - ki_{cave}[T1R_{surf}] + kr_{cave}[T1R_{cave}] - ki_{EE}[T1R_{surf}] + kr_{EE}[T1R_{EE}]$$
$$- k_{LRC}[TGF-\beta][T1R_{surf}][T2R_{surf}] + kr_{cave}[LRC_{cave}] + kr_{EE}[LRC_{EE}] \tag{A10}$$

$$\frac{d[T1R_{cave}]}{dt} = ki_{cave}[T1R_{surf}] - kr_{cave}[T1R_{cave}] \tag{A11}$$

$$\frac{d[T1R_{EE}]}{dt} = ki_{EE}[T1R_{surf}] - kr_{EE}[T1R_{EE}] - k_{deg}^{T1R}[T1R_{EE}] \tag{A12}$$

$$\frac{d[T2R_{surf}]}{dt} = v_{T2R} - ki_{cave}[T2R_{surf}] + kr_{cave}[T2R_{cave}] - ki_{EE}[T2R_{surf}] + kr_{EE}[T2R_{EE}]$$
$$- k_{LRC}[TGF-\beta][T1R_{surf}][T2R_{surf}] + kr_{cave}[LRC_{cave}] + kr_{EE}[LRC_{EE}] \tag{A13}$$

$$\frac{d[T2R_{cave}]}{dt} = ki_{cave}[T2R_{surf}] - kr_{cave}[T2R_{cave}] \tag{A14}$$

$$\frac{d[T2R_{EE}]}{dt} = ki_{EE}[T2R_{surf}] - kr_{EE}[T2R_{EE}] - k_{deg}^{T2R}[T2R_{EE}] \tag{A15}$$

$$\frac{d[LRC_{surf}]}{dt} = k_{LRC}[TGF-\beta][T1R_{surf}][T2R_{surf}] - ki_{cave}[LRC_{surf}] - ki_{EE}[LRC_{surf}] \tag{A16}$$

$$\frac{d[LRC_{cave}]}{dt} = ki_{cave}[LRC_{surf}] - kr_{cave}[LRC_{cave}] - k_{lid}[LRC_{cave}][Smads\_complex_n] \tag{A17}$$

$$\frac{d[LRC_{EE}]}{dt} = ki_{EE}[LRC_{surf}] - kr_{EE}[LRC_{EE}] - k_{cd}[LRC_{EE}] \tag{A18}$$

$$\frac{d[Smad2_c]}{dt} = \frac{V_n}{V_c}k_{exp}^{Smad2}[Smad2_n] - k_{imp}^{Smad2}[Smad2_c] -$$
$$k_{Smads\_complex}[Smad2_c][Smad4_c][LRC_{EE}]\frac{K_I}{K_I+[PSmadL]} \tag{A19}$$

$$\frac{d[Smad2_n]}{dt} = \frac{V_c}{V_n} k_{imp}^{Smad2}[Smad2_c] - k_{exp}^{Smad2}[Smad2_n] +$$

$$k_{diss}^{Smads\_complex}[Smads\_complex_n] \frac{K_{III}}{K_{III} + [PSmadL]} \tag{A20}$$

$$\frac{d[Smad4_c]}{dt} = \frac{V_n}{V_c} k_{exp}^{Smad4}[Smad4_n] - k_{imp}^{Smad4}[Smad4_c] -$$

$$k_{Smads\_complex}[Smad2_c][Smad4_c][LRC_{EE}] \frac{K_I}{K_I + [PSmadL]} \tag{A21}$$

$$\frac{d[Smad4_n]}{dt} = \frac{V_c}{V_n} k_{imp}^{Smad4}[Smad4_c] - k_{exp}^{Smad4}[Smad4_n] +$$

$$k_{diss}^{Smads\_complex}[Smads\_complex_n] \frac{K_{III}}{K_{III} + [PSmadL]} \tag{A22}$$

$$\frac{d[Smads\_complex_c]}{dt} = k_{Smads\_complex}[Smad2_c][Smad4_c][LRC_{EE}] \frac{K_I}{K_I + [PSmadL]}$$

$$- k_{imp}^{Smads\_complex}[Smads\_complex_c] \tag{A23}$$

$$\frac{d[Smads\_complex_n]}{dt} = \frac{V_c}{V_n} k_{imp}^{Smads\_complex}[Smads\_complex_c] -$$

$$k_{diss}^{Smads\_complex}[Smads\_complex_n] \frac{K_{III}}{K_{III} + [PSmadL]} \tag{A24}$$

$$\frac{d[TGF-\beta]}{dt} = \frac{V_c}{V_{extra}} \left(kr_{cave}[LRC_{cave}] + kr_{EE}[LRC_{EE}] - k_{LRC}[TGF-\beta][T1R_{surf}][T2R_{surf}] \right) \tag{A25}$$

This model consists of 24 states and 44 parameters. The nominal values of the Michaelis constants K_1–K_{10} are given in nM.

$$K_1 = 10, K_2 = 8, K_3 = K_4 = K_5 = K_6 = K_7 = K_8 = K_9 = K_{10} = 15 \tag{A26}$$

The catalytic rate constants k_3, k_4, k_7, k_8 are given in min^{-1}

$$k_3 = k_4 = k_7 = k_8 = 1.5 \tag{A27}$$

The maximal enzyme rates V_2, V_5, V_6, V_9, V_{10} are expressed in nM·min^{-1}

$$V_2 = 15, V_5 = V_6 = 45, V_9 = V_{10} = 30 \tag{A28}$$

The nominal values and biological interpretations of the other parameters are shown in Table A1. The initial conditions and biological interpretations of the state variables are listed in Table A2.

Table A1. Parameter values in the model.

Parameter	Biological Meaning	Value
v_{T1R}	type I receptor production rate constant	0.0103 nM·min^{-1}
v_{T2R}	type II receptor production rate constant	0.02869 nM·min^{-1}
ki_{EE}	internalization rate constant of receptor from cell surface to early endosome	0.33 min^{-1}
kr_{EE}	recycling rate constant of receptor from early endosome to cell surface	0.033 min^{-1}
ki_{cave}	internalization rate constant of receptor from cell surface to caveolar lipid-raft	0.33 min^{-1}
kr_{cave}	recycling rate constant of receptor from caveolar lipid-raft to cell surface	0.03742 min^{-1}
k_{cd}	constitutive degradation rate constant for ligand-receptor complex in early endosome	0.005 min^{-1}
k_{LCR}	ligand-receptor complex formation rate constant from TGF-β and receptors	2197 N·m^2·min^{-1}
k_{lid}	ligand induced degradation rate constant for ligand-receptor complex in caveolar lipid-raft	0.02609 min^{-1}
k_{deg}^{T1R}	constitutive degradation rate constant for type I receptor in early endosome	0.005 min^{-1}
k_{deg}^{T2R}	constitutive degradation rate constant for type II receptor in early endosome	0.025 min^{-1}
k_{imp}^{Smad2}	nuclear import rate constant for Smad2	0.16 min^{-1}
k_{exp}^{Smad2}	nuclear export rate constant for Smad2	1 min^{-1}
k_{imp}^{Smad4}	nuclear import rate constant for Smad4	0.08 min^{-1}
k_{exp}^{Smad4}	nuclear export rate constant for Smad4	0.5 min^{-1}
$k_{Smads\_complex}$	formation rate constant for the phosphorylated Smad complex	6.85 × 10^{-5} n·M^2·min^{-1}
$k_{imp}^{Smads\_complex}$	nuclear import rate constant for the phosphorylated Smad complex	0.16 min^{-1}
$k_{diss}^{Smads\_complex}$	dissociation rate constant for the nuclear phosphorylated Smad complex	0.1174 min^{-1}
V_C/V_n	ratio of cytoplasmic to nuclear volume	3
V_0/V_{extra}	ratio of cytoplasmic volume to the average extracellular medium volume per cell	0.001
V_0	the concentration of activated receptor MKKKK at $t = t_0$	0.18 nM·s^{-1}
γ	the characteristic time of activated receptor MKKKK	0.004
t_0	the delayed time of receptor response to extracellular stimuli	0
K_I	interaction strength	5
K_{III}	interaction strength	5

Table A2. Initial conditions of state variables in the model.

Variable	Biological Meaning	Initial Value (nM)
$MKKK$	MAPK kinase kinase	100
$MKKK_P$	Singly phosphorylated MKKK	0
MKK	MAPK kinase	300
MKK_P	Singly phosphorylated MKK	0
MKK_{PP}	doubly phosphorylated MKK	0
$MAPK$	MAP kinase	300
$MAPK_P$	Singly phosphorylated MAPK	0
$MAPK_{PP}$	doubly phosphorylated MAPK	0
$T1R_{surf}$	type I receptor at cell surface	0.237
$T1R_{cave}$	type I receptor in caveolar lipid-raft	2.092
$T1R_{EE}$	type I receptor in early endosome	1.148
$T2R_{surf}$	type II receptor at cell surface	0.202
$T2R_{cave}$	type II receptor in caveolar lipid-raft	1.778
$T2R_{EE}$	type II receptor in early endosome	1.148
LRC_{surf}	ligand-receptor complex at cell surface	0
LRC_{cave}	ligand-receptor complex in caveolar lipid-raft	0
LRC_{EE}	ligand-receptor complex in early endosome	0
$Smad2_c$	Smad2 in the cytoplasm	492.61
$Smad2_n$	Smad2 in the nucleus	236.45
$Smad4_c$	Smad4 in the cytoplasm	1149.4
$Smad4_n$	Smad4 in the nucleus	551.72
$Smads\_complex_c$	Smad complex in the cytoplasm	0
$Smads\_complex_n$	Smad complex in the nucleus	0
TGF-β	TGF-β in the extracellular medium	0.080

Conflicts of Interest

The authors declare no conflict of interest.

References

1. Hough, C.; Radu, M.; Doré, J.J. Tgf-β induced Erk phosphorylation of smad linker region regulates smad signaling. *PLoS One* **2012**, *7*, e42513.
2. Derynck, R.; Zhang, Y.E. Smad-dependent and Smad-independent pathways in TGF-β family signalling. *Nature* **2003**, *425*, 577–584.
3. Kyprianou, N.; Isaacs, J.T. Expression of transforming growth factor-β in the rat ventral prostate during castration-induced programmed cell death. *Mol. Endocrinol.* **1989**, *3*, 1515–1522.
4. Guo, Y.; Kyprianou, N. Restoration of transforming growth factor *β* signaling pathway in human prostate cancer cells suppresses tumorigenicity via induction of caspase-1-mediated apoptosis. *Cancer Res.* **1999**, *59*, 1366–1371.
5. Zhu, B.; Fukada, K.; Zhu, H.; Kyprianou, N. Prohibitin and cofilin are intracellular effectors of transforming growth factor β signaling in human prostate cancer cells. *Cancer Res.* **2006**, *66*, 8640–8647.

6. Zi, Z.; Klipp, E. Constraint-based modeling and kinetic analysis of the Smad dependent TGF-β signaling pathway. *PLoS One* **2007**, *2*, e936.

7. Tian, X.-J.; Zhang, H.; Xing, J. Coupled reversible and irreversible bistable switches underlying TGFβ-induced epithelial to mesenchymal transition. *Biophys. J.* **2013**, *105*, 1079–1089.

8. Javelaud, D.; Mauviel, A. Crosstalk mechanisms between the mitogen-activated protein kinase pathways and Smad signaling downstream of TGF-β: Implications for carcinogenesis. *Oncogene* **2005**, *24*, 5742–5750.

9. Inoki, K.; Haneda, M.; Ishida, T.; Mori, H.; Maeda, S.; Koya, D.; Sugimoto, T.; Kikkawa, R. Role of mitogen-activated protein kinases as downstream effectors of transforming growth factor-β; in mesangial cells. *Kidney Int.* **2000**, *58*, S76–S80.

10. Guo, X.; Wang, X.-F. Signaling cross-talk between TGF-β/BMP and other pathways. *Cell Res.* **2008**, *19*, 71–88.

11. Hayashida, T.; Decaestecker, M.; Schnaper, H.W. Cross-talk between ERK MAP kinase and Smad signaling pathways enhances TGF-β-dependent responses in human mesangial cells. *FASEB J.* **2003**, *17*, 1576–1578.

12. Funaba, M.; Zimmerman, C.M.; Mathews, L.S. Modulation of Smad2-mediated signaling by extracellular signal-regulated kinase. *J. Biol. Chem.* **2002**, *277*, 41361–41368.

13. Burch, M.L.; Yang, S.N.; Ballinger, M.L.; Getachew, R.; Osman, N.; Little, P.J. TGF-β stimulates biglycan synthesis via p38 and ERK phosphorylation of the linker region of Smad2. *Cell. Mol. Life Sci.* **2010**, *67*, 2077–2090.

14. Kretzschmar, M.; Doody, J.; Timokhina, I.; Massague, J. A mechanism of repression of TGF-β/Smad signaling by oncogenic Ras. *Genes Dev.* **1999**, *13*, 804–816.

15. Kretzschmar, M.; Doody, J.; Massagu, J. Opposing BMP and EGF signalling pathways converge on the TGF-β family mediator Smad1. *Nature* **1997**, *389*, 618–622.

16. Oft, M.; Peli, J.; Rudaz, C.; Schwarz, H.; Beug, H.; Reichman, E. TGF-β1 and Ha-Ras collaborate in modulating the phenotypic plasticity and invasiveness of epithelial tumor cells. *Genes Dev.* **1996**, *10*, 2462–2477.

17. Calonge, M.A.J.; Massagué, J. Smad4/DPC4 silencing and hyperactive Ras jointly disrupt transforming growth factor-β antiproliferative responses in colon cancer cells. *J. Biol. Chem.* **1999**, *274*, 33637–33643.

18. Massagué, J.; Chen, Y.-G. Controlling TGF-β signaling. *Genes Dev.* **2000**, *14*, 627–644.

19. Chapnick, D.A.; Warner, L.; Bernet, J.; Rao, T.; Liu, X. Partners in crime: The TGF-β and MAPK pathways in cancer progression. *Cell Biosci.* **2011**, doi:10.1186/2045-3701-1-42.

20. Heinrich, R.; Neel, B.G.; Rapoport, T.A. Mathematical models of protein kinase signal transduction. *Mol. Cell* **2002**, *9*, 957–970.

21. Saez-Rodriguez, J.; Kremling, A.; Conzelmann, H.; Bettenbrock, K. Modular analysis of signal transduction networks. *Control Syst. IEEE* **2004**, *24*, 35–52.

22. Kholodenko, B.N. Negative feedback and ultrasensitivity can bring about oscillations in the mitogen-activated protein kinase cascades. *Eur. J. Biochem.* **2000**, *267*, 1583–1588.

23. Schoeberl, B.; Eichler-Jonsson, C.; Gilles, E.D.; Müller, G. Computational modeling of the dynamics of the MAP kinase cascade activated by surface and internalized EGF receptors. *Nat. Biotechnol.* **2002**, *20*, 370–375.

24. Hartsough, M.T.; Mulder, K.M. Transforming growth factor β activation of p44mapk in proliferating cultures of epithelial cells. *J. Biol. Chem.* **1995**, *270*, 7117–7124.

25. Zhang, Y.E. Non-Smad pathways in TGF-β signaling. *Cell Res.* **2008**, *19*, 128–139.

26. Gui, T.; Sun, Y.; Shimokado, A.; Muragaki, Y. The roles of mitogen-activated protein kinase pathways in TGF-*β*-induced epithelial-mesenchymal transition. *J. Signal Transduct.* **2012**, *2012*, doi:10.1155/2012/289243.

27. Blanchette, F.; Rivard, N.; Rudd, P.; Grondin, F.; Attisano, L.; Dubois, C.M. Cross-talk between the p42/p44 MAP kinase and Smad pathways in transforming growth factor β1-induced furin gene transactivation. *J. Biol. Chem.* **2001**, *276*, 33986–33994.

28. Pessah, M.; Prunier, C.; Marais, J.; Ferrand, N.; Mazars, A.; Lallemand, F.; Gauthier, J.M.; Atfi, A. c-jun interacts with the corepressor TG-interacting factor (TGIF) to suppress Smad2 transcriptional activity. *Proc. Natl. Acad. Sci. USA* **2001**, *98*, 6198–6203.

29. Suzuki, K.; Wilkes, M.C.; Garamszegi, N.; Edens, M.; Leof, E.B. Transforming growth factor β signaling via Ras in mesenchymal cells requires p21-activated kinase 2 for extracellular signal-regulated kinase-dependent transcriptional responses. *Cancer Res.* **2007**, *67*, 3673–3682.

30. Liberati, N.T.; Datto, M.B.; Frederick, J.P.; Shen, X.; Wong, C.; Rougier-Chapman, E.M.; Wang, X.F. Smads bind directly to the Jun family of AP-1 transcription factors. *Proc. Natl. Acad. Sci. USA* **1999**, *96*, 4844–4849.

31. Massagué, J. How cells read TGF-β signals. *Nat. Rev. Mol. Cell Biol.* **2000**, *1*, 169–178.

32. Chu, Y.; Hahn, J. Parameter set selection via clustering of parameters into pairwise indistinguishable groups of parameters. *Ind. Eng. Chem. Res.* **2008**, *48*, 6000–6009.

33. Dai, W.; Bansal, L.; Hahn, J.; Word, D. Parameter set selection for dynamic systems under uncertainty via dynamic optimization and hierarchical clustering. *AIChE J.* **2014**, *60*, 181–192.

34. Morris, M.D. Factorial sampling plans for preliminary computational experiments. *Technometrics* **1991**, *33*, 161–174.

35. Martin, M.M.; Buckenberger, J.A.; Jiang, J.; Malana, G.E.; Knoell, D.L.; Feldman, D.S.; Elton, T.S. TGF-β1 stimulates human AT1 receptor expression in lung fibroblasts by cross talk between the Smad, p38 MAPK, JNK, and PI3K signaling pathways. *Am. J. Physiol. Lung Cell. Mol. Physiol.* **2007**, *293*, L790–L799.

Resolving Early Signaling Events in T-Cell Activation Leading to IL-2 and FOXP3 Transcription

Jeffrey P. Perley, Judith Mikolajczak, Gregery T. Buzzard, Marietta L. Harrison and Ann E. Rundell

Abstract: Signal intensity and feedback regulation are known to be major factors in the signaling events stemming from the T-cell receptor (TCR) and its various coreceptors, but the exact nature of these relationships remains in question. We present a mathematical model of the complex signaling network involved in T-cell activation with cross-talk between the Erk, calcium, PKCθ and mTOR signaling pathways. The model parameters are adjusted to fit new and published data on TCR trafficking, Zap70, calcium, Erk and IκBα signaling. The regulation of the early signaling events by phosphatases, CD45 and SHP1, and the TCR dynamics are critical to determining the behavior of the model. Additional model corroboration is provided through quantitative and qualitative agreement with experimental data collected under different stimulating and knockout conditions. The resulting model is analyzed to investigate how signal intensity and feedback regulation affect TCR- and coreceptor-mediated signal transduction and their downstream transcriptional profiles to predict the outcome for a variety of stimulatory and knockdown experiments. Analysis of the model shows that: (1) SHP1 negative feedback is necessary for preventing hyperactivity in TCR signaling; (2) CD45 is required for TCR signaling, but also partially suppresses it at high expression levels; and (3) elevated FOXP3 and reduced IL-2 signaling, an expression profile often associated with T regulatory cells (Tregs), is observed when the system is subjected to weak TCR and CD28 costimulation or a severe reduction in CD45 activity.

Reprinted from *Processes*. Cite as: Perley, J.P.; Mikolajczak, J.; Buzzard, G.T.; Harriso, M.L. Resolving Early Signaling Events in T-Cell Activation Leading to IL-2 and FOXP3 Transcription. *Processes* **2014**, *2*, 867–900.

1. Introduction

The actions of CD4+ T-cells are controlled by the stimulatory signals and cytokine milieu in their surrounding tissue environment [1,2]. Extracellular signals naturally drive CD4+ T-cells to activate and assume one of many important immunological roles, each characterized by a distinct profile of signaling events and cytokine secretion. These signals are processed by the intracellular signaling pathways to alter the transcription profiles that direct T-cell activation and differentiation. To explore this process from a quantitative perspective, a mathematical model is constructed from the known molecular interactions and regulations.

This model encompasses the signal transduction events relating to the process of directing gene expression that ultimately determine the actions or immunological role of the T-cell. This process begins at the engagement of the T-cell receptor (TCR) to cognate peptide-bound major histocompatibility complexes (pMHC) on the surface of antigen-presenting cells (APCs) [3–5].

TCR engagement triggers a number of early phosphorylation and dephosphorylation events. The CD45 phosphatase primes the CD4 coreceptor-associated 56-kD lymphocyte-specific tyrosine kinase (Lck), a member of the sarcoma (Src) family of kinases (SFK) [6,7], by removing the inhibitory phosphate group from the Y505 residue [8,9]. Primed Lck is activated by autophosphorylation at Y394, mediated by receptor clustering [9]. Activated Lck phosphorylates the immunoreceptor tyrosine-based activation motifs (ITAMs) on the TCR-associated CD3 ζ-chain. The zeta-chain-associated protein kinase 70 (Zap70) binds with high affinity to doubly phosphorylated ζ-chain ITAMs through its two Src homology 2 (SH2) domains and is activated by Lck-mediated phosphorylation of its Y319 and Y493 residues [10–12]. Zap70 facilitates the activation of phospholipase C-γ1 (PLCγ1), which hydrolyzes phosphatidylinositol 4,5-bisphosphate (PIP$_2$) to second messengers diacylglycerol (DAG) and inositol 1,4,5-trisphosphate (IP$_3$) [13–15]. IP$_3$ migrates to the cytoplasm to regulate intracellular calcium (Ca^{2+}) flux from the endoplasmic reticulum [13,16,17], a key modulator of nuclear factor of activated T-cells (NFAT) [18]. While IP$_3$ migrates to the cytoplasm, DAG operates within the plane of the cell membrane to activate two major signaling molecules: rat sarcoma (Ras) and protein kinase Cθ (PKCθ) [19]. Ras triggers the extracellular signal-regulated kinase (Erk) cascade that results in the activation of activator protein 1 (AP1) [20], while PKCθ activates AP1 and the nuclear factor κB (NFκB) pathway [21–24]. Upon entering the nucleus, transcription factors, such as NFAT, AP1, NFκB and others, interact to support cooperative regulation of interleukin-2 (IL-2) and other genes involved in T-cell activation [25,26].

A critical second signal driving T-cell activation and fate determination is achieved by incorporating CD28-mediated costimulation [27] and its downstream signaling events into the model. In spite of the ability of the TCR to upregulate factors necessary for full activation, TCR-antigen recognition ultimately leads to anergy or a state of hyporesponsiveness in the absence of a second signal [28]. Alternatively, TCR signaling with additional costimulation of CD28 on the surface of T-cells greatly enhances IL-2 production, T-cell proliferation and the prevention of anergy. This synergistic effect is believed to be mediated through PKCθ and Akt (also known as protein kinase B), which recruit and regulate several signaling molecules, including NFκB [23,29] and mammalian target for rapamycin (mTOR) [30]. The serine/threonine protein kinase mTOR is another important player in CD4+ T-cell activation. mTOR complex 1 (mTORC1) has many known functions, including the ability to control the expression of forkhead box P3 (FOXP3), a master transcriptional regulator in the development of T regulatory cells (Tregs) [30,31]. While much less is known about mTORC2, it was recently shown to promote PKC, Akt and NFκB signaling and to regulate T helper type 1 (Th1) and Th2 commitment [32].

A number of mathematical models exist that span the entirety or portions of these T-cell signaling events. The models presented in [33,34] span these events and are able to capture many dominant behaviors. However, these discrete-time logical models do not support a detailed analysis of the interactions in a continuous state and parameter space and the ability to predict responses to analog knockdown or mutation scenarios. Other models are very detailed, but only consider portions of the system, such as the MAPK [35] and NFκB [36,37] pathways. Some of these models consider every hypothesized molecular interaction or complex formation. This level of detail is prohibitive to complete for the entire signaling network, predominantly due to computational tractability, but also due to uncertainties in the existence or rates of these events. A few models set out to address the role of location as a regulatory event, analyzing T-cell signaling events within a spatial context [38,39]. These models tend to be limited in scope in order to offset the added computational expense of such an analysis. Our model differs from past approaches in that we consider a broad scope of early TCR and CD28 signaling leading to gene transcription in continuous time, consider only the major known phosphorylation, dephosphorylation, association and dissociation events and address location through compartments and not as a partial differential equation with diffusion.

Here, we present a mathematical model of T-cell activation that focuses on the early steps in TCR trafficking and signal amplification by kinases, phosphatases and other second messengers, as well as CD28 coreceptor signaling and gene transcription of IL-2 and FOXP3. Sections 2 and 3 describe the formulation of the model and its fitting to training data. Model simulations and corroborating experimental evidence are provided in Section 4, demonstrating the model's ability to reproduce many known experimental results. This section also describes predictions made by the model about the roles of phosphatases CD45 and SHP1 in T-cell activation. Finally, conclusions are summarized in Section 5.

2. Materials and Methods

This section provides an overview of the proposed T-cell activation model and details the methods and materials employed in its development. These include the computational methods and experimental datasets used for model parameter identification.

2.1. The Mathematical Model: Overview and Scope

The mathematical model reflects the current understanding of the intracellular signaling events leading to T-cell activation. The reaction scheme, as shown in Figure 1, encompasses the primary events that occur upon TCR engagement and CD28 costimulation. Appendix 1 describes each signaling element and assigns a symbol and initial value for use in constructing the mathematical model. Appendix 1 also lists the rate equations, parameter values and summarizes the reactions primarily in terms of protein binding, phosphate group transfers and kinase and phosphatase activation. The part of the model representing the TCR-mediated Erk pathway is based on the ordinary differential equation (ODE) model developed by Zheng [40] with updates in Perley *et al.* [41]. The Zheng model takes into account the major events of early T-cell signaling leading to the activation of the protein tyrosine kinases Zap70 and SFK, the downstream Erk pathway, as well as negative feedback regulation by phosphatase SHP1.

Figure 1. Reaction network of the T-cell activation model. Solid arrows denote reactions for which the forward and reverse directions are indicated; dashed arrows connect reactions (either forward or reverse) with their catalysts. Colored arrows denote reactions that are catalyzed by specific species as indicated. Symbols and reactions are described in Appendix 1.

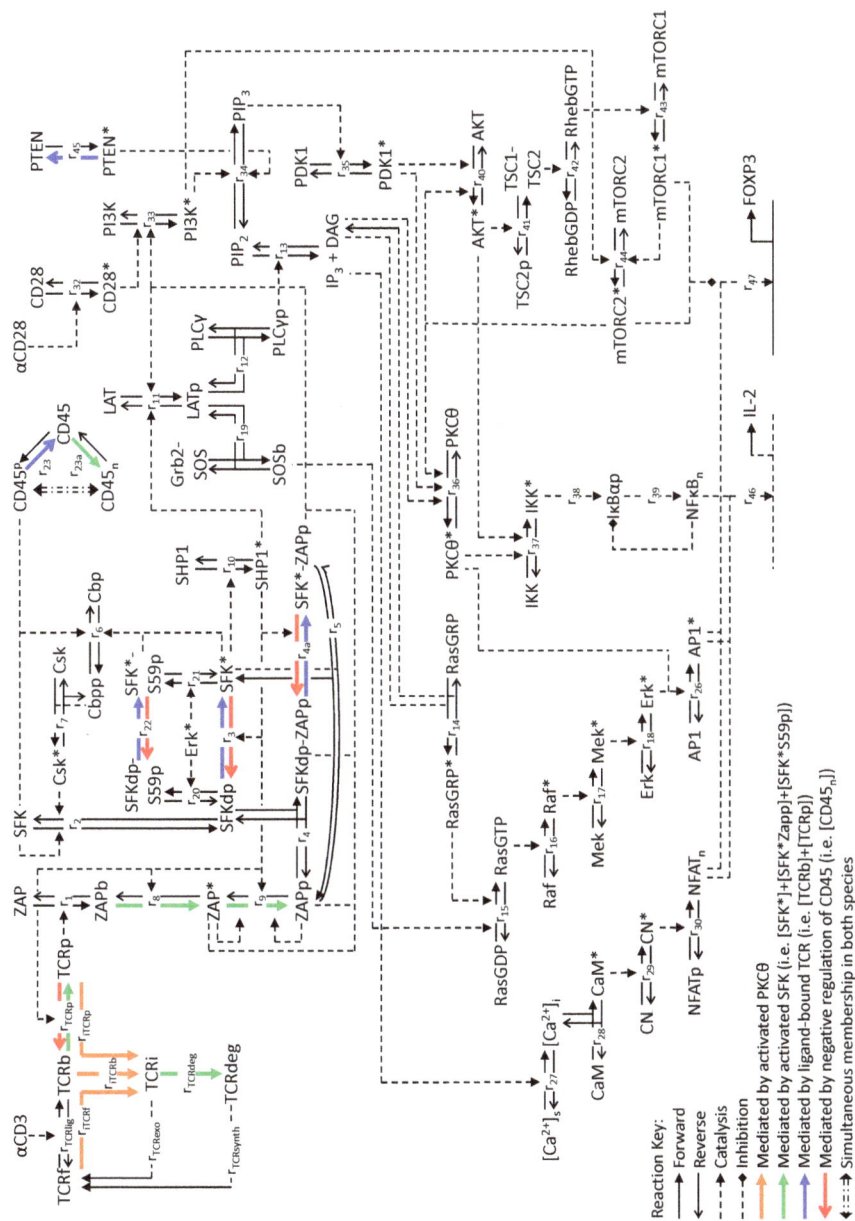

The Zheng model was extended to specifically incorporate TCR trafficking, tuning of CD45 and SHP1 phosphatase activity, refinement of Erk signaling with AP1 activation and the addition of the calcium, NFκB and mTOR pathways leading to IL-2 and FOXP3 transcription. These changes are further detailed in Section 3. Molecular association and complex dissociation are designed to obey the law of mass action, *i.e.*, assuming that the rate of a reaction is proportional to the product of reactant concentrations. Most enzyme-catalyzed reactions utilize second-order kinetics derived from a simplification of the Michaelis–Menten enzyme kinetics by assuming that the substrate concentration is limiting. Such a simplification contributes to reducing the complexity of the nonlinearities while retaining the structure of the pathway connectivity. For reactions catalyzed by enzymes not explicitly included in this model, the expression further reduces to first-order kinetics, where the enzyme concentration becomes a constant. In addition, single molecular transitions are also governed by first-order kinetics. Alternatively, enzyme-catalyzed reactions in which cooperative binding between substrates plays a significant role utilize the Hill equation. Scenarios characterized by these equations include the cooperative transcriptional regulation of IL-2 and FOXP3 and the activation of certain enzymes. The model is also designed to mimic the cell response to stimulation by αCD3, αCD28, phorbol 12-myristate 13-acetate (PMA) and the calcium ionophore ionomycin, as well as mimic perturbations in kinase and phosphatase activity by various small molecule inhibitors. The complete model consists of 48 ODEs with 154 reaction parameters and simulated in MATLAB using the ode15s solver.

2.2. Global Sensitivity Analysis

Due to the size and complexity of the model, it is necessary to reduce the dimension of the uncertain parameter space to facilitate an effective and computationally tractable analysis of the model. Parameter sensitivity analysis provides a powerful tool for analyzing mathematical models of complex biological systems and can be used to facilitate parameter estimation by identifying and ranking the various contributors of parametric uncertainty in the model. Parameters with the lowest sensitivity ranks, *i.e.*, parameters whose uncertainty causes the least amount of variability in the output space, can be neglected during the model fitting process and fixed at their initial guesses. For the purpose of this study, a variance-based sensitivity analysis is used to determine how the model output changes over the global parameter space:

$$S_{i,j} = \frac{\text{var}(\text{E}[y_i|\theta_j])}{\text{var}(\text{E}[y_i])} \tag{1}$$

where $S_{i,j}$ is the sensitivity index of the $i-th$ output to the $j-th$ parameter. The sensitivity indices are numerically computed using Sobol's method by the efficient sparse-grid-based algorithm proposed by Buzzard [42,43]. Parameters having little effect on model outputs upon perturbation are determined to be insignificant and fixed at their nominal values. Significant parameters are retained for parameter identification.

2.3. Parameter Identification

The ability of the model simulated with a given parameter vector to reproduce available experimental data is quantified by the objective function:

$$J(\theta) = \log_{10}\left(1 + \sum_{i=1}^{N_s} e_{t_i}^T W_{t_i} e_{t_i}\right) \text{ with } e_{t_i} = C(t_i)[y(t_i) - \hat{y}(t_i)] \tag{2}$$

where e_{t_i} is the error at each sample time point, W_{t_i} is the inverse of the measurement error covariance matrix at time t_i, $y(t_i)$ and $\hat{y}(t_i)$ are the simulated model outputs and the mean values of the available data at time t_i, respectively, $C(t_i)$ is a binary matrix that indicates which outputs are measured at time t_i and N_s is the total number of sampled time points. This cost function is computed in log space in order to compress the range of and smooth the function, making its approximation with sparse-grid interpolation more effective. The "1" in the cost function is added to avoid the singularity of log at zero.

Specific parameter values that are capable of producing data-consistent model simulations are extracted directly from the literature whenever possible. For those not found in the literature, they are either derived using known conditions or constraints of the system or estimated from experimental data using a combination of manual and automated calibration. First, the starting guess values are roughly estimated by back-of-the-envelope calculations and manual tuning to within a reasonable level of accuracy. This step is used to establish our parameter search space and its bounds, defined to be a hypercube spanning the nominal parameter vector by an order of magnitude both above and below. Next, Equation (2) is evaluated at a number of samples from the parameter hypercube, generated using Latin hypercube sampling in log space. Finally, starting from the parameter vector with the lowest sampled cost, MATLAB's nonlinear constrained optimization solver, fmincon, is employed to identify the parameter vector that minimizes Equation (2) constrained within the hypercube. To increase the chance of identifying the global minimizer, multiple starting points were considered that span the region of low cost. This procedure of automated parameter identification is aided by the use of sparse-grid interpolation, a powerful tool for generating fast-evaluating approximations of mathematical models [44].

2.4. Experimental Datasets for Model Development

Experimental datasets for the purposes of model development are compiled from the published literature or generated from experiments performed in our lab specifically for this manuscript. For datasets compiled from the published literature, only those that are commensurate with our model in terms of cell type, stimuli, output species and time scale are considered. Table 1 summarizes the attributes of the compiled datasets used to train the model. These datasets consist of dynamics (*i.e.*, time courses), dose response experiments (*i.e.*, multiple doses of an input or stimulus) and input response experiments (*i.e.*, single doses of different input or stimuli combinations). In order to compare the model and observed behaviors directly, simulations are generated to mimic the experimental conditions used in our experimental set-ups (see Appendices 3 and 4) or as indicated in the corresponding source articles. Model simulations are shown with the corresponding data and error bars whenever possible; however, for cases in which data could not be visualized on the same plot, the reader is referred to the original source.

Table 1. Description of experimental datasets used for model development.

Species	Variable(s)	Experiment Type	Cell Type	Figure	Source(s)
TCR	αCD3	Dynamic dose response	Jurkat	2	[45,46]
Zap70	αCD3	Dose response	Jurkat	4	[40]
Erk	Sanguinarine, U0126	Dynamic dose response	Jurkat	5	[41]
Ca^{2+}	αCD3	Dynamics	Jurkat	6	Appendix 3
IκBα	αCD3+αCD28	Dynamics	Jurkat	7	Appendix 4
Akt, PKCθ	αCD3, αCD28	Input response	CD4+ T-cells (murine)	8	[32]

3. Model Development

The following subsections describe the process of formulating the proposed T-cell activation model as shown in Figure 1 from the original ODE model presented by Zheng [40]. These changes include the incorporation of TCR trafficking mechanisms, tuning of CD45 and SHP1 phosphatase activity, refinement of Erk signaling with AP1 activation and the addition of the calcium, NFκB and mTOR pathways leading to IL-2 and FOXP3 transcription. These changes are further detailed in the following subsections.

3.1. Modeling TCR Trafficking

The TCR expression level at the cell surface is the result of a dynamic equilibrium maintained by the membrane expression of newly synthesized TCR, receptor internalization, recycling to the cell surface and degradation [46,47]. As depicted in Figure 1, the TCR complex is represented in the model by five states: free TCR (TCR_f), ligand-bound TCR (TCR_b), phosphorylated ligand-bound TCR (TCR_p), internalized TCR (TCR_i) and degraded TCR (TCR_{deg}). TCR trafficking is modeled by allowing TCR complexes from all surface states (TCR_f, TCR_b and TCR_p) to be internalized [38]. Internalized TCR subunits can be degraded or recycled back to the unbound state. Newly synthesized TCR contributes to the free receptor state with a rate proportional to degraded TCR.

Early studies demonstrated that the TCR is a constitutively cycling receptor [46,47]. Thus, at steady state, a certain amount of TCR is endocytosed, while at the same time, an equal amount of TCR is exocytosed. The associated model rate parameters are set by the following observations. Several studies seem to agree that the constitutive endocytic rate constant for the TCR in resting T-cells is ~0.01 min^{-1}, meaning that ~1% of the cell surface-expressed TCR is internalized each minute [46].

TCR ligation induces downregulation and degradation of the TCR in a dose-dependent manner [45,46]. In theory, TCR downregulation can be accomplished by an increase in the endocytic rate constant, a decrease in the exocytic rate constant or a combination of both. However, most studies found that TCR downregulation is caused by an increase in the endocytic rate constant to ~0.038 min^{-1} after TCR triggering [46], whereas the exocytic rate was constant. In addition, TCR degradation was also found to accelerate after TCR ligation. The TCR subunits in non-stimulated Jurkat cells were observed to degrade with rate constants of ~0.0011 min^{-1}, resulting in a half-life of ~10.5 h. Triggering of the TCR by anti-TCR Abs resulted in a three-fold increase in the degradation rate constants to ~0.0033 min^{-1}, resulting in a half-life of ~3.5 h [45]. This process was modeled by making the receptor internalization and degradation rate parameters functions related to T-cell activation. These mechanisms were mathematically encoded in the model using:

$$k = k_{min} + \frac{X^n}{X^n + K^n}(k_{max} - k_{min}) \tag{3}$$

Equation (3) ensures a continuous transition between the constitutive endocytosis or degradation rates (k_{min}) and the ligand-induced rates (k_{max}) using a Hill equation, where X is the induction substrate concentration, K is the substrate concentration of the transition midpoint and n is the Hill coefficient denoting positive ($n > 1$) or negative ($n < 1$) cooperativity. Since the TCR cycling pathway is dependent on the CD3γ di-leucine-based motif and is activated by PKC [46,47], for receptor cycling, we set $k = k_{int}$, $X = [PKC\theta]$, $n = 2$, and K is set to 5% of total PKCθ with $k_{min} = 0.01$ min^{-1} and $k_{max} = 0.038$ min^{-1}. Since the activation of Lck and Zap70 leads to recruitment of Cbl and ubiquitination (Ub) of the CD3 and ζ chains to induce degradation of the TCR in the lysosomes [46], for receptor degradation, we set $k = k_{deg}$, $X = [SFKact] + [SFKactZapp] + [SFKactS59p]$, $n = 2$, and K is set to 5% of total SFK with $k_{min} = 0.0011$ min^{-1} and $k_{max} = 0.0033$ min^{-1}.

At equilibrium without any stimulus, most studies also seem to agree that the pool of recycling TCR was distributed with approximately 75%–85% at the cell surface and 15%–25% inside the cells. Assuming the minimum constitutive rates for internalization and degradation with 80% surface TCR and 20% inside the cell, of which half are being degraded, we can compute the constitutive exocytosis rate as $k_{exo} = 0.0789$ min^{-1} and the constitutive synthesis rate as $k_{synth} = k_{deg_{min}}$ to achieve the observed dynamic equilibrium in resting T-cells.

The resulting simulated surface and degraded TCR dynamics are consistent with expectations at equilibrium and after receptor engagement with a ligand. The dose response curves in Figure 2A,B show good agreement with the behaviors reported by Menne et al. [47], Geisler et al. [46] and von Essen et al. [45].

Figure 2. Resting and ligand-induced (**A**) surface TCR expression and internalization and (**B**) degradation activity in response to varying doses of αCD3 stimulation (data shown in [45]).

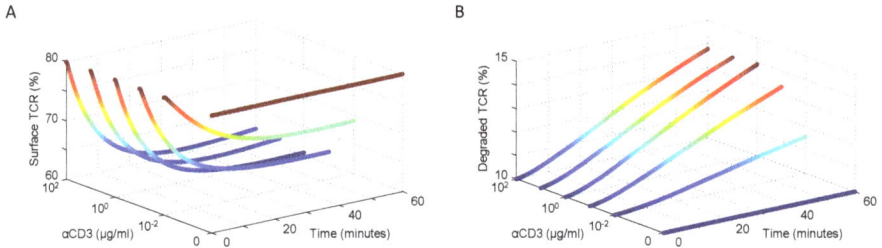

3.2. Tuning the Roles of CD45 and SHP1

The protein tyrosine phosphatases, CD45 and SHP1, are central players in signal amplification following antigen recognition by the TCR and the activation of many downstream second messenger and signaling molecules. CD45 is a leukocyte-specific transmembrane glycoprotein and a receptor-like protein tyrosine phosphatase (PTP) [8]. The SFK member Lck is the best characterized CD45 substrate in T-cells [7]. Lck exists in dynamic equilibrium with three main sub-populations: (1) open and activated ($SFKact$ in the model); (2) open and not activated ("primed") ($SFKdp$); (3) closed and not activated (SFK) [9]. Phosphorylation of SFKs at the negative regulatory site (Y505 in Lck) by the kinase Csk results in an intramolecular interaction with the SH2 domain, creating a folded inactive conformation [48]. Dephosphorylation at this site by CD45 opens up the molecule, creating a "primed" molecule. Clustering of these primed SFKs results in the transphosphorylation of the activation loop (Y394 in Lck), which displaces it from the catalytic site and creates an active kinase by allowing substrate access. Dephosphorylation at this site by CD45 or other PTPs, such as SHP1, downregulates SFK activity and returns them to the primed state [9]. Thus, CD45 functions as both a positive and negative regulator of the T-cell antigen receptor and in setting the threshold of activation.

The dual role of CD45 as both positive and negative regulator of T-cell activation is modulated in part by the distribution and movement of CD45 and its substrates proximal to the receptor complex inside and outside lipid rafts. Lck inside lipid rafts has been reported to be hyperphosphorylated and less active when CD45 is excluded [9]. The engineered inclusion of CD45 to the lipid domains also decreases TCR signaling, consistent with its ability to downregulate signaling by dephosphorylating the positive regulatory site at later time points [9]. These observations show that the localization of both CD45 and the SFKs can affect the phosphorylation state of SFKs. It is believed that upon receptor-mediated clustering, the SFKs are activated and often relocalize to lipid rafts, where the concentration of CD45 is much lower and the kinases can experience sustained signaling. The later recruitment of CD45 to these domains will then primarily dephosphorylate the activation site and downregulate activity [9].

The model is augmented to include both the positive and negative regulatory roles of CD45, as well as spatial localization effects, on T-cell activation with two explicit states: $CD45^p$ and $CD45_n$ (see Figure 3). The first, $CD45^p$, reflects its ability to activate substrates immediately following TCR triggering and promote signaling. Initially, this state exists at a high level with its major role being to dephosphorylate negative regulatory sites and "prime" SFKs (R_2). Following TCR engagement, the receptor cluster formation (represented in simulation by ligand-bound TCR) separates CD45 and its substrate. This causes $CD45^p$ to enter an ineffective state, an implicitly modeled state called $CD45$ (R_{23}), and decreases its ability to activate its substrate. The second state, $CD45_n$, represents the negative regulatory function of CD45 on SFK. Initially, this state is minimally active with most existing in the inactive state, $CD45$. As TCR signaling progresses (represented in simulation by activated SFK), CD45 is recruited back to the lipid rafts (R_{23a}), where its ability to dephosphorylate activated SFKs is the dominant role (R_3, R_{4a}, R_{22} and R_{TCRp}). Since CD45 can possess both positive and negative roles simultaneously, the species $CD45^p$ and $CD45_n$ are not mutually exclusive; however, the intersection between these two sets, a CD45-related state possessing both roles ($CD45_n^p$), is not explicitly modeled or tracked. $CD45_{tr}$, representing completely inactive CD45 that is translocated away from its substrates, is also not explicitly modeled.

Figure 3. Representation of the phosphatase CD45 in the model. The model states $CD45^p$ and $CD45_n$ are represented by the regions outlined in blue and red, respectively. $CD45_n^p$, which is the intersection between the two modeled states outlined in purple, has both positive and negative roles, but is not explicitly modeled. $CD45_{tr}$ represents completely inactive CD45 that is translocated away from its substrates and is also not explicitly modeled.

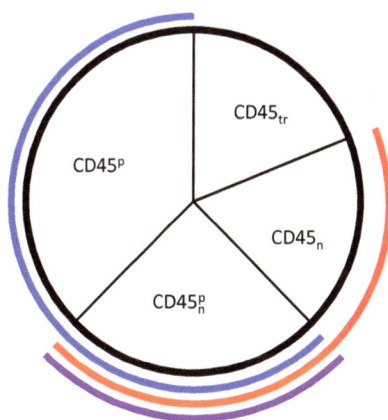

In contrast to the dual role of CD45, Src homology region 2 domain-containing phosphatase-1 (SHP1) functions primarily as a negative regulator of TCR signal transduction. SHP1 associates with and negatively regulates the Syk family kinase Zap70 upon T-cell activation (R_8 and R_9), thereby suppressing TCR signaling [49]. SHP1 also forms a negative feedback loop that is composed of (1) SHP1 phosphorylation by activated Lck, (2) binding of phospho-SHP1 to Lck and (3) Lck inactivation by SHP1-mediated dephosphorylation (R_3) [50]. Erk, on the other hand, antagonizes SHP1 activity by modifying Lck (S59 phosphorylated by MAPK/Erk), which interferes with SHP1 recruitment and Lck inactivation [50,51]. These events are captured by partitioning the SFK states with S59 phosphorylation (*i.e.*, $SFKdpS59p$ and $SFKactS59p$) from those without (R_{20} and R_{21}).

Parameters for this updated model structure, which include reactions R_1-R_{10} and R_{20}-R_{23a}, are trained to recapitulate the behaviors and roles of CD45 and SHP1 in early TCR signaling. To effectively isolate the early TCR signaling module from downstream feedback, the Erk positive feedback loop is suppressed with the presence of the Mek1/2 inhibitor U0126. Figure 4 shows the changes in the Zap70-Y319 phosphorylation level with different doses of αCD3 stimulation in the presence of U0126. The results show that with increasing doses of stimulation, the response of Zap70 increases slowly initially, then rapidly between 2 and 5 μg/mL of αCD3 and eventually saturates and even decreases slightly at the highest stimulus level (100 μg/mL). The dose response curve demonstrates the existence of a threshold αCD3 stimulation concentration. Below the threshold level, CD45 is unable to activate enough SFK and indirectly Zap70 to overcome the negative regulation by SHP1 and CD45 itself to sustain signaling. Once the threshold is crossed, the negative feedback barrier caused by SHP1 is overcome, and Zap70 signaling increases rapidly and eventually saturates at a high stimulus level. The Zap70 dose response simulations show good agreement with the observations originally reported by Zheng [40].

Figure 4. Zap70-Y319 phosphorylation in response to various doses of αCD3 stimulation in the absence of Erk feedback. Jurkat cells were incubated in the presence of a Mek1/2 inhibitor (2 μg/mL U0126) and stimulated with αCD3 at the indicated concentrations. Samples were taken 5 min post-stimulation and analyzed by western blot. The data shown are the means and standard errors from at least three independent experiments (data source: [40]).

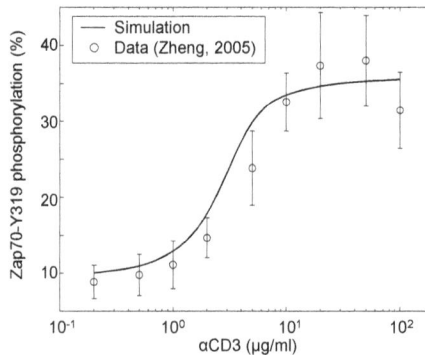

3.3. Tuning Erk Signaling

The Erk signaling cascade plays an important role in IL-2 activation via the transcription factor, AP1. Activated Zap70 mobilizes a number of adapter proteins to the receptor complex, including linker of activated T-cells (LAT) and growth factor receptor-bound protein 2 (Grb2) [52]. Son of Sevenless (SOS), a Ras guanine exchange factor (GEF), forms a complex with Grb2 and LAT to facilitate the activation of Ras in T-cells. LAT also facilitates the activation of Ras through the RAS guanyl nucleotide-releasing protein-1 (RasGRP1), which is itself activated by DAG [19]. Ras triggers the extracellular signal-regulated kinase (Erk) cascade that results in the activation of AP1 [20]. As mentioned in Section 3.2, Erk also promotes its own activation by phosphorylating Lck at S59 and antagonizing SHP1 activity [50,51].

These reactions (R_{11}-R_{19}) are modeled using the first- and second-order mass action kinetic equations of the base model presented by Zheng [40]. The kinetic parameters in this module are fitted to Erk phosphorylation data presented by Perley *et al.* [41]. Figure 5A,B shows the changes in the Erk phosphorylation with different doses of the MKP inhibitor sanguinarine and Mek1/2 inhibitor U0126. For each experiment, Jurkat cells were stimulated with 10 µg/mL αCD3, followed by a dose of inhibitor once the indicated amount of time had past. Samples were taken before and after the inhibitor dose administrations to show how the system changes over time with varying degrees of inhibition. At the lowest concentrations (where inhibition is negligible), the transient nature of Erk phosphorylation can be seen by the decreasing level over time. As the inhibitor concentrations increase, a threshold is crossed, resulting in more forceful inhibition: 10 µg/mL for sanguinarine and 1 µg/mL for U0126. The phospho-Erk simulations show good agreement with the time course and dose response observations originally reported by Perley *et al.* [41].

Figure 5. Erk activation in response to various doses of (**A**) MKP inhibitor sanguinarine and (**B**) Mek1/2 inhibitor U0126. Doses of sanguinarine and U0126 were administered at 15 and 6 min post-stimulation (10 µg/mL αCD3), respectively. Samples of phospho-Erk were taken at indicated times relative to the inhibitor doses and measured via western blot. The data shown are the means and standard errors from at least three independent experiments (data source: [41]).

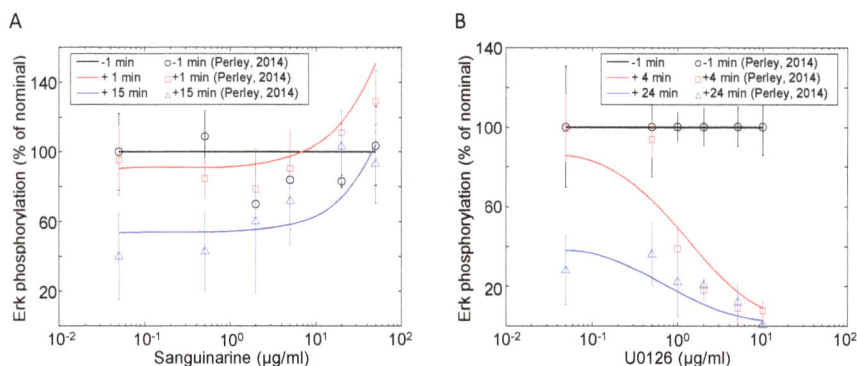

3.4. Modeling Calcium Signaling

Calcium is an important second messenger involved in a number of cellular processes, including the activation of NFAT and IL-2. PLCγ1, activated by Zap70 via LAT, hydrolyzes PIP_2 to DAG and IP_3 (R_{13}) [13–15]. IP_3 migrates to the cytoplasm and interacts with IP_3 receptors on the endoplasmic reticulum, inducing the release of stored calcium (Ca^{2+}) (R_{27}) [13,16,17]. Calmodulin (CaM), a small calcium-sensing protein and signal transducer, binds released Ca^{2+} ions (R_{28}) and undergoes a conformational change to activate the calcium-dependent serine-threonine phosphatase calcineurin (CN) (R_{29}) [53]. Activated CN dephosphorylates the nuclear factor of activated T-cell (NFAT) proteins, exposing their nuclear-localization signal (NLS) and inducing nuclear translocation (R_{30}) [18].

The model is modified to recapitulate calcium signaling with these reactions (R_{13}, R_{27}-R_{30}), modeled using first- and second-order mass action kinetic expressions similar to that of the Erk signaling module. The kinetic parameters in this module are fitted to Ca^{2+} flux data for which the experimental methodology is presented in Appendix 3. Figure 6 shows the experimental observations of intracellular calcium in Jurkat cells stimulated with 10 µg/mL αCD3 and the corresponding model simulation. Upon TCR stimulation, calcium ions are rapidly released from intracellular stores, peaking after 2 min before returning to a slightly elevated level. It is evident from the plot that the model readily captures the rapid transience of intracellular calcium flux.

Figure 6. Intracellular calcium release in response to stimulation with 10 µg/mL αCD3. The data shown are the means and standard errors from 12 independent experiments.

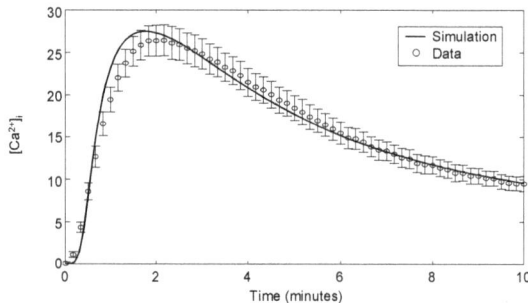

3.5. Modeling CD28 Costimulation

In spite of the ability of the TCR to upregulate many factors necessary for full activation, TCR-antigen recognition ultimately leads to anergy, or a state of hyporesponsiveness, in the absence of a second signal [28]. By contrast, signaling via CD28 on the surface of T-cells, in addition to TCR signaling, greatly enhances IL-2 production, T-cell proliferation and the prevention of anergy. As a result, CD28-mediated costimulation provides a critical second signal in T-cell activation and fate determination [27]. CD28 signaling is believed to be mediated by phosphoinositide 3-kinase (PI_3K)

(R_{33}) [54]. PI$_3$K phosphorylates PIP$_2$ to become phosphatidylinositol 3,4,5-trisphosphate (PIP$_3$), an activity that is directly antagonized by phosphatase and tensin homolog (PTEN) (R_{34}). PIP$_3$ serves as pleckstrin homology (PH) domain membrane anchors for 3-phosphoinositide-dependent kinase-1 (PDK1) (R_{35}). PDK1 activation leads to the membrane recruitment of PKCθ (R_{36}) and protein kinase B (PKB, or Akt) (R_{40}) to regulate several signaling pathways, including NFκB [23,29] and mTOR [30].

The model is augmented to simulate the effects of CD28 costimulation on T-cell activation. As with the other signaling modules, these reactions (R_{32}-R_{35}) are modeled primarily using first- and second-order mass action kinetic expressions. The corresponding kinetic parameters are tuned after accounting for downstream signaling events, such as NFκB and mTOR signaling, which are described in the following sections.

3.6. Modeling NFκB Signaling

Several lines of evidence indicate that the NFκB pathway is perhaps the most relevant biochemical or transcriptional target for the costimulatory activity of CD28 [29]. TCR- and CD28-mediated induction of the NFκB signaling pathway intersect at the central regulator, PKCθ (via mechanisms described in Section 3.5). PKCθ induces NFκB signaling by activating IκB kinase (IKK) (R_{37}) [21], which phosphorylates the inhibitor IκB (at positions S32 and S36). This triggers the rapid polyubiquitination (at positions K21 and K22) and proteolysis of IκB in the 26S proteasome complex (R_{38}). IκB degradation exposes the nuclear localization signal of NFκB, allowing its rapid translocation into the nucleus (R_{39}) [23,24].

Physiologically, the defining characteristic of IκBα is its ability to regulate rapid, but transient, induction of NFκB activity, owing to the participation of IκBα in an autoregulatory feedback loop. That is, the activation of NFκB causes the upregulation of transcription of IκBα, which, in turn, serves to shut off its own nuclear localization signal [36,37]. This upregulation occurs due to the presence of κB sites in the IκBα promoter. Thus, IκBα is thought to maintain the transient effect of inducing agents on the transcription of NFκB responsive genes [23]. It was demonstrated that *in vivo* degradation of IκBα is required for the appearance of NFκB in the nucleus. In addition, some investigators were able to demonstrate by co-immunoprecipitating rel proteins with IκBα, from stimulated cells treated with proteasome inhibitors to block the degradation of phosphorylated IκBα, that IκBα undergoes degradation mediated by the 26S proteasome after phosphorylation, but before dissociation [23].

We model the autoregulatory relationship between IκBα and NFκB (R_{38} and R_{39}) as a two-component negative feedback system:

$$\begin{aligned}
[p\dot{I\kappa B\alpha}] &= [IKK] - \alpha\,[pI\kappa B\alpha] - \beta\,[NF\kappa B] \\
[\dot{NF\kappa B}] &= \gamma\,[pI\kappa B\alpha] - \delta\,[NF\kappa B]
\end{aligned} \qquad (4)$$

The dynamic behavior of the negative feedback depends on the relative efficiency of the feedback regulation (α and β) regulating oscillation persistence *versus* self-regulation (γ and δ), causing oscillation damping. The output, $NF\kappa B$, can range from persistent oscillations (high feedback efficiency and no damping, $\alpha = \delta = 0$) to gradual rising to a plateau level (low feedback efficiency and high damping). Figure 7 demonstrates the phosphorylation of $I\kappa B\alpha$ by IKK, subsequent rapid degradation by the 26S proteasome and synthesis of $I\kappa B\alpha$ promoted by nuclear NFκB. The corresponding NFκB signal is also depicted. Model parameters are fitted to phospho-$I\kappa B\alpha$ data for which the experimental methodology is specified in Appendix 4. In our case, the fitted parameters correspond to damped oscillations (intermediate feedback efficiency and intermediate damping). Although we note that the sample size is small, the model simulation does show good agreement with the trend seen in the data.

Figure 7. $I\kappa B\alpha$ phosphorylation following 2 μg/mL αCD3 and αCD28 costimulation. The data shown are quantified western blot data from one experiment.

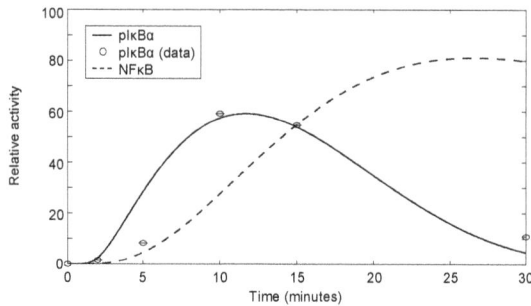

3.7. Modeling mTOR Signaling

Mammalian target of rapamycin (mTOR) is an evolutionary conserved serine/threonine protein kinase that is well known for its ability to control T-cell activation and differentiation. mTOR is necessary for TCR-induced signaling to drive differentiation into Th1, Th17 or Th2 effector types *in vitro* or *in vivo* under polarizing conditions. On the other hand, mTOR-deficiency drives CD4+ T-cells to the generation of FOXP3+ T-cells, even under normal activating conditions [30]. The ability of mTOR to regulate the expression of FOXP3 makes it a key player in the development of Tregs [30,31].

TCR and CD28 costimulation leads to the membrane recruitment of Akt (via mechanisms described in Section 3.5), where it is phosphorylated (at position T308) by PDK1 (R_{40}). Activated Akt phosphorylates tuberous sclerosis complex 2 (TSC2) in an inhibitory manner, yielding a separation of the TSC1/TSC2 complex (R_{41}). This causes Ras homolog enriched in brain (Rheb) to lose its GTPase-activating protein (GAP) activity (R_{42}). The resulting accumulation of Rheb-GTP promotes mTOR complex 1 (mTORC1) function (R_{43}), which acts to downregulate FOXP3 expression (R_{47}) [30]. By contrast, mTORC2 appears to be involved in cross-talk between Akt, FOXP3, PKCθ and NFκB. mTORC2 was shown to promote phosphorylation of Akt (R_{40}) and PKC (R_{36}), Akt activity, nuclear NFκB and to regulate Th1 and Th2 commitment in response to T-cell activation [32].

The model is augmented with the described reactions (R_{40}-R_{47}), which consist primarily of first- and second-order mass action kinetic expressions. Hill equations are employed in cases, such as PTEN deactivation by TCR triggering (R_{45}), TSC2 phosphorylation (R_{41}) and IL-2 (R_{46}) and FOXP3 (R_{47}) transcription, to better describe apparent cooperativity between transcriptions factors [26,31]. Kinetic parameters are chosen to reproduce qualitative observations of Akt and PKC phosphorylation in response to αCD3 and αCD28 costimulation originally reported by Lee *et al.* [32], the results of which are shown in Figure 8. In the model, as with the data, while αCD3 and αCD28 are each capable of inducing Akt and PKCθ phosphorylation, both are required to induce full activation of these pathways.

Figure 8. T-cell signaling in response to doses of αCD3 and αCD28 as measured by (**A**) P-Akt and (**B**) P-PKCθ. CD4+ T-cells were stimulated (40 min) with 0.5 mg/mL plate-bound αCD3, 2.5 mg/mL of soluble αCD28 or both. Bar graphs quantify phosphorylation of Akt and PKC, with each sample normalized to the level of unphosphorylated protein in one experiment representative of three replicates (data source: [32]).

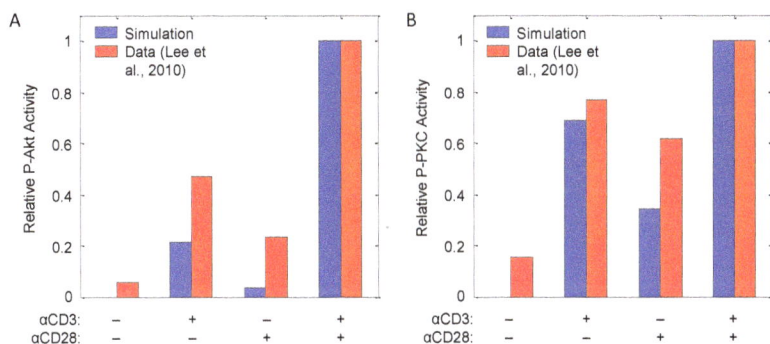

4. Results and Discussion

The following results and discussion subsections are divided into two main groups. First, the complete and calibrated model, as described in Section 3, is evaluated against additional data not used in the model tuning process. This experimental evidence and corroboration of the model's predictive accuracy is described in Section 4.1. The sections that follow then describe predictions made using the model. As mentioned in the introduction, these will mainly focus on the roles of signal strength and feedback loops on the regulation of T-cell activation.

4.1. Model Corroboration

4.1.1. Experimental Datasets for Model Corroboration

Experimental datasets for the purposes of model corroboration are compiled from the published literature. Only those that are commensurate with our model in terms of cell type, stimuli, output species and time scale are considered. Table 2 summarizes the attributes of the compiled dataset. Each dataset is used to analyze the model's ability to recapitulate the dynamics of the biological system as indicated. These datasets consist of dynamics (*i.e.*, time courses), dose responses (*i.e.*, multiple doses of an input or stimulus), input responses (*i.e.*, single doses of different input or stimuli combinations) and knockdown or knockout (*i.e.*, activity of a particular species is reduced or eliminated) experiments. In order to compare the model and observed behaviors directly, simulations are generated to mimic the experimental conditions used in the published experimental set-ups. Model simulations are shown with the corresponding data and error bars whenever possible.

Table 2. Description of experimental datasets used for model corroboration.

Species	Variable(s)	Experiment Type	Cell Type	Figure	Source
Ca^{2+}	αCD3, αCD28, PMA, ionomycin	Dynamic input response	Jurkat	9	[27]
NFAT	Cyclosporin A	Dose response	Jurkat	9	[53]
NFAT, AP1, NFκB	αCD3, αCD28, PMA	Input response	Jurkat	10	[27]
Lck, Erk	CD45	Knockdown	Murine DPthymocytes	11	[55]
Ca^{2+}	CD45	Knockout	Murine DP thymocytes	11	[56]
IL-2	SHP1, PMA, aTCR, ionomycin	Input response, Knockdown	Jurkat	12	[49]

4.1.2. Corroboration of Signaling Events and Transcription Factor Activation

In order to corroborate the model, we perform several *in silico* experiments and compare the simulations to published experimental results (described in Table 2). Figure 9A shows the changes in the intracellular calcium release in response to different combinations of stimuli. As shown in the figure, only αCD3 and ionomycin are capable of inducing calcium signaling, with ionomycin (administered with PMA) being the stronger of the two inducers. For the most part, the model simulations are in good qualitative agreement with the observations reported by Smeets *et al.* [27]; however, there are two noticeable discrepancies. First, these data show that the model slightly overestimates the system's calcium sensitivity to ionomycin stimulation relative to that of αCD3. Since the model is shown to fit calcium signaling quite well (see Figure 6), more information on the effects of ionomycin would likely resolve this issue. Second, the model's predicted calcium response demonstrates a slight sensitivity to PMA, a trend that is not observed in the data. This model behavior is likely caused by the presence of the positive feedback loop mediated by Erk. PMA is considered a DAG substitute, activating the substrates of DAG, such as RasGRP, which leads to the activation of Erk. Erk, in turn, promotes signaling by preventing SFK from activating the SHP1 negative feedback loop. In the model, the effect of elevated SFK facilitates further signaling through LAT, PLCγ, IP$_3$ and eventually calcium. This may be evidence that the model, particularly the positive feedback loop, may be more sensitive than these experimental data suggest.

Figure 9B shows changes in nuclear translocation of NFAT, a downstream target of calcium signaling, with varying doses of the calcineurin inhibitor cyclosporin A (CsA). Model simulations show excellent agreement with the observations reported by Clipstone *et al.* [53] in terms of half-maximal inhibitory concentration (IC_{50}) at ~3.5 ng/mL CsA. However, the model does appear to be slightly less sensitive to changes in inhibitor concentrations, as the slopes of the two curves are slightly different.

Figure 10 shows the activity of transcription factors NFAT, NFκB and AP1, as well as IL-2 synthesis in response to different combinations of stimuli. While there are a few obvious quantitative discrepancies between the model simulations and the data presented by Smeets *et al.*, there is good qualitative agreement among them in the input/output relationships. Nuclear localization of NFAT is only achieved when stimulating with combinations involving αCD3 (Figure 10A). This is because NFAT translocation requires strong calcium signaling, which αCD28 and PMA stimulation alone could not induce, consistent with [17] and Figure 9A. On the other hand, both αCD3 and PMA are sufficient to induce nuclear localization of NFκB and AP1 (Figure 10B,C). This is promoted by αCD28, but a combination of αCD3 and PMA have the greatest influence on translocation. IL-2 transcription in response to these same stimuli is shown in Figure 10D. The results indicate that TCR signaling is necessary for IL-2 transcription, but not sufficient. Coupling with CD28 coreceptor signaling causes a moderate increase in IL-2 transcription, but coupling with PMA results in the greatest observed increase, consistent with [17,27]. These experimental results corroborate the model and suggest that it is capable of accurately recapitulating the transcription profiles and, indirectly, the upstream signaling pathways, for a wide variety of stimuli.

Figure 9. Intracellular calcium and NFAT signaling in response to various combinations of stimuli. (**A**) Jurkat cells were stimulated as indicated (αCD3, αCD28, ionomycin: 1 µg/mL; PMA: 10 ng/mL), and intracellular Ca^{2+} release was monitored over time (data sampled from [27]). Solid lines represent corresponding model simulations. (**B**) NFAT activity in response to 10 µg/mL αCD3 stimulation and varying doses of calcineurin inhibitor cyclosporin A (CsA). The data are measured at 30 min post-stimulation (data sampled from [53]).

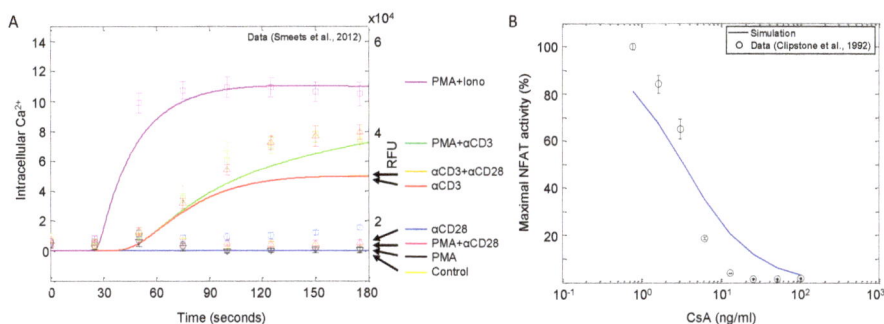

Figure 10. Activity of transcription factors (**A**) NFAT, (**B**) NFκB and (**C**) AP1, as well as (**D**) synthesis of IL-2 in response to various combinations of stimuli (αCD3, αCD28: 1 µg/mL; PMA: 10 ng/mL). Activity was measured 15 min post-stimulation (data sampled from [27]). Left and right axes correspond to model simulations and measured relative absorbance values, respectively.

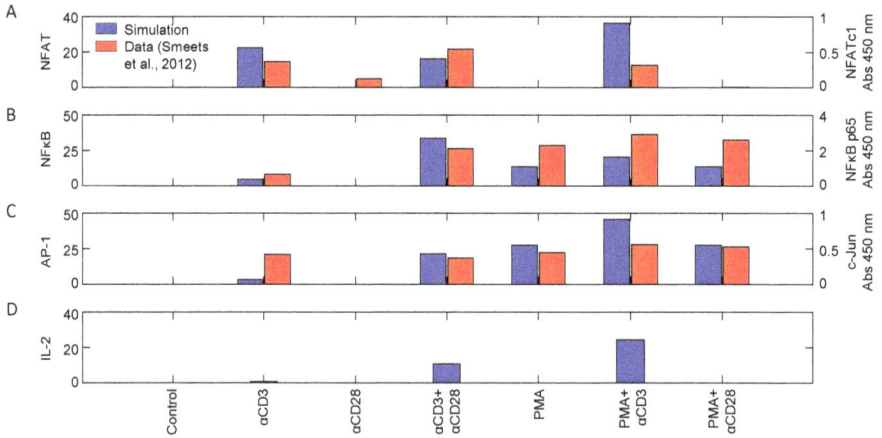

4.1.3. Corroboration of CD45 Activity

In this model, CD45 exists initially as fully active (proximal to the SFKs), then is deactivated (translocates away from the SFKs) by the formation of the TCR complex. At later time points, CD45 is reintroduced to the receptor cluster with an increased negative regulatory role. This results in dephosphorylating active SFK and phosphorylated ligand-bound TCR, thus terminating the TCR signal. Figure 11A depicts Lck-Y505 phosphorylation as a function of CD45 activity after 15 min of stimulation. At normal expression levels (*i.e.*, 100% of WT), Y505 phosphorylation is reduced by approximately 75%. Phosphorylation at Y505 is inversely related to CD45 activity, which is corroborated by the observations of McNeill *et al.* [55]. The model is able to approximate the data from 100% WT activity down through 5% WT activity very well. Only when CD45 is completely suppressed do the model and data diverge, although the trend is still present.

In CD45-null thymocytes, the p56Lck and p59Fyn tyrosine kinases are hyperphosphorylated, and p56Lck is found in its inactive conformation [56]. Both basal and TCR-stimulated tyrosine phosphorylation of TCRζ and CD3ε are also much reduced. These defects are associated with the failure of Zap70 kinase recruitment to the TCRζ chain; however, TCR-induced signaling is not entirely ablated. Figure 11B shows Erk phosphorylation as a function of αCD3 stimulation and CD45 activity. It is clearly evident that pErk increases with CD45 activity; however, above ~50%, there is qualitative switch, and pErk begins to decrease as CD45 expression approaches wild-type levels, which is also corroborated by observations by McNeill *et al.* [55].

Figure 11. Downstream TCR signaling in response to CD45 knockdown. (**A**) Model simulations of Lck phosphorylation at the negative regulatory residue Y505 as a function of CD45 activity (data sampled from [55]). (**B**) Model simulations of Erk phosphorylation 3 min after stimulation with 0 or 50 µg/mL αCD3 as a function of CD45 activity (data sampled from [55]). (**C**) Model simulations of intracellular calcium release over time in wild-type (WT) and CD45 knockdown mutant (5% of WT activity) stimulated with 10 µg/mL αCD3 at 90 s as a function of CD45 activity (data sampled from [56]).

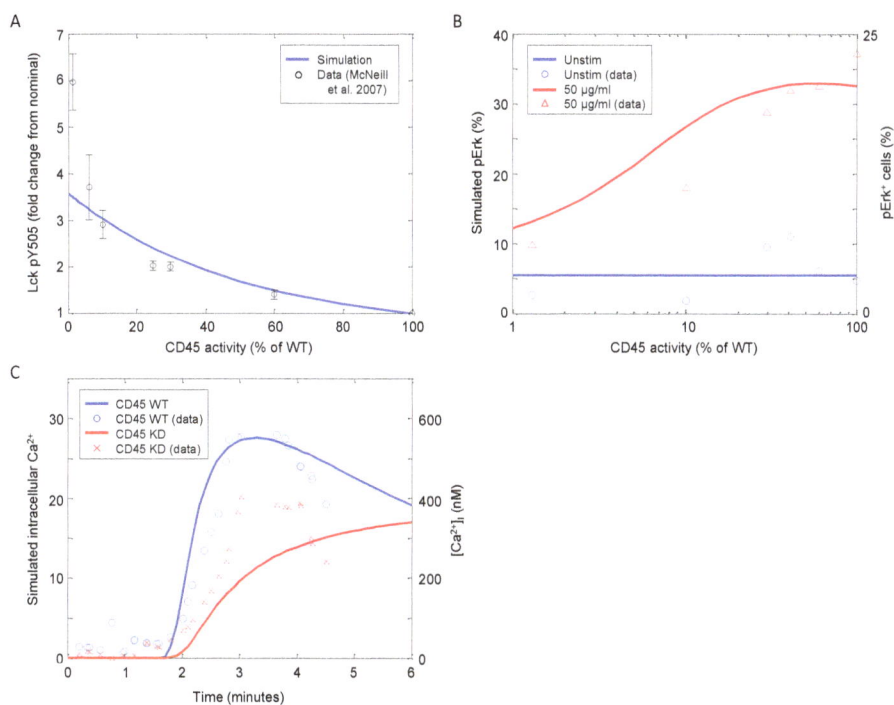

Furthermore, significant inositol phosphate and calcium signals are observed in CD45-null thymocytes. In our simulation, the CD45 defect is approximated with a knockdown to 5% of WT activity as complete CD45 knockout suppressed all signaling. Although greatly reduced from the nominal system, calcium signaling does not appear to be insignificant in CD45-defective CD4+ T-cells (Figure 11C). The molecular analysis presented by Stone *et al.* suggests that the threshold for TCR signal transduction is greatly increased in CD45-null T-cells, thus explaining the profound defects in thymic development [56].

4.1.4. Corroboration of SHP1 Activity

SHP1 functions as a negative regulator by deactivating Zap70 and SFK upon T-cell activation [49]. SHP1 also forms a negative feedback loop that is composed of SHP1 phosphorylation by activated Lck, binding of phospho-SHP1 to Lck and Lck inactivation by SHP1-mediated dephosphorylation [50]. To evaluate the ability of the model to capture the role of SHP1 feedback on T-cell activation, we simulate the SHP1 knockdown experiment presented by [49]. Figure 12 shows the stimulation of the IL-2 reporter with different combinations of stimuli and two different levels of SHP1 activity. Model-simulated results and data are depicted by bars and dots, respectively. In the wild-type system, neither PMA nor αCD3 alone are sufficient to upregulate IL-2 transcription. However, the combination of the two is able to generate a substantial signal. Pairing PMA with ionomycin produces the maximally-observed signal. In the SHP1-knockdown system, IL-2 synthesis is drastically increased as a result of αCD3 and PMA + αCD3 stimulation, thus demonstrating the lack of inhibition. PMA + ionomycin stimulation is not affected by SHP1 knockdown because SHP1 functions upstream of their substrates: RasGRP1, PKCθ and calcium. The experimental results show mostly good agreement with the model, suggesting that the model is able to adequately capture the reported behavior of SHP1.

Figure 12. IL-2 reporter stimulation in response to SHP1 knockdown. The model simulates the wild-type and SHP1 knockdown mutant (C453S mutation, resulting in catalytically inactive SHP1) stimulated with 10 μg/mL αCD3. IL-2 reporter stimulation was measured 2 h post-stimulation (data sampled from [49]).

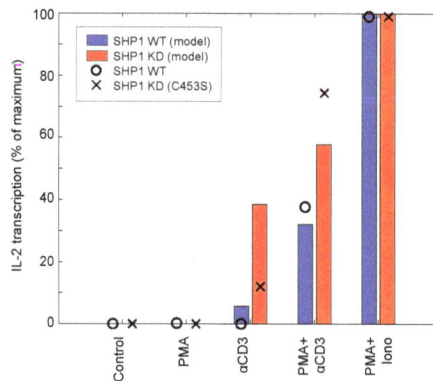

4.2. Weak CD28 Costimulation Predicted to Elevate FOXP3 Transcription

TCR signal intensity and costimulation is known to differentially affect the activation of T-cells. We study this by performing a two-way dose response experiment, measuring the effects of αCD3 and αCD28 on IL-2 and FOXP3 transcription. Figure 13A,B shows IL-2 and FOXP3 transcription, respectively, following 30 min of stimulation by various combinations of αCD3 and αCD28. At trivial doses of both stimuli, neither species is active, indicating that the cell is at rest. At high doses of both stimuli, IL-2 is the dominant species, indicating that the cell is fully active. However, when costimulation through CD28 is relatively weak, the balance of dominance shifts toward FOXP3, while leaving IL-2 at a much reduced level. Indeed, this behavior is corroborated by the observations by Kretschmer *et al.* [57] that weak TCR signals and limited costimulation have been linked to FOXP3 induction and the development of the regulatory phenotype in CD4+ T-cells.

Figure 13. (**A**) IL-2 and (**B**) FOXP3 transcription in response to stimulation by combinations of αCD3 and αCD28. Results show outputs 30 min after stimulation.

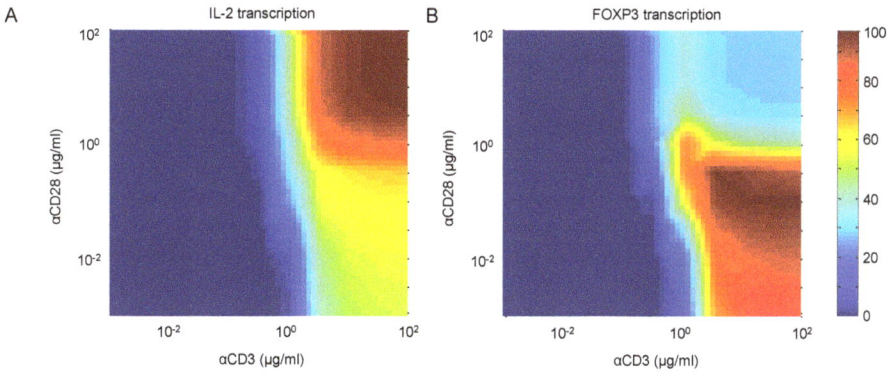

4.3. Reduced CD45 Activity Predicted to Elevate FOXP3 Transcription

We investigate the roles of CD45 and SHP1 on T-cell activation by simulating a two-way knockdown response experiment and measuring the effects on a small number of important outputs. Figure 14A–F show the model response to various CD45- and SHP1-knockdown scenarios. SHP1 clearly has the effect of downregulating most species playing a role in T-cell activation. As SHP1 activity is reduced (*i.e.*, moving top to bottom on the right axis in each plot), calcium, Erk, PKCθ and IL-2 are all upregulated. This is due to the loss of their transient behaviors and constitutive activation at elevated levels. For CD45, on the other hand, the analysis demonstrates a dual role as both activator and inhibitor of T-cell activation. As CD45 activity is reduced (*i.e.*, moving right to left on the top axis in each plot), TCR signaling increases until CD45 activity reaches ~1%–10% of the wild-type, then decreases to complete inactivation as CD45 is completely suppressed. This result indicates that CD45 is required for TCR signaling, without which the pool of SFKs would remain in the closed and inactive state; however, the highest level of activation is not actually achieved for wild-type CD45 expression, but rather at a reduced rate between approximately 1%–10% of the wild-type with intact SHP1 activity. Stronger TCR signaling can be achieved at full CD45 activity by downregulating SHP1.

It is evident that the system is quite sensitive to CD45 activity, particularly at the lower end of expression. By contrast, SHP1 is much more effective closer to wild-type levels; however, its activity is heavily dependent on the presence or absence of CD45. The results suggest that there exists a distinct threshold of CD45 activity (~1% of WT) that may be critical to determining the outcome of TCR-mediated activation. Below this threshold, the system is presumably unable to activate with every state remaining at baseline. However, at the threshold and above, the signaling molecules relating to Th development become active and the system is rescued. Recently,

McNeill *et al.* reported that only 3% of normal CD45 activity is sufficient to reconstitute CD45-deficient mice with normal numbers of mature T-cells, biased toward CD4+ T-cell lineage commitment [55].

Figure 14. Model response to various CD45- and SHP1-knockdown scenarios. (**A**) Intracellular calcium; (**B**) Erk; (**C**) PKCθ; (**D**) Akt; (**E**) IL-2; and (**F**) FOXP3 transcription factor activation in systems with various levels of CD45 and SHP1 downregulation. Note that downregulation is shown as percentages of wild-type activity (*i.e.*, 100% corresponds to normal function, 0% corresponds to full knockout). Results show outputs 30 min after stimulation by 10 µg/mL αCD3 and αCD28.

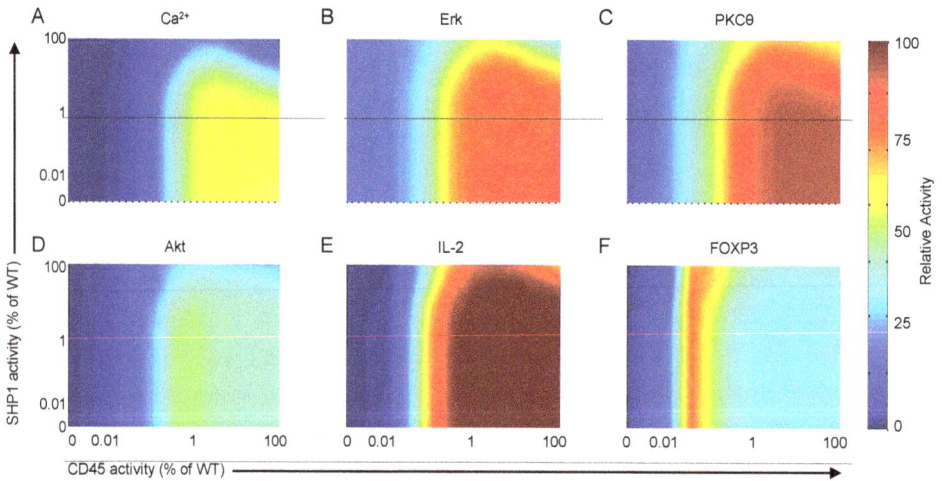

CD45 also appears to have an effect on the activation of FOXP3. At total CD45 suppression, the cell remains unresponsive, even at full stimulation. For wild-type activity, TCR signaling promotes mTORC1 activation, thereby also suppressing FOXP3. However, a reduction in CD45 activity to around the aforementioned threshold actually induces a substantial upregulation in FOXP3 expression while IL-2 remains low. This may be due to the transcriptional regulators of FOXP3 becoming sufficiently induced, while signaling remains too low to promote mTOR-mediated inhibition. This result suggests that CD45 downregulation may lead to weak IL-2 induction and increased FOXP3 expression, which is often observed in Treg lineage commitment.

5. Conclusions

Signal intensity and feedback regulation are known to be major determinants of the signaling profile for the TCR and its various coreceptors. While the exact nature of these relationships remains under investigation, it is believed to involve a complex signaling network with cross-talk between the calcium, Erk, PKCθ and mTOR signaling pathways.

In this manuscript, we present a mathematical model encompassing the signal transduction events relating to the process of TCR-mediated cell activation and gene expression. The model is able to reproduce key behaviors in: (1) ligand-induced TCR trafficking, synthesis and degradation; (2) early kinase and phosphatase interactions between SFK, Zap70, CD45 and SHP1; (3) CD28 costimulation; and (4) downstream signal transduction pathways leading to IL-2 and FOXP3 synthesis, including calcium, Erk, PKCθ and mTOR.

In addition to corroborating many experimentally-observed behaviors, the model is able to provide insight into the positive and negative regulatory roles of the phosphatases CD45 and SHP1 during T-cell activation. Analysis of the model demonstrates that: (1) SHP1 negative feedback is necessary for preventing hyperactivity in TCR signaling; (2) CD45 is required for TCR signaling, but also partially suppresses it at high activity levels; and (3) elevated FOXP3 and reduced IL-2 signaling, an expression profile often observed in developing Tregs, can be achieved either by weak TCR and CD28 stimulation or a severe reduction in CD45 activity. However, we do note that further investigation and experimental evidence is required in order to corroborate these predictions.

While the proposed model is a demonstrably powerful tool for predicting many events involved in CD4+ T-cell activation, there are certain caveats that need to be considered to accurately define the model's scope and capability. First, as a finite mathematical approximation, this model is inherently an abstraction of the biological reality. As such, the model does not attempt to explain every possible mechanism involved in the processes in question. For example, the Foxo family protein, Foxo1, is thought to bind to the FOXP3 locus and induce FOXP3 gene transcription. Foxo1 activity is subject to modulation by Akt kinase signaling, and Tregs have dampened Akt signaling in response to TCR stimulation compared with conventional T-cells. However, it is not within the scope of this work to test and corroborate every one of these mechanisms. Second, the purpose of the model is to simulate the dynamics of a small number of key species involved in T-cell activation or, more specifically, and clinically relevant, activation of human primary CD4+ T lymphocytes. Due to practical limitations on the availability of experimental data, however, the model is partially calibrated with datasets from alternate sources, including Jurkat cells. While Jurkat cells are an immortalized cell line of human T lymphocytes, they share many similarities with their primary counterparts and are very useful in studying T-cell signaling and IL-2 production. However, we do recognize that this constitutes an amalgamation of data from a variety of sources and that further investigation using primary T-cells is highly desirable.

As demonstrated in this manuscript, the mathematical model captures the key events in TCR and CD28 co-mediated signal transduction events leading to IL-2 and FOXP3 activation. As such, the model enables researchers to study these processes with a combination of broad scope, using a large-scale highly-connected network and with quantitative detail and accuracy not allowed by comparable models. In the future, we plan to use this mathematical model to design informative and hypothesis-driven experiments to refine our understanding of the dynamical nature of CD4+ T-cell activation. Our goal is to use such a model as a foundation for the design of strategies to drive and control T-cell activation and differentiation for therapeutic gain.

74

Acknowledgments

This research was funded by a grant awarded by the National Science Foundation, No. DMS-09002677. The authors would also like to thank current and former members of the Rundell research group for their helpful discussions and expertise.

Author Contributions

Jeffrey Perley performed the literature review, developed the model and contributed to the analysis. Judith Mikolajczak performed the experiments and collected the data used to develop and tune the model. Marietta Harrison and Gregery Buzzard contributed to designing and supervising the experimental and mathematical modeling aspects, respectively, and Ann Rundell oversaw all aspects of the research.

Appendix 1: Model Equations

Table A1 presents a comprehensive list of all species illustrated in Figure 1 that were chosen to participate in the model. The list also includes the rate equation (further defined in Table A2), biological meaning, the initial and total quantities for each state in units of molecules and the source (if applicable) providing these values or used to compute them. Stars (*) denote active forms.

Table A1. Summary of model states.

State	Rate Equation	Biological Meaning	Initial	Total	Source(s)
TCR_b	$R_{TCR_{lig}} - R_{TCR_p} - R_{iTCR_b}$	Ligand-bound TCR	0	2×10^5	[40]
TCR_p	$R_{TCR_p} - R_{iTCR_p}$	Phosphorylated TCR-ζ chain	0	2×10^5	[40]
TCR_i	$R_{iTCR_b} + R_{iTCR_p} + R_{iTCR_f} - R_{TCR_{exo}} - R_{TCR_{deg}}$	Internalized TCR	2×10^4	2×10^5	[40,46]
TCR_{deg}	$R_{TCR_{deg}} - R_{TCR_{synth}}$	Degraded TCR	2×10^4	2×10^5	[40,46]
Zapb	$R_1 - R_8$	Protein tyrosine kinase Zap70 bound to the phosphorylated TCR-ζ chain	0	9.3×10^4	[40]
Zap*	$R_8 - R_9$	Activated Zap70 (phosphorylated at Y493 in the activation loop)	0	9.3×10^4	[40]
Zapp	$R_9 - R_4 + R_5$	Doubly phosphorylated Zap70	9×10^3	9.3×10^4	[40]
SFKdp	$R_2 - R_3 - R_4 - R_{20}$	Src family kinases (including Lck and Fyn) with dephosphorylated inhibitory site (Y505 on Lck)	100	1×10^5	[40]
SFKdp-Zapp	$R_4 - R_{4a}$	Dephosphorylated SFK bound to pY319 of Zap70	0	9.3×10^4	[40]
SFKdpS59p	$R_{20} - R_{22}$	Dephosphorylated SFK phosphorylated at serine-59 by activated Erk	0	1×10^5	[40]
SFK*	$R_3 + R_5 - R_{21}$	Free fully activated SFK	0	1×10^5	[40]
SFK*-Zapp	$R_{4a} - R_5$	Fully activated SFK bound to pY319 of Zap70	0	9.3×10^4	[40]
SFK*S59p	$R_{21} + R_{22}$	Fully activated SFK phosphorylated at serine-59 by activated Erk	0	1×10^5	[40]
$CD45^P$	$-R_{23}$	Positive regulatory role of transmembrane tyrosine phosphatase CD45	1×10^5	1×10^5	[40]
$CD45_n$*	R_{23a}	Negative regulatory role of CD45	0	1×10^5	[40]
Cbpp	$R_6 - R_7$	Phosphorylated transmembrane scaffold protein Cbp (also known as PAG)	50	5×10^4	[40]
Csk*	R_7	Membrane-localized protein tyrosine kinase Csk recruited by Cbpp	2.5×10^3	5×10^4	[40]
SHP1*	R_{10}	Tyrosine phosphatase SHP1 recruited to the membrane and activated	0	1×10^6	[40]
LATp	$R_{11} - R_{12} - R_{19}$	Phosphorylated transmembrane protein LAT at tyrosine residues	0	5×10^4	[40]
SOSb	R_{19}	LATp-bound scaffold protein Grb2 and guanine nucleotide exchange factor SOS	0	5×10^4	[40]
PLCγp	R_{12}	Activated phospholipase Cγ and bound to LATp	0	5×10^4	[40]
DAG	$R_{13} - R_{13a} - R_{14}$	Diacylglycerol	0	1×10^7	[40]
IP_3	$R_{13} - R_{13b}$	Inositol 1,4,5-triphosphate	0	1×10^7	[40]
RasGRP*	R_{14}	Activated Ras guanine nucleotide releasing protein (RasGRP)	0	1×10^5	[40]
RasGTP	R_{15}	Guanine triphosphate (GTP)-bound Ras protein	0	1×10^7	[40]
Raf*	R_{16}	Phosphorylated and activated mitogen-activated protein (MAP) kinase kinase kinase Raf	0	4×10^4	[40]
Mek*	R_{17}	Phosphorylated and activated MAP kinase kinase Mek	0	2×10^7	[40]
Erk*	R_{18}	Phosphorylated and activated MAP kinase (MAPK) Erk	2.0960×10^6	2×10^7	[40]
AP1*	R_{26}	Activated transcription factor activator protein 1	0	2×10^7	Derived
Ca^{2+}	$R_{27} - R_{28}$	Cytoplasmic calcium ions released from intracellular stores (endoplasmic reticulum)	3.011×10^4	1×10^8	[58]
CaM*	R_{28}	Calcium-binding protein calmodulin bound to calcium	0	1×10^6	Derived
CN*	R_{29}	Activated calcium-dependent serine-threonine phosphatase calcineurin	0	1×10^6	Derived
NFATn	R_{30}	Dephosphorylated NFAT with unobstructed nuclear localization signal	0	1×10^6	Derived

Table A1. *Cont.*

State	Rate Equation	Biological Meaning	Initial	Total	Source(s)
CD28*	R_{32}	Ligand-bound and activated CD28 coreceptor	0	2×10^5	Derived
PI3K*	R_{33}	Activated phosphoinositide 3-kinase related kinase (PI3K)	0	1×10^4	Derived
PIP$_3$	R_{34}	Phosphatidylinositol 3,4,5-trisphosphate	0	1×10^7	Derived
PDK1*	R_{35}	Activated 3-phosphoinositide-dependent kinase-1 (PDK1)	0	1×10^4	Derived
PKCθ*	$R_{36} + R_{14a}$	Active protein kinase C-θ	0	1×10^4	Derived
IKK*	R_{37}	Activated IκB kinase	0	1×10^4	Derived
IκBαp	R_{38}	Phosphorylated IκB marked for proteasomal degradation	0	1×10^4	Derived
NFκBn	R_{39}	Nuclear NFκB	0	1×10^4	Derived
AKT*	R_{40}	Activated serine-threonine kinase Akt, also known as protein kinase B (PKB)	0	1×10^4	Derived
TSC1-TSC2	$-R_{41}$	GTPase-activating protein (GAP) consisting of tuberous sclerosis complex 1 (TSC1) and TSC2	1×10^4	1×10^4	Derived
RhebGTP	$-R_{42}$	GTP-bound Ras homolog enriched in brain (Rheb) GTPase	5×10^3	1×10^4	Derived
mTORC1*	R_{43}	Activated mammalian target of rapamycin (mTOR) complex 1	0	1×10^4	Derived
mTORC2*	R_{44}	Activated mTORC2	0	1×10^4	Derived
PTEN*	R_{45}	Activated phosphatase and tensin homolog (PTEN)	1×10^4	1×10^4	Derived
IL2	R_{46}	Interleukin-2, a cytokine marking T-cell activation	0	1×10^4	Derived
FOXP3	R_{47}	Forkhead box P3, regulator of regulatory T-cell development and function	0	1×10^4	Derived

Table A2 presents a comprehensive list of all reaction equations used to model the biochemical reactions of this system (illustrated in Figure 1). These expressions form the basis for the rate equations of the ODE-based model presented in Table A1. The list includes the reaction identifier, mathematical equation and biological meaning for each reaction.

Table A2. Summary of model equations.

Reaction		Equation	Biological Meaning
$R_{TCR_{lig}}$	=	$k_{f,r_{00}} * TCR_{lig} * TCR_f - k_{r,r_{00}} * TCR_b$	Association/dissociation of ligand and TCR complex
R_{TCR_p}	=	$(k_{f1,r_0} * (SFKact + SFKactS59p) + k_{f2,r_0} * SFKactZapp) *$ $TCR_b - (k_{r1,r_0} * SHP1act + k_{r3,r_0} * CD45_n + k_{r2,r_0})TCR_p$	SFK-mediated phosphorylation and SHP1/CD45-mediated dephosphorylation of ligand-bound TCR complex
R_{iTCR_f}	=	$k_{int} * TCR_f$	Internalization of free TCR
R_{iTCR_b}	=	$k_{int} * TCR_b$	Internalization of ligand-bound TCR
R_{iTCR_p}	=	$k_{int} * TCR_p$	Internalization of phosphorylated TCR
$R_{TCR_{exo}}$	=	$k_{exo} * TCR_i$	Exocytosis of internalized TCR
$R_{TCR_{deg}}$	=	$k_{deg} * TCR_i$	Degradation of internalized TCR
$R_{TCR_{synth}}$	=	$k_{synth} * TCR_{deg}$	Synthesis of new TCR
R_1	=	$k_{f,r_1} * (2 * TCRp - Zapb - Zapact - (Zapp - Zapp_0) -$ $SrcbactZapp - SrcdpZapp) * Zap - k_{r,r_1} * Zapb$	Association/dissociation of phosphorylated TCR complex and Zap70
R_2	=	$k_{f,r_2} * CD45^p * SFK - k_{r,r_2} * Cskact * SFKdp$	SFK dephosphorylation by CD45 and re-phosphorylation by Csk* at the inhibitory site (Y505 in Lck, Y528 in Fyn)
R_3	=	$(k_{f,r_3} * (TCR_b + TCR_p)) * SFKdp - (k_{r1,r_3} * SHP1act + k_{r3,r_3} *$ $CD45_n + k_{r2,r_3}) * SFKact$	SFK phosphorylation at the activation loop (Y394 in Lck, Y417 in Fyn) by autophosphorylation (or by another kinase) and dephosphorylation by SHP1
R_4	=	$k_{f,r_4} * (Zapp - Zapp_0) * SFKdp - k_{r,r_4} * SFKdpZapp$	Association/dissociation of Zapp and SFKdp
R_{4a}	=	$k_{f,r_4} * (Zapp - Zapp_0) * SFKdp - k_{r,r_4} * SFKdpZapp$	TCR-mediated phosphorylation of Zapp-bound SFKdp
R_5	=	$k_{f,r_5} * SFKactZapp - k_{r,r_5} * SFKact * (Zapp - Zapp_0)$	Dissociation/association of activated SFK and Zapp
R_6	=	$(k_{f1,r_6} * (SFKact + SFKactS59p) + k_{f2,r_6}) * Cbp - k_{r,r_6} *$ $CD45^p * Cbpp$	Cbp phosphorylation by activated SFK (or other kinases) and dephosphorylation by CD45
R_7	=	$k_{f,r_7} * Cbpp * Csk - k_{r,r_7} * Cskact$	Cbpp-mediated activation of Csk
R_8	=	$(k_{f1,r_8} * (SFKact + SFKactS59p) + k_{f2,r_8} * SFKactZapp) *$ $Zapb - (k_{r1,r_8} * SHP1act + k_{r2,r_8}) * Zapact$	Zap70 phosphorylation at the activation loop (Y493) by activated SFK (SFK* and SFK*-Zapp) and dephosphorylation by PTPs including SHP1
R_9	=	$(k_{f1,r_9} * (Zapact + (Zapp - Zapp_0) + SFKactZapp + SFKdpZapp) +$ $k_{f2,r_9} * (SFKact + SFKactS59p) + k_{f3,r_9} * SFKactZapp) *$ $Zapact - (k_{r1,r_9} * SHP1act + k_{r2,r_9}) * (Zapp - Zapp_0)$	Additional Zap70 phosphorylation at Y319 by activated SFK and Zap70 and dephosphorylation by PTPs including SHP1
R_{10}	=	$k_{f,r_{10}} * SFKact * SHP1 - k_{r,r_{10}} * SHP1act$	SHP1 activation (by SFK*) and deactivation
R_{11}	=	$k_{f,r_{11}} * (Zapact + (Zapp - Zapp_0) + SFKactZapp + SFKdpZapp) *$ $LAT - (k_{r1,r_{11}} * SHP1act + k_{r2,r_{11}}) * LATp$	LAT phosphorylation by activated Zap70 and dephosphorylation SHP1
R_{12}	=	$k_{f,r_{12}} * LATp * PLCg - k_{r,r_{12}} * PLCgp$	PLCγ phosphorylation by LATp
R_{13}	=	$k_{f,r_{13}} * PLCgp * PIP2$	PLCγ-mediated hydrolysis of PIP$_2$ to IP$_3$ and DAG
R_{13a}	=	$k_{r,r_{13a}} * DAG$	Degradation of DAG
R_{13b}	=	$k_{r,r_{13b}} * IP3$	Degradation of IP$_3$
R_{14}	=	$k_{f,r_{14}} * DAG * RasGRP - k_{r,r_{14}} * RasGRPact$	DAG-mediated activation of RasGRP
R_{14a}	=	$k_{f,r_{14a}} * DAG * PKC\theta$	DAG-mediated activation of PKCθ
R_{15}	=	$(k_{f1,r_{15}} * RasGRPact + k_{f2,r_{15}} * SOSb) * RasGDP - k_{r,r_{15}} * RasGTP$	RasGRP- and Grb2SOS-mediated activation of Ras
R_{16}	=	$k_{f,r_{16}} * RasGTP * (Raf) - k_{r,r_{16}} * Rafp$	Ras-mediated activation of Raf
R_{17}	=	$k_{f,r_{17}} * Rafp * Mek - k_{r,r_{17}} * Mekp$	Rafp-mediated activation of Mek
R_{18}	=	$k_{f,r_{18}} * Mekp * Erk - k_{r,r_{18}} * (Erkp - Erkp_0)$	Mekp-mediated activation of Erk

Table A2. *Cont.*

Reaction		Equation	Biological Meaning
R_{19}	=	$k_{f,r_{19}} * LATp * Grb2SOS - k_{r,r_{19}} * SOSb$	LATp-mediated association and activation of the Grb2-SOS complex
R_{20}	=	$k_{f,r_{20}} * SFKdp * (Erkp - Erkp_0) - k_{r,r_{20}} * SFKdpS59p$	Erkp-mediated phosphorylation of SFKdp at serine-59
R_{21}	=	$k_{f,r_{21}} * SFKact * (Erkp - Erkp_0) - k_{r,r_{21}} * SFKactS59p$	Erkp-mediated phosphorylation of SFKact at serine-59
R_{22}	=	$(k_{f,r_3} * (TCR_b + TCR_p)) * SFKdpS59p - (k_{r1,r_3} * SHP1act + k_{r3,r_3} * CD45_n + k_{r2,r_3}) * SFKactS59p$	TCR-mediated activation and SHP1-mediated deactivation of SFK-s59p
R_{23}	=	$k_{f,r_{23}} * (TCRb + TCRp) * CD45^P - k_{r,r_{23}} * (CD45_{tot} - CD45^P)$	Positive regulatory role of CD45 and translocation caused by receptor cluster formation
R_{23a}	=	$k_{f,r_{23a}} * (SFKact + SFKactZapp + SFKactS59p) * CD45_n - k_{r,r_{23a}} * (CD45_{tot} - CD45_n)$	Negative regulatory role of CD45 and recruitment to receptor cluster
R_{26}	=	$k_{f,r_{26}} * (Erkp - Erkp_0) * PKC\theta act * AP1 - k_{r,r_{26}} * AP1act$	Erkp- and PKCθ-mediated activation of AP1
R_{27}	=	$k_{f,r_{27}} * IP3 * Ca_s - k_{r,r_{27}} * (Ca - Ca_0)$	IP$_3$-induced calcium release into the cytoplasm
R_{28}	=	$k_{f,r_{28}} * (Ca - Ca_0) * CaM - k_{r,r_{28}} * CaMact$	Association/dissociation of calcium and calmodulin
R_{29}	=	$k_{f,r_{29}} * CaMact * CN - k_{r,r_{29}} * CNact$	Calmodulin-mediated activation of calcineurin
R_{30}	=	$k_{f,r_{30}} * CNact * NFATp - k_{r,r_{30}} * NFATn$	Calcineurin-mediated dephosphorylation and nuclear translocation of NFAT
R_{32}	=	$k_{f,r_{32}} * CD28_{lig} * CD28_f - k_{r,r_{32}} * CD28act$	Association/dissociation of ligand and CD28
R_{33}	=	$k_{f,r_{33}} * CD28act * (Zapact + (Zapp - Zapp0) + SFKdpZapp + SFKactZapp) * PI3K - k_{r,r_{33}} * PI3Kact$	PI3K activation by CD28 and Zap70 and deactivation
R_{34}	=	$k_{f,r_{34}} * PI3Kact * PIP2 - (k_{r1,r_{34}} * PTENact + k_{r2,r_{34}}) * PIP3$	PI3K-mediated phosphorylation of PIP$_2$ and PTEN-mediated dephosphorylation of PIP$_3$
R_{35}	=	$k_{f,r_{35}} * PIP3 * PDK1 - k_{r,r_{35}} * PDK1act$	PIP$_3$-mediated activation of PDK1
R_{36}	=	$(k_{f1,r_{36}} * PDK1act + k_{f2,r_{36}} * mTORC2act) * PKC\theta - k_{r,r_{36}} * PKC\theta act$	Activation of PKCθ mediated by PDK1, DAG, and mTORC2
R_{37}	=	$(k_{f1,r_{37}} * PKC\theta act + k_{f2,r_{37}} * Aktp) * IKK - k_{r,r_{37}} * IKKact$	PKCθ- and AKT-mediated activation of IKK
R_{38}	=	$k_{f,r_{38}} * IKKact - k_{r1,r_{38}} * IkBp - k_{r2,r_{38}} * NFkBn$	IKK-mediated phosphorylation and 26S proteasome-mediated degradation of IκBα; NFκB-induced synthesis of new IκBα
R_{39}	=	$k_{f,r_{39}} * IkBp - k_{r,r_{39}} * NFkBn$	Activation and nuclear translocation of NFκB
R_{40}	=	$(k_{f,r_{40}} * PDK1act + k_{f2,r_{40}} * mTORC2act + k_{f3,r_{40}} * PDK1act * mTORC2act) * Akt - k_{r,r_{40}} * Aktp$	PDK1- and mTORC2-mediated phosphorylation of AKT
R_{41}	=	$k_{f,r_{41}} * \dfrac{Aktp^{n_{r41}}}{Aktp^{n_{r41}} + k_{r_{41}}^{n_{r41}}} * TSC - k_{r,r_{41}} * TSC2p$	AKT-mediated phosphorylation, dissociation, and deactivation of TSC
R_{42}	=	$k_{f,r_{42}} * TSC * RhebGTP - k_{r,r_{42}} * RhebGDP$	GAP activity of TSC on Rheb
R_{43}	=	$k_{f,r_{43}} * (RhebGTP - RhebGTP_0) * mTORC1 - k_{r,r_{43}} * mTORC1act$	RhebGTP-mediated activation of mTORC1
R_{44}	=	$k_{f,r_{44}} * PI3Kact * mTORC2 - (k_{r1,r_{44}} * mTORC1act + k_{r2,r_{44}}) * mTORC2act$	PI3K-mediated activation and mTORC1-mediated inhibition of mTORC2
R_{45}	=	$(k_{f1,r_{45}} * FOXP3 + k_{f2,r_{45}}) * PTEN - k_{r,r_{45}} * \dfrac{(TCR_b + TCR_p)^{n_{r45}}}{(TCR_b + TCR_p)^{n_{r45}} + k_{r_{45}}^{n_{r45}}} * PTENact$	TCR-mediated inhibition and FOXP3-mediated activation of PTEN
R_{46}	=	$k_{f,r_{46}} * \dfrac{AP1act^{n_{1,r46}}}{AP1act^{n_{1,r46}} + k_{1,r46}} * \dfrac{NFAT^{n_{2,r46}}}{NFAT^{n_{2,r46}} + k_{2,r46}} * \dfrac{NFkBn^{n_{3,r46}}}{NFkBn^{n_{3,r46}} + k_{3,r46}} - (k_{r1,r_{46}} * FOXP3 + k_{r2,r_{46}}) * IL2$	AP1, NFAT, and NFkB regulate transcription of IL-2; FOXP3-mediated inhibition of IL-2
R_{47}	=	$k_{f,r_{47}} * \dfrac{AP1act^{n_{1,r47}}}{AP1act^{n_{1,r47}} + k_{1,r47}} * \dfrac{NFAT^{n_{2,r47}}}{NFAT^{n_{2,r47}} + k_{2,r47}} - (k_{r1,r_{47}} * (mTORC1 * mTORC2) + k_{r2,r_{47}}) * FOXP3$	AP1 and NFAT regulate transcription of FOXP3; mTOR-mediated inhibition of FOXP3

Table A3 presents a comprehensive list of all reaction parameters included in the model. The list also includes the biological meaning, value, 95% confidence interval (if applicable), units and source (if applicable) for each parameter. For parameters not provided by external sources, they are either estimated from data or explicitly derived to satisfy a condition of the model, for example, to ensure equilibrium when the system should be at rest. Confidence intervals are estimated for the model calibration data (Table 1) using the likelihood ratio method [59].

Table A3. Summary of model parameters.

Parameter	Biological Meaning	Value	95% CI	Units	Source
$k_{f,r_{00}}$	Association rate of ligand and TCR complex	0.0900	[0.0558, 0.1231]	$(mol·min)^{-1}$	Fitted
$k_{r,r_{00}}$	Dissociation rate of ligand and TCR complex	3×10^{-4}	$[1.2346 \times 10^{-4}, 5.5632 \times 10^{-4}]$	min^{-1}	Fitted
k_{f1,r_0}	Phosphorylation rate of ligand-bound TCR mediated by SFK* and SFK*-S59p	0.3000	[0.0475, 3.8961]	$(mol·min)^{-1}$	[40]
k_{f2,r_0}	Phosphorylation rate of ligand-bound TCR mediated by SFK*-Zapp	$1.13 \cdot k_{f1,r_0}$		$(mol·min)^{-1}$	[40]
k_{r1,r_0}	Dephosphorylation rate of ligand-bound TCR mediated by activated SHP1	0.0022	[0.0015, 0.0221]	$(mol·min)^{-1}$	[40]
k_{r2,r_0}	Constitutive dephosphorylation rate of ligand-bound TCR	16.1100	[1.0731, 61.7493]	min^{-1}	[40]

Table A3. *Cont.*

Parameter	Biological Meaning	Value	95% CI	Units	Source
k_{r3,r_0}	Dephosphorylation rate of ligand-bound TCR mediated by activated CD45	0.0300	$[8.1854 \times 10^{-4}, 0.1574]$	$(\text{mol} \cdot \text{min})^{-1}$	Fitted
$k_{int_{min}}$	Resting TCR internalization rate	0.0100		min^{-1}	[46]
$k_{int_{max}}$	Maximum induced TCR internalization rate	0.0380		min^{-1}	[46]
n_{int}	TCR internalization Hill coefficient	2		unitless	Derived
K_{int}	Enzyme quantity producing half-maximum TCR internalization rate	$0.05 \cdot [\text{PKC}\theta]_{total}$		mol	Derived
k_{exo}	Constitutive TCR exocytosis rate	0.0789		min^{-1}	Derived
$k_{deg_{min}}$	Resting TCR degradation rate	0.0011		min^{-1}	[45]
$k_{deg_{max}}$	Maximum induced TCR degradation rate	0.0033		min^{-1}	[45]
n_{deg}	TCR degradation Hill coefficient	2		unitless	Derived
K_{deg}	Enzyme quantity producing half-maximum TCR internalization rate	$0.05 \cdot [\text{SFK}]_{total}$		mol	Derived
k_{synth}	TCR synthesis rate	$k_{deg_{min}}$		min^{-1}	Derived
k_{f,r_1}	Association rate of Zap70 to phosphorylated TCRζ-chain	6×10^{-4}	$[5.4721 \times 10^{-5}, 9.6525 \times 10^{-4}]$	$(\text{mol} \cdot \text{min})^{-1}$	Fitted
k_{r,r_1}	Dissociation rate of Zap70 to phosphorylated TCRζ-chain	1.2600	$[0.4913, 41.7225]$	min^{-1}	Fitted
k_{f,r_2}	Dephosphorylation rate of SFK at the inhibitory site by CD45	3×10^{-6}	$[2.0755 \times 10^{-6}, 3.6068 \times 10^{-6}]$	$(\text{mol} \cdot \text{min})^{-1}$	Fitted
k_{r,r_2}	Phosphorylation rate of SFK at the inhibitory site by Csk	0.1199		$(\text{mol} \cdot \text{min})^{-1}$	Derived
k_{f,r_3}	Phosphorylation rate of SFKdp at the activation site mediated by TCRb and TCRp	13.7700	$[1.3177, 75.4525]$	$(\text{mol} \cdot \text{min})^{-1}$	Fitted
k_{r1,r_3}	Dephosphorylation rate of SFK* at the activation site by activated SHP1	k_{r1,r_0}		$(\text{mol} \cdot \text{min})^{-1}$	[40]
k_{r2,r_3}	Constitutive dephosphorylation rate of SFK* at the activation site	k_{r2,r_0}		min^{-1}	[40]
k_{r3,r_3}	Dephosphorylation rate of SFK* at the activation site by activated CD45	k_{r3,r_0}		$(\text{mol} \cdot \text{min})^{-1}$	Derived
k_{f,r_4}	Association rate of SFKdp to Zapp	0.0217	$[2.169 \times 10^{-4}, 2.1690]$	$(\text{mol} \cdot \text{min})^{-1}$	Fitted
k_{r,r_4}	Dissociation rate of SFKdp to Zapp	0.0025	$[2.415 \times 10^{-4}, 0.0151]$	min^{-1}	Fitted
$k_{f,r_{4a}}$	Phosphorylation rate of Zapp-bound SFKdp at the activation site mediated by TCRb and TCRp	k_{f,r_3}		$(\text{mol} \cdot \text{min})^{-1}$	[40]
$k_{r1,r_{4a}}$	Dephosphorylation rate of Zapp-bound SFK* at the activation site by activated SHP1	0.0068	$[0.0012, 0.0641]$	$(\text{mol} \cdot \text{min})^{-1}$	Fitted
$k_{r2,r_{4a}}$	Constitutive dephosphorylation rate of Zapp-bound SFK* at the activation site	13.4040	$[1.3409, 84.0744]$	min^{-1}	Fitted
$k_{r3,r_{4a}}$	Dephosphorylation rate of Zapp-bound SFK* at the activation site by activated CD45	0.0300	$[0.0019, 0.3542]$	$(\text{mol} \cdot \text{min})^{-1}$	Fitted
k_{f,r_5}	Dissociation rate of SFK* to Zapp	70.7880	$[68.7880, 102.3198]$	min^{-1}	Fitted
k_{r,r_5}	Association rate of SFK* to Zapp	k_{f,r_4}		$(\text{mol} \cdot \text{min})^{-1}$	[40]
k_{f1,r_6}	Phosphorylation rate of Cbp by SFK*	1.788×10^{-6}	$[1.0338 \times 10^{-6}, 1.788 \times 10^{-5}]$	$(\text{mol} \cdot \text{min})^{-1}$	Fitted
k_{f2,r_6}	Constitutive phosphorylation rate of Cbp	0.0207	$[0.0127, 0.1200]$	min^{-1}	Fitted
k_{r,r_6}	Dephosphorylation rate of Cbpp by CD45	1.9644×10^{-4}		$(\text{mol} \cdot \text{min})^{-1}$	Derived
k_{f,r_7}	Association rate of Csk to Cbpp	6.984×10^{-4}	$[1.2137 \times 10^{-5}, 0.0070]$	$(\text{mol} \cdot \text{min})^{-1}$	Fitted
k_{r,r_7}	Dissociation rate of Csk to Cbpp	0.6635		min^{-1}	Derived
k_{f1,r_8}	Phosphorylation rate of bound Zap by SFK* and SFK*S59p	0.0021	$[0.0015, 0.0022]$	$(\text{mol} \cdot \text{min})^{-1}$	Fitted
k_{f2,r_8}	Phosphorylation rate of bound Zap by SFK*-Zapp	$1.13 \cdot k_{f1,r_8}$		$(\text{mol} \cdot \text{min})^{-1}$	[40]
k_{r1,r_8}	Dephosphorylation rate of Zap* by activated SHP1	k_{r1,r_0}		$(\text{mol} \cdot \text{min})^{-1}$	[40]
k_{r2,r_8}	Constitutive dephosphorylation rate of Zap*	k_{r2,r_0}		min^{-1}	[40]
k_{f1,r_9}	Phosphorylation rate of Zap* by free and bound Zapp	3×10^{-4}	$[3 \times 10^{-6}, 3.6068 \times 10^{-4}]$	$(\text{mol} \cdot \text{min})^{-1}$	Fitted
k_{f2,r_9}	Phosphorylation rate of Zap* by SFK* and SFK*S59p	k_{f1,r_8}		$(\text{mol} \cdot \text{min})^{-1}$	[40]
k_{f3,r_9}	Phosphorylation rate of Zap* by SFK*-Zapp	$1.13 \cdot k_{f1,r_8}$		$(\text{mol} \cdot \text{min})^{-1}$	[40]
k_{r1,r_9}	Dephosphorylation rate of Zapp by activated SHP1	k_{r1,r_0}		$(\text{mol} \cdot \text{min})^{-1}$	[40]
k_{r2,r_9}	Constitutive dephosphorylation rate of Zapp	k_{r2,r_0}		min^{-1}	[40]
$k_{f,r_{10}}$	Activation rate of SHP1 by SFK*	8.19×10^{-6}	$[8.1121 \times 10^{-6}, 8.2192 \times 10^{-6}]$	$(\text{mol} \cdot \text{min})^{-1}$	Fitted
$k_{r,r_{10}}$	Deactivation rate of SHP1	0.3660	$[0.3450, 0.3861]$	min^{-1}	Fitted
$k_{f1,r_{11}}$	Phosphorylation rate of LAT by activated Zap	3×10^{-4}	$[2.5835 \times 10^{-4}, 3.9477 \times 10^{-4}]$	$(\text{mol} \cdot \text{min})^{-1}$	Fitted
$k_{r1,r_{11}}$	Dephosphorylation rate of LAT by activated SHP1	0.0020	$[0.0017, 0.0025]$	$(\text{mol} \cdot \text{min})^{-1}$	Fitted
$k_{r2,r_{11}}$	Constitutive dephosphorylation rate of SHP1	90	$[87.9332, 130.0896]$	min^{-1}	Fitted
$k_{f,r_{12}}$	Association rate of PLCγ and LATp (PLCγ is immediately phosphorylated)	0.0030	$[0.0021, 0.0033]$	$(\text{mol} \cdot \text{min})^{-1}$	Fitted
$k_{r,r_{12}}$	Lumped rate of dephosphorylation of PLCγ and its dissociation from LATp	300	$[295.1992, 302.0020]$	min^{-1}	Fitted
$k_{f,r_{13}}$	PIP$_2$ hydrolysis rate catalyzed by PLCγp	3×10^{-8}	$[2.9199 \times 10^{-8}, 3.0011 \times 10^{-8}]$	$(\text{mol} \cdot \text{min})^{-1}$	Fitted
$k_{r,r_{13a}}$	DAG degradation rate	2.4311	$[1.3990, 2.9228]$	min^{-1}	Fitted
$k_{r,r_{13b}}$	IP$_3$ degradation rate	$k_{r,r_{13a}}$		min^{-1}	Derived
$k_{f,r_{14}}$	Activation rate of RasGRP by DAG	0.0030	$[0.0025, 0.0052]$	$(\text{mol} \cdot \text{min})^{-1}$	Fitted
$k_{r,r_{14}}$	Inactivation rate of RasGRP	30	$[17.2632, 36.0679]$	min^{-1}	Fitted
$k_{f,r_{14a}}$	Activation rate of PKCθ by DAG	3×10^{-4}	$[2.7665 \times 10^{-5}, 3.7263 \times 10^{-4}]$	$(\text{mol} \cdot \text{min})^{-1}$	Fitted
$k_{f1,r_{15}}$	Rate of Ras guanine nucleotide exchange catalyzed by activated RasGRP	1.2×10^{-5}	$[8.3020 \times 10^{-6}, 1.7345 \times 10^{-5}]$	$(\text{mol} \cdot \text{min})^{-1}$	Fitted
$k_{f2,r_{15}}$	Rate of Ras guanine nucleotide exchange catalyzed by recruited SOS	1.2×10^{-6}	$[1.2111 \times 10^{-7}, 8.3020 \times 10^{-6}]$	$(\text{mol} \cdot \text{min})^{-1}$	Fitted
$k_{r,r_{15}}$	Constitutive rate of RasGTP hydrolysis to RasGDP	30	$[24.9529, 43.3632]$	min^{-1}	Derived
$k_{f,r_{16}}$	Activation rate of Raf by RasGTP	2.4×10^{-4}	$[1.6604 \times 10^{-4}, 2.8854 \times 10^{-4}]$	$(\text{mol} \cdot \text{min})^{-1}$	Fitted
$k_{r,r_{16}}$	Constitutive rate of Raf inactivation by phosphatase	30	$[24.9529, 43.3632]$	min^{-1}	Derived
$k_{f,r_{17}}$	Activation rate of Mek by activated Raf	0.0030	$[0.0025, 0.0036]$	$(\text{mol} \cdot \text{min})^{-1}$	Fitted

Table A3. *Cont.*

Parameter	Biological Meaning	Value	95% CI	Units	Source
$k_{r,r_{17}}$	Constitutive rate of Mek inactivation by phosphatase	30	[24.9529, 36.0679]	min^{-1}	Derived
$k_{f,r_{18}}$	Activation rate of Erk by activated Mek	3×10^{-6}	[2.4953×10^{-6}, 3.6068×10^{-6}]	$(mol \cdot min)^{-1}$	Fitted
$k_{r,r_{18}}$	Constitutive rate of Erk inactivation by phosphatase	30	[24.9529, 36.0679]	min^{-1}	Derived
$k_{f,r_{19}}$	Association rate of Grb-SOS complex to LATp	6×10^{-4}	[5.7393×10^{-6}, 9.151×10^{-4}]	$(mol \cdot min)^{-1}$	Fitted
$k_{r,r_{19}}$	Dissociation rate of Grb-SOS complex from LATp	30	[3.3632, 100]	min^{-1}	Derived
$k_{f,r_{20}}$	Phosphorylation rate of SFKdp at S59 by activated Erk	3×10^{-5}	[6.3433×10^{-7}, 0.0011]	$(mol \cdot min)^{-1}$	Fitted
$k_{r,r_{20}}$	Constitutive dephosphorylation rate of SFKdp at S59	30	[4.5543, 80.0101]	min^{-1}	Fitted
$k_{f,r_{21}}$	Phosphorylation rate of SFK* at S59 by activated Erk	$k_{f,r_{20}}$		$(mol \cdot min)^{-1}$	[40]
$k_{r,r_{21}}$	Constitutive dephosphorylation rate of SFK* at S59	$k_{r,r_{20}}$		min^{-1}	[40]
$k_{f,r_{22}}$	Phosphorylation rate of SFKdpS59p at the activation site mediated by TCRb and TCRp	k_{f,r_3}		$(mol \cdot min)^{-1}$	Derived
$k_{r1,r_{22}}$	Dephosphorylation rate of SFK*S59p at the activation site by activated SHP1	k_{r1,r_3}		$(mol \cdot min)^{-1}$	[40]
$k_{r2,r_{22}}$	Constitutive dephosphorylation rate of SFK*S59p at the activation site	k_{r2,r_3}		min^{-1}	[40]
$k_{r3,r_{22}}$	Dephosphorylation rate of SFK*S59p at the activation site by activated CD45	k_{r3,r_3}		$(mol \cdot min)^{-1}$	Derived
$k_{f,r_{23}}$	Translocation rate of CD45 mediated by receptor complex	3×10^{-7}	[6.5488×10^{-8}, 6.2679×10^{-7}]	$(mol \cdot min)^{-1}$	Fitted
$k_{r,r_{23}}$	Constitutive return rate of CD45	0.0030	[9.1223×10^{-5}, 0.1009]	min^{-1}	Fitted
$k_{f,r_{23a}}$	Activation rate of CD45 negative regulator by SFK*	6×10^{-7}	[4.151×10^{-7}, 3.98×10^{-5}]	$(mol \cdot min)^{-1}$	Fitted
$k_{r,r_{23a}}$	Constitutive deactivation rate of CD45 negative regulator	0.0030	[9.2531×10^{-5}, 0.0961]	min^{-1}	Fitted
$k_{f,r_{26}}$	Activation rate of AP1 by Erk* and PKCθ*	3×10^{-9}	[3.3439×10^{-10}, 7.9843×10^{-8}]	$(mol^2 \cdot min)^{-1}$	Fitted
$k_{r,r_{26}}$	Constitutive deactivation rate of AP1*	30	[23.1193, 43.8583]	min^{-1}	Fitted
$k_{f,r_{27}}$	Release rate of calcium stored in the endoplasmic reticulum by IP3	0.0300	[0.0212, 0.0380]	$(mol \cdot min)^{-1}$	Fitted
$k_{r,r_{27}}$	Constitutive calcium uptake rate	30	[19.0031, 54.1565]	min^{-1}	Fitted
$k_{f,r_{28}}$	Association rate of calmodulin to calcium	1.5×10^{-8}	[1.5267×10^{-9}, 1.0877×10^{-6}]	$(mol \cdot min)^{-1}$	Fitted
$k_{r,r_{28}}$	Dissociation rate of calmodulin from calcium	0.3000	[0.0182, 3.2856]	min^{-1}	Fitted
$k_{f,r_{29}}$	Activation rate of calcineurin by calmodulin*	1.5×10^{-7}	[5.8843×10^{-8}, 8.6424×10^{-6}]	$(mol \cdot min)^{-1}$	Fitted
$k_{r,r_{29}}$	Constitutive deactivation rate of calcineurin*	0.3000	[0.0943, 19.1092]	min^{-1}	Fitted
$k_{f,r_{30}}$	Activation rate of NFAT (NFAT immediately translocates to nucleus)	1.5×10^{-6}	[1.0869×10^{-6}, 1.9341×10^{-6}]	$(mol \cdot min)^{-1}$	Fitted
$k_{r,r_{30}}$	Constitutive deactivation rate of NFAT	0.3000	[0.0210, 0.3433]	min^{-1}	Fitted
$k_{f,r_{32}}$	Association rate of ligand to CD28 coreceptor	0.0300	[0.0199, 0.0360]	$(mol \cdot min)^{-1}$	Fitted
$k_{r,r_{32}}$	Dissociation rate of ligand from CD28 coreceptor	6×10^{-5}	[1.8795×10^{-5}, 7.4673×10^{-5}]	min^{-1}	Fitted
$k_{f,r_{33}}$	Activation rate of PI3K by ligand-bound CD28 and activated Zap	3×10^{-9}	[3.9548×10^{-10}, 5.1943×10^{-9}]	$(mol^2 \cdot min)^{-1}$	Fitted
$k_{r,r_{33}}$	Constitutive deactivation rate of PI3K	30	[15.3560, 37.5731]	min^{-1}	Fitted
$k_{f,r_{34}}$	Phosphorylation rate of PIP2 by PI3K*	3	[0.3923, 5.7161]	$(mol \cdot min)^{-1}$	Fitted
$k_{r1,r_{34}}$	Dephosphorylation rate of PIP3 by PTEN*	30	[17.4422, 65.3412]	$(mol \cdot min)^{-1}$	Fitted
$k_{r2,r_{34}}$	Constitutive dephosphorylation rate of PIP3	1×10^{-10}	[5.1834×10^{-11}, 2.7164×10^{-10}]	min^{-1}	Fitted
$k_{f,r_{35}}$	Activation rate of PDK1 by PIP3	3×10^{-5}	[9.3939×10^{-6}, 5.7164×10^{-5}]	$(mol \cdot min)^{-1}$	Fitted
$k_{r,r_{35}}$	Constitutive deactivation rate of PDK1*	30	[5.4422, 40.4623]	min^{-1}	Fitted
$k_{f1,r_{36}}$	Activation rate of PKCθ by PDK1*	3×10^{-6}	[8.4379×10^{-7}, 1.0892×10^{-5}]	$(mol \cdot min)^{-1}$	Fitted
$k_{f2,r_{36}}$	Activation rate of PKCθ by mTORC2*	3×10^{-5}	[8.0982×10^{-7}, 9.9339×10^{-5}]	$(mol \cdot min)^{-1}$	Fitted
$k_{r,r_{36}}$	Constitutive deactivation rate of PKCθ*	0.3000	[0.1754, 1.3217]	min^{-1}	Fitted
$k_{f1,r_{37}}$	Activation rate of IKK by PKCθ*	0.0015	[9.0321×10^{-4}, 0.0065]	$(mol \cdot min)^{-1}$	Fitted
$k_{f2,r_{37}}$	Activation rate of IKK by Aktp	0.0030	[0.0019, 0.0046]	$(mol \cdot min)^{-1}$	Fitted
$k_{r,r_{37}}$	Constitutive deactivation rate of IKK*	15	[5.3394, 48.7211]	min^{-1}	Fitted
$k_{f,r_{38}}$	Phosphorylation rate of IκBα by IKK*	0.4500	[0.4109, 0.5512]	min^{-1}	Fitted
$k_{r1,r_{38}}$	Proteasomal degradation rate of pIκBα	0.1500	[0.1093, 0.1978]	min^{-1}	Fitted
$k_{r2,r_{38}}$	Deactivation rate of IκBα by NFκB	0.1500	[0.0936, 0.2380]	min^{-1}	Fitted
$k_{f,r_{39}}$	Activation rate of NFκB by IκBα deactivation	0.1500	[0.0994, 0.2121]	min^{-1}	Fitted
$k_{r,r_{39}}$	Constitutive deactivation rate of NFκB	0.0150	[0.0124, 0.0272]	min^{-1}	Fitted
$k_{f1,r_{40}}$	Phosphorylation rate of Akt by PDK1*	1×10^{-10}	[7.9346×10^{-11}, 2.7832×10^{-10}]	$(mol \cdot min)^{-1}$	Fitted
$k_{f2,r_{40}}$	Phosphorylation rate of Akt by mTORC2*	1×10^{-10}	[9.3976×10^{-11}, 1.7832×10^{-10}]	$(mol \cdot min)^{-1}$	Fitted
$k_{f3,r_{40}}$	Phosphorylation rate of Akt by PDK1* and mTORC2*	1×10^{-8}	[6.4853×10^{-9}, 5.7164×10^{-8}]	$(mol^2 \cdot min)^{-1}$	Fitted
$k_{r,r_{40}}$	Dephosphorylation rate of Aktp by phosphatases	1	[0.5422, 7.0427]	min^{-1}	Derived
$k_{f,r_{41}}$	Phosphorylation rate of the TSC2 subunit of the TSC1-TSC2 complex (complex immediately dissociates)	10	[6.8726, 30.3895]	min^{-1}	Fitted
$k_{r,r_{41}}$	Dephosphorylation rate of TSC2p and its association with TSC1	1	[0.3783, 8.2627]	min^{-1}	Derived
$n_{r_{41}}$	TSC2 phosphorylation Hill coefficient	5		unitless	Derived
$k_{r_{41}}$	Enzyme quantity producing half-maximum TSC2 phosphorylation rate	$0.15 \cdot [Akt]_{total}$		mol	Fitted
$k_{f,r_{42}}$	Rate of Rheb guanine nucleotide exchange catalyzed by TSC1-TSC2 complex	1×10^{-4}	[3.6763×10^{-5}, 9.9339×10^{-4}]	$(mol \cdot min)^{-1}$	Derived
$k_{r,r_{42}}$	Rate of RhebGDP to RhebGTP exchange	1	[0.3020, 10.5309]	min^{-1}	Derived
$k_{f,r_{43}}$	Activation rate of mTORC1 by RhebGTP	1×10^{-4}	[6.3984×10^{-5}, 0.0047]	$(mol \cdot min)^{-1}$	Fitted
$k_{r,r_{43}}$	Constitutive deactivation rate of mTORC1*	1	[0.2653, 4.7164]	min^{-1}	Derived
$k_{f,r_{44}}$	Activation rate of mTORC2 by PI3K*	0.0030	[2.7360e-4, 0.0993]	$(mol \cdot min)^{-1}$	Fitted
$k_{r1,r_{44}}$	Deactivation rate of mTORC2* by mTORC1	0.0030	[0.0023, 0.0102]	$(mol \cdot min)^{-1}$	Fitted
$k_{r2,r_{44}}$	Constitutive deactivation rate of mTORC2*	3	[1.3631, 3.8319]	min^{-1}	Fitted
$k_{f1,r_{45}}$	Activation rate of PTEN by FOXP3*	1×10^{-4}	[1.8647×10^{-5}, 0.0010]	$(mol \cdot min)^{-1}$	Derived
$k_{f2,r_{45}}$	Constitutive activation of PTEN	1	[0.1925, 18.9529]	min^{-1}	Derived
$k_{r,r_{45}}$	Deactivation of PTEN* by TCR complex	20	[11.4372, 106.1915]	min^{-1}	Fitted
$n_{r_{45}}$	PTEN* deactivation Hill coefficient	10		unitless	Derived

Table A3. *Cont.*

Parameter	Biological Meaning	Value	95% CI	Units	Source
$k_{r_{45}}$	Ligand-bound TCR quantity producing half-maximum PTEN* deactivation	$0.75 \cdot [\text{TCR}]_{total}$		mol	Derived
$k_{f,r_{46}}$	Activation rate of IL-2 activity	1×10^4	$[574.7247, 1.8543 \times 10^5]$	$\text{mol} \cdot \text{min}^{-1}$	Fitted
$n_{1,r_{46}}$	Hill coefficient for AP1-induced IL-2 activity	2		unitless	Derived
$n_{2,r_{46}}$	Hill coefficient for NFAT-induced IL-2 activity	2		unitless	Derived
$n_{3,r_{46}}$	Hill coefficient for NFκB-induced IL-2 activity	2		unitless	Derived
$k_{1,r_{46}}$	AP1 quantity producing half-maximum IL-2 activity	$0.1 \cdot [\text{AP1}]_{total}$		mol	Derived
$k_{2,r_{46}}$	NFAT quantity producing half-maximum IL-2 activity	$0.3 \cdot [\text{NFAT}]_{total}$		mol	Derived
$k_{3,r_{46}}$	NFκB quantity producing half-maximum IL-2 activity	$0.1 \cdot [\text{NFκB}]_{total}$		mol	Derived
$k_{r1,r_{46}}$	Deactivation rate of IL-2 activity by FOXP3	1×10^{-4}	$[3.4974 \times 10^{-5}, 0.0093]$	$(\text{mol} \cdot \text{min})^{-1}$	Fitted
$k_{r2,r_{46}}$	Constitutive deactivation rate of IL-2	1	$[0.1832, 8.9847]$	min^{-1}	Derived
$k_{f,r_{47}}$	Activation rate of FOXP3 activity	1×10^4	$[748.5767, 2.4329 \times 10^4]$	$\text{mol} \cdot \text{min}^{-1}$	Fitted
$n_{1,r_{47}}$	Hill coefficient for AP1-induced FOXP3 activity	2		unitless	Derived
$n_{2,r_{47}}$	Hill coefficient for NFAT-induced FOXP3 activity	2		unitless	Derived
$k_{1,r_{47}}$	AP1 quantity producing half-maximum FOXP3 activity	$0.1 \cdot [\text{AP1}]_{total}$		mol	Derived
$k_{2,r_{47}}$	NFAT quantity producing half-maximum FOXP3 activity	$0.1 \cdot [\text{NFAT}]_{total}$		mol	Derived
$k_{r1,r_{47}}$	Deactivation rate of FOXP3 activity by mTOR	2×10^{-7}	$[7.3598 \times 10^{-8}, 8.9275 \times 10^{-6}]$	$(\text{mol}^{\cdot}2\text{min})^{-1}$	Fitted
$k_{r2,r_{47}}$	Constitutive deactivation rate of FOXP3 activity	1	$[0.3685, 4.7328]$	min^{-1}	Derived

Appendix 2: Parameter Sensitivity Analysis

Figure A1 shows the results of the parameter sensitivity analysis used to determine which parameters should be considered for parameter identification. The sensitivity indices are standardized by rows, enabling the parameters to be ranked for each model state independently. For a given model state, a value shown in red denotes that the state is relatively sensitive to perturbations in the corresponding parameter. On the other hand, a value shown in green implies that the particular state is insensitive to the parameter in question. As shown in Figure A1, it can be common for certain states to be only sensitive to a fraction of the parameters. This blocking structure is leveraged during the parameter identification process to reduce the dimension of the uncertain parameter space when calibrating individual modules.

Appendix 3: Calcium Flux Measurements

Jurkat T leukemia cells (Jurkat clone E6.1; ATCC) were harvested in log phase growth ($4–8 \times 10^5$ cell/mL), washed once with Hank's Balanced Salt Solution (HBSS) without phenol red. Phenol red-free, calcium containing HBSS (Lonza) was used for the remainder of the experiment. Cells were resuspended at 2×10^6 cells/mL in HBSS supplemented with 2.5 mM probenecid (unloaded control) or in HBSS with 1 μM Fluo-4 AM, 2.5 mM probenecid and 10 μL/mL PowerLoad (Invitrogen, Carlsbad, USA). Cells were incubated for 30 min at room temperature in the dark. Following incubation, cells were washed twice with 1 mL HBSS with 2.5 mM probenecid and then resuspended in 100 μL HBSS plus 2.5 mM probenecid. Cells were loaded into a black-walled, clear bottom 96-well plate. Wells with cells were surrounded two wells deep with 100 μL water each as a temperature buffer. The plate was covered in foil, and cells were allowed to settle for 5 min at 37 °C.

80

Figure A1. Parameter sensitivity indices. Values are standardized by rows. Red = most sensitive; Green = least sensitive.

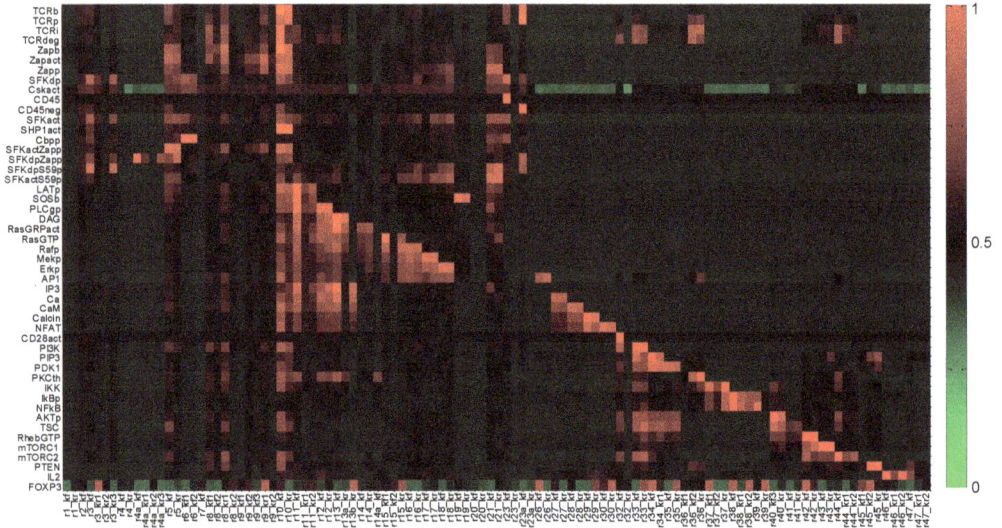

Fluorescence was measured on a BioTek plate reader (BioTek Instruments, Winooski, VT, USA) using the following parameters: excitation wavelength, 495 nm; emission, 520 nm; temperature, 37 °C. Prior to stimulation, a background reading was taken. The plate was ejected from the plate reader and 10 µg/mL αCD3ε or 1 µg/mL ionomycin (positive control) were added to the wells using a repeat dispense pipette. Following the addition of stimulants, a kinetic measurement run was done over 5–10 min, measuring fluorescence every 10 s. From the obtained data, the background reading was subtracted.

Appendix 4: Phospho-IκBα Measurements

Jurkat cells were grown in RPMI 1640 (Sigma-Aldrich, St. Louis, MO, USA) supplemented with 7.5% heat-inactivated fetal bovine serum (Bio-West, Logan, UT, USA), 1 mM sodium pyruvate (Life Technologies, Carlsbad, CA, USA), 12.5 mM HEPES pH 7.4 (Sigma-Aldrich, St. Louis, MO, USA), 12 µM sodium bicarbonate (Sigma-Aldrich, St. Louis, MO, USA) 50 µM 2-mercaptoethanol (Sigma-Aldrich, St. Louis, MO, USA), 50 µg/mL streptomycin and 50 units/mL penicillin in an incubator at 37 °C in humidified air containing 5% carbon dioxide. Cells were harvested in log-phase growth at a density of 1×10^7 cells per treatment. The treatments were stimulated with either 2 µg/mL αCD3 or 2 µg/mL αCD3 plus αCD28 in a 37 °C water bath for up to 30 min. Samples of 1×10^6 cells were taken at the indicated time points and lysed in 1% NP40 lysis buffer (1% NP40, 25 mM Tris, pH 7.4, 150 mM NaCl, 5 mM EDTA, 1 mM NaV, 10 mM NaF, 10 µg/mL each of aprotinin and leupeptin) for 15 min on ice. Lysates were centrifuged for 5 min at $18,000 \times$ g at 4 °C. The supernatant was added to the same volume of 2X protein solubilizing mixture (PSM, 25% (w/v)

sucrose, 2.5% (w/v) sodium dodecyl sulfate, 25 mM Tris, 2.5 mM EDTA, 0.05% bromophenol blue) and boiled for five min. Proteins were separated via SDS-PAGE, blotted for α-tubulin, phospho-IκB (S32/36) and IκB (Cell Signaling Technology, Beverly, MA, USA). IRDye 800 and 680 secondary anti-mouse and anti-rabbit antibodies (Li-Cor Biosciences, Lincoln, OR, USA) were used for signal detection using an Odyssey infrared scanner. Blot images were analyzed using ImageJ to produce quantitative data for model comparison.

Conflicts of Interest

The authors declare no conflict of interest.

References

1. Yamane, H.; Paul, W.E. Cytokines of the gamma(c) family control CD4(+) T-cell differentiation and function. *Nat. Immunol.* **2012**, *13*, 1037–1044.
2. Zhu, J.F.; Paul, W.E. Heterogeneity and plasticity of T helper cells. *Cell Res.* **2010**, *20*, 4–12.
3. Doherty, P.C.; Zinkernagel, R.M. Biological role for major histocompatibility antigens. *Lancet* **1975**, *1*, 1406–1409.
4. Marrack, P.; Hannum, C.; Harris, M.; Haskins, K.; Kubo, R.; Pigeon, M.; Shimonkevitz, R.; White, J.; Kappler, J. Antigen-specific, major histocompatibility complex-restricted T-cell receptors. *Immunol. Rev.* **1983**, *76*, 131–145.
5. Weiss, A.; Imboden, J.; Hardy, K.; Manger, B.; Terhorst, C.; Stobo, J. The role of the T3-antigen receptor complex in T-cell activation. *Annu. Rev. Immunol.* **1986**, *4*, 593–619.
6. Veillette, A.; Bookman, M.A.; Horak, E.M.; Bolen, J.B. The CD4 and CD8 T-cell surface-antigens are associated with the internal membrane tyrosine-protein kinase p56Lck. *Cell* **1988**, *55*, 301–308.
7. Palacios, E.H.; Weiss, A. Function of the Src-family kinases, Lck and Fyn, in T-cell development and activation. *Oncogene* **2004**, *23*, 7990–8000.
8. Hermiston, M.L.; Xu, Z.; Weiss, A. CD45: A critical regulator of signaling thresholds in immune cells. *Annu. Rev. Immunol.* **2003**, *21*, 107–137.
9. Saunders, A.E.; Johnson, P. Modulation of immune cell signalling by the leukocyte common tyrosine phosphatase, CD45. *Cell. Signal.* **2010**, *22*, 339–348.
10. Chan, A.C.; Iwashima, M.; Turck, C.W.; Weiss, A. ZAP-70—A 70kd protein-tyrosine kinase that associates with the TCR zeta-chain. *Cell* **1992**, *71*, 649–662.
11. Iwashima, M.; Irving, B.A.; Vanoers, N.S.C.; Chan, A.C.; Weiss, A. Sequential interactions of the TCR with 2 distinct cytoplasmic tyrosine kinases. *Science* **1994**, *263*, 1136–1139.
12. Szabo, M.; Czompoly, T.; Kvell, K.; Talaber, G.; Bartis, D.; Nemeth, P.; Berki, T.; Boldizsar, F. Fine-tuning of proximal TCR signaling by ZAP-70 tyrosine residues in JurkaT-cells. *Int. Immunol.* **2012**, *24*, 79–87.
13. Berridge, M.J.; Irvine, R.F. Inositol trisphosphate, a novel 2nd messenger in cellular signal transduction. *Nature* **1984**, *312*, 315–321.

14. Nishibe, S.; Wahl, M.I.; Hernandez-Sotomayor, S.M.T.; Tonks, N.K.; Rhee, S.G.; Carpenter, G. Increase of the catalytic activity of phospholipase C-gamma-1 by tyrosine phosphorylation. *Science* **1990**, *250*, 1253–1256.

15. Zhang, W.G.; Sloan-Lancaster, J.; Kitchen, J.; Trible, R.P.; Samelson, L.E. LAT: The ZAP-70 tyrosine kinase substrate that links T-cell receptor to cellular activation. *Cell* **1998**, *92*, 83–92.

16. Imboden, J.B.; Stobo, J.D. Transmembrane signaling by the T-cell antigen receptor—Perturbation of the T3-antigen receptor complex generates inositol phosphates and releases calcium-ions from intracellular stores. *J. Exp. Med.* **1985**, *161*, 446–456.

17. Imboden, J.B.; Weiss, A.; Stobo, J.D. The antigen receptor on a human T-cell line initiates activation by increasing cytoplasmic free calcium. *J. Immunol.* **1985**, *134*, 663–665.

18. Macian, F. NFAT proteins: Key regulators of T-cell development and function. *Nat. Rev. Immunol.* **2005**, *5*, 472–484.

19. Roose, J.P.; Mollenauer, M.; Gupta, V.A.; Stone, J.; Weiss, A. A diacylglycerol-protein kinase C-RasGRP1 pathway directs Ras activation upon antigen receptor stimulation of T-cells. *Mol. Cell. Biol.* **2005**, *25*, 4426–4441.

20. Genot, E.; Cantrell, D.A. Ras regulation and function in lymphocytes. *Curr. Opin. Immunol.* **2000**, *12*, 289–294.

21. Isakov, N.; Altman, A. Protein kinase C theta in T-cell activation. *Annu. Rev. Immunol.* **2002**, *20*, 761–794.

22. Baeuerle, P.A.; Baltimore, D. I-kappa-B—A specific inhibitor of the NF-kappa-B transcription factor. *Science* **1988**, *242*, 540–546.

23. Ghosh, S.; May, M.J.; Kopp, E.B. NF-kappa B and rel proteins: Evolutionarily conserved mediators of immune responses. *Annu. Rev. Immunol.* **1998**, *16*, 225–260.

24. Hacker, H.; Karin, M. Regulation and function of IKK and IKK-related kinases. *Science's STKE* **2006**, *2006*, re13.

25. Jain, J.N.; McCaffrey, P.G.; Miner, Z.; Kerppola, T.K.; Lambert, J.N.; Verdine, G.L.; Curran, T.; Rao, A. The T-cell transcription factor NFAT(p) is a substrate for calcineurin and interacts with Fos and Jun. *Nature* **1993**, *365*, 352–355.

26. Jain, J.; Loh, C.; Rao, A. Transcriptional regulation of the IL-2 gene. *Curr. Opin. Immunol.* **1995**, *7*, 333–342.

27. Smeets, R.L.; Fleuren, W.W.M.; He, X.; Vink, P.M.; Wijnands, F.; Gorecka, M.; Klop, H.; Bauerschmidt, S.; Garritsen, A.; Koenen, H.J.P.M.; *et al.* Molecular pathway profiling of T lymphocyte signal transduction pathways; Th1 and Th2 genomic fingerprints are defined by TCR and CD28-mediated signaling. *BMC Immunol.* **2012**, *13*, doi:10.1186/1471-2172-13-12.

28. Powell, J.D.; Ragheb, J.A.; Kitagawa-Sakakida, S.; Schwartz, R.H. Molecular regulation of interleukin-2 expression by CD28 co-stimulation and anergy. *Immunol. Rev.* **1998**, *165*, 287–300.

29. Kane, L.P.; Lin, J.; Weiss, A. It's all Rel-ative: NF-kappa B and CD28 costimulation of T-cell activation. *Trends Immunol.* **2002**, *23*, 413–420.

30. Powell, J.D.; Pollizzi, K.N.; Heikamp, E.B.; Horton, M.R. Regulation of Immune Responses by mTOR. *Annu. Rev. Immunol.* **2012**, *30*, 39–68.

31. Huehn, J.; Polansky, J.K.; Hamann, A. Epigenetic control of FOXP3 expression: The key to a stable regulatory T-cell lineage? *Nat. Rev. Immunol.* **2009**, *9*, 83–89.

32. Lee, K.; Gudapati, P.; Dragovic, S.; Spencer, C.; Joyce, S.; Killeen, N.; Magnuson, M.A.; Boothby, M. Mammalian Target of Rapamycin Protein Complex 2 Regulates Differentiation of Th1 and Th2 Cell Subsets via Distinct Signaling Pathways. *Immunity* **2010**, *32*, 743–753.

33. Saez-Rodriguez, J.; Simeoni, L.; Lindquist, J.A.; Hemenway, R.; Bommhardt, U.; Arndt, B.; Haus, U.U.; Weismantel, R.; Gilles, E.D.; Klamt, S.; *et al*. A logical model provides insights into T-cell receptor signaling. *PLoS Comput. Biol.* **2007**, *3*, 1580–1590.

34. Miskov-Zivanov, N.; Turner, M.S.; Kane, L.P.; Morel, P.A.; Faeder, J.R. The Duration of T-cell Stimulation Is a Critical Determinant of Cell Fate and Plasticity. *Sci. Signal.* **2013**, *6*, ra97, doi:10.1126/scisignal.2004217.

35. Levchenko, A.; Bruck, J.; Sternberg, P.W. Scaffold proteins may biphasically affect the levels of mitogen-activated protein kinase signaling and reduce its threshold properties. *Proc. Natl. Acad. Sci. USA* **2000**, *97*, 5818–5823.

36. Hoffmann, A.; Levchenko, A.; Scott, M.L.; Baltimore, D. The I kappa B-NF-kappa B signaling module: Temporal control and selective gene activation. *Science* **2002**, *298*, 1241–1245.

37. O'Dea, E.L.; Barken, D.; Peralta, R.Q.; Tran, K.T.; Werner, S.L.; Kearns, J.D.; Levchenko, A.; Hoffmann, A. A homeostatic model of I kappa B metabolism to control constitutive NF-kappa B activity. *Mol. Syst. Biol.* **2007**, *3*, doi:10.1038/msb4100148.

38. Coombs, D.; Kalergis, A.M.; Nathenson, S.G.; Wofsy, C.; Goldstein, B. Activated TCRs remain marked for internalization after dissociation from pMHC. *Nat. Immunol.* **2002**, *3*, 926–931.

39. Wylie, D.C.; Hori, Y.; Dinner, A.R.; Chakraborty, A.K. A hybrid deterministic-stochastic algorithm for modeling cell signaling dynamics in spatially inhomogeneous environments and under the influence of external fields. *J. Phys. Chem. B* **2006**, *110*, 12749–12765.

40. Zheng, Y. A Systems Biology Study to Delineate the T-cell Receptor-Activated Erk-MAPK Signaling Pathway. Ph.D. Thesis, Purdue University, West Lafayette, IN, USA, 2005.

41. Perley, J.P.; Mikolajczak, J.; Harrison, M.L.; Buzzard, G.T.; Rundell, A.E. Multiple Model-Informed Open-Loop Control of Uncertain Intracellular Signaling Dynamics. *PLoS Comput. Biol.* **2014**, *10*, e1003546.

42. Buzzard, G.T.; Xiu, D.B. Variance-Based Global Sensitivity Analysis via Sparse-Grid Interpolation and Cubature. *Commun. Comput. Phys.* **2011**, *9*, 542–567.

43. Buzzard, G.T. Global sensitivity analysis using sparse grid interpolation and polynomial chaos. *Reliab. Eng. Syst. Saf.* **2012**, *107*, 82–89.

44. Donahue, M.M.; Buzzard, G.T.; Rundell, A.E. Robust Parameter Identification with Adaptive Sparse Grid-based Optimization for Nonlinear Systems Biology Models. In Proceedings of the IEEE 2009 American Control Conference, St. Louis, MO, USA, 10–12 June 2009; IEEE: New York, NY, USA, 2009; Volumes 1–9, pp. 5055–5060.

45. Von Essen, M.; Bonefeld, C.M.; Siersma, V.; Rasmussen, A.B.; Lauritsen, J.P.H.; Nielsen, B.L.; Geisler, C. Constitutive and ligand-induced TCR degradation. *J. Immunol.* **2004**, *173*, 384–393.

46. Geisler, C. TCR trafficking in resting and stimulated T-cells. *Crit. Rev. Immunol.* **2004**, *24*, 67–85.

47. Menne, C.; Sorensen, T.M.; Siersma, V.; von Essen, M.; Odum, N.; Geisler, C. Endo- and exocytic rate constants for spontaneous and protein kinase C-activated T-cell receptor cycling. *Eur. J. Immunol.* **2002**, *32*, 616–626.

48. Okada, M.; Nada, S.; Yamanashi, Y.; Yamamoto, T.; Nakagawa, H. Csk—A protein-tyrosine kinase involved in regulation of src family kinases. *J. Biol. Chem.* **1991**, *266*, 24249–24252.

49. Plas, D.R.; Johnson, R.; Pingel, J.T.; Matthews, R.J.; Dalton, M.; Roy, G.; Chan, A.C.; Thomas, M.L. Direct regulation of ZAP-70 by SHP-1 in T-cell antigen receptor signaling. *Science* **1996**, *272*, 1173–1176.

50. Stefanova, I.; Hemmer, B.; Vergelli, M.; Martin, R.; Biddison, W.E.; Germain, R.N. TCR ligand discrimination is enforced by competing ERK positive and SHP-1 negative feedback pathways. *Nat. Immunol.* **2003**, *4*, 248–254.

51. Lipniacki, T.; Hat, B.; Faeder, J.R.; Hlavacek, W.S. Stochastic effects and bistability in T-cell receptor signaling. *J. Theor. Biol.* **2008**, *254*, 110–122.

52. Buday, L.; Egan, S.E.; Viciana, P.R.; Cantrell, D.A.; Downward, J. A complex of Grb2 adapter protein, SOS exchange factor, and a 36-kDa membrane-bound tyrosine phosphoprotein is implicated in Ras activation in T-cells. *J. Biol. Chem.* **1994**, *269*, 9019–9023.

53. Clipstone, N.A.; Crabtree, G.R. Identification of calcineurin as a key signaling enzyme in lyphocyte-T activation. *Nature* **1992**, *357*, 695–697.

54. Rudd, C.E.; Taylor, A.; Schneider, H. CD28 and CTLA-4 coreceptor expression and signal transduction. *Immunol. Rev.* **2009**, *229*, 12–26.

55. McNeill, L.; Salmond, R.J.; Cooper, J.C.; Carret, C.K.; Cassady-Cain, R.L.; Roche-Molina, M.; Tandon, P.; Holmes, N.; Alexander, D.R. The differential regulation of lck kinase phosphorylation sites by CD45 is critical for T-cell receptor signaling responses. *Immunity* **2007**, *27*, 425–437.

56. Stone, J.D.; Conroy, L.A.; Byth, K.F.; Hederer, R.A.; Howlett, S.; Takemoto, Y.; Holmes, N.; Alexander, D.R. Aberrant TCR-Mediated signaling in CD45-null thymocytes involves dysfunctional regulation of Lck, Fyn, TCR-zeta and ZAP-70. *J. Immunol.* **1997**, *158*, 5773–5782.

57. Kretschmer, K.; Apostolou, I.; Hawiger, D.; Khazaie, K.; Nussenzweig, M.C.; von Boehmer, H. Inducing and expanding regulatory T-cell populations by foreign antigen. *Nat. Immunol.* **2005**, *6*, 1219–1227.

58. Lewis, R.S. Calcium signaling mechanisms in T lymphocytes. *Annu. Rev. Immunol.* **2001**, *19*, 497–521.

59. Dogan, G. Confidence Interval Estimation in System Dynamics Models: Bootstrapping vs. Likelihood Ratio Method. In Proceedings of the 22nd International Conference of the System Dynamics Society, Oxford, UK, 2004.

Mathematical Modeling of Pro- and Anti-Inflammatory Signaling in Macrophages

Shreya Maiti, Wei Dai, Robert C. Alaniz, Juergen Hahn and Arul Jayaraman

Abstract: Inflammation is a beneficial mechanism that is usually triggered by injury or infection and is designed to return the body to homeostasis. However, uncontrolled or sustained inflammation can be deleterious and has been shown to be involved in the etiology of several diseases, including inflammatory bowel disorder and asthma. Therefore, effective anti-inflammatory signaling is important in the maintenance of homeostasis in the body. However, the inter-play between pro- and anti-inflammatory signaling is not fully understood. In the present study, we develop a mathematical model to describe integrated pro- and anti-inflammatory signaling in macrophages. The model incorporates the feedback effects of *de novo* synthesized pro-inflammatory (tumor necrosis factor α; TNF-α) and anti-inflammatory (interleukin-10; IL-10) cytokines on the activation of the transcription factor nuclear factor κB (NF-κB) under continuous lipopolysaccharide (LPS) stimulation (mimicking bacterial infection). In the model, IL-10 upregulates its own production (positive feedback) and also downregulates TNF-α production through NF-κB (negative feedback). In addition, TNF-α upregulates its own production through NF-κB (positive feedback). Eight model parameters are selected for estimation involving sensitivity analysis and clustering techniques. We validate the mathematical model predictions by measuring phosphorylated NF-κB, *de novo* synthesized TNF-α and IL-10 in RAW 264.7 macrophages exposed to LPS. This integrated model represents a first step towards modeling the interaction between pro- and anti-inflammatory signaling.

Reprinted from *Processes*. Cite as: Maiti, S.; Dai, W.; Alaniz, R.; Hahn, J.; Jayaraman, A. Mathematical Modeling of Pro- and Anti-Inflammatory Signaling in Macrophages. *Processes* **2015**, *3*, 1-18.

1. Introduction

Inflammation is a beneficial self-defense mechanism that is initiated by the body to eliminate pathogens and prevent the spread of infection [1]. The inflammatory responses to pathogens and other inflammatory stimuli are mediated by innate (dendritic cells and macrophages) and adaptive immune cells (T-cells and B-cells) [2]. Immune cells have transmembrane receptors called Toll-like receptors (TLR) that recognize foreign molecules based on pathogen-associated molecular patterns (PAMPs), such as flagellin of bacterial flagella [3], lipopolysaccharide (LPS) of Gram-negative bacteria and peptidoglycan of Gram-positive bacteria [4]. Recognition of PAMPs by immune cells (such as macrophages) triggers the production and secretion of pro-inflammatory cytokines, which leads to the recruitment of phagocytic cells, such as neutrophils [5], for eliminating pathogens. While inflammation is a beneficial body response, unabated (chronic) inflammation is deleterious, as it can result in immune cells attacking other host cells. Chronic inflammation has been shown to be involved in the etiology of several diseases, including inflammatory bowel disease (IBD) [6] and asthma [7]. Chronic inflammation can also arise in the absence of pathogen infection. Since the mucosal immune cells in the gastro-intestinal (GI) tract are in close proximity with intestinal microbiota [8], any alteration in the intestinal microbial community (*i.e.*, dysbiosis) can also lead to uncontrolled pro-inflammatory responses. This sustained inflammation in the absence of any infection has been shown to result in ulcerative colitis or Crohn's disease [6].

Nuclear factor-κB (NF-κB) is an important transcription factor that plays a pivotal role in mediating inflammatory responses in immune cells, such as macrophages [9]. NF-κB is made up of two subunits, p50 and p65 [10], and is sequestered as an inactive complex in the cytosol by an inhibitor protein, IκBα [9]. When macrophages detect the presence of bacteria (by detecting LPS) through their cell surface receptor, TLR4, an LPS-TLR4 complex is formed that triggers the activation of IκBα kinase (IKK), resulting in phosphorylation of IκBα-NFκB and subsequent ubiquitination and degradation of IκBα [9]. NF-κB, which is catalytically released from the inactive IκBα-NFκB complex, translocates into the nucleus and binds to response elements in the promoter region of its target genes to activate their transcription [9]. Several target genes with functions in inflammation and immune regulation have been identified for NF-κB [11], of which TNF-α and IL-10 are the most prominent pro- and anti-inflammatory cytokines, respectively [12–14]. In addition to TNF-α and IL-10, other NF-κB responsive genes that have significant NF-κB regulatory functions are IκBα (sequesters free NF-κB) [15] and A20 (inactivates IKK) [16]. However, NF-κB is not the only transcription factor that regulates IL-10 and TNF-α signaling and often acts in concert with other transcription factors. For example, signal transducer and activator of transcription 3 (STAT3) is a well-studied transcription factor involved in IL-10 signaling [17,18]. STAT3 not only regulates transcription of IL-10, but is itself activated by IL-10 [19] and LPS [20] in a feedback manner. The effects of pro-inflammatory cytokines, such as TNF-α and IL-1β, are countered by signaling initiated by anti-inflammatory cytokines. IL-10 is a potent anti-inflammatory cytokine and suppresses the production of pro-inflammatory cytokines, like TNF-α [21], by downregulating NF-κB through inhibition of IKK activation and suppression of free phosphorylated NF-κB translocation from cytosol to nucleus [22,23].

Several computational models of inflammatory signaling have been previously developed. These include a model for the IL-6 signal transduction pathway by Singh *et al.* [24], the TNF-α signaling pathway by Huang *et al.* [25], Lipniacki *et al.* [26], Rangamani *et al.* [27] and Hoffmann *et al.* [28]. A characteristic feature of these models is that they describe the dynamics of signaling initiated by a single pro-inflammatory cytokine. Moya *et al.* [29] developed a mathematical model to represent interactions between IL-6 (pro-inflammatory) and IL-10 (anti-inflammatory) in hepatocytes when both of these cytokines were used as stimuli to the cells. The current work describes an interplay between *de novo* synthesized pro-inflammatory (TNF-α) and anti-inflammatory (IL-10) cytokines in macrophages exposed to LPS (Figure 1). Since the inter-play between the pro- and anti-inflammatory signaling in macrophages is poorly understood, our integrated model represents a first step towards modeling the interaction between pro- and anti-inflammatory signaling mediators that is important in inflammation and maintaining homeostasis.

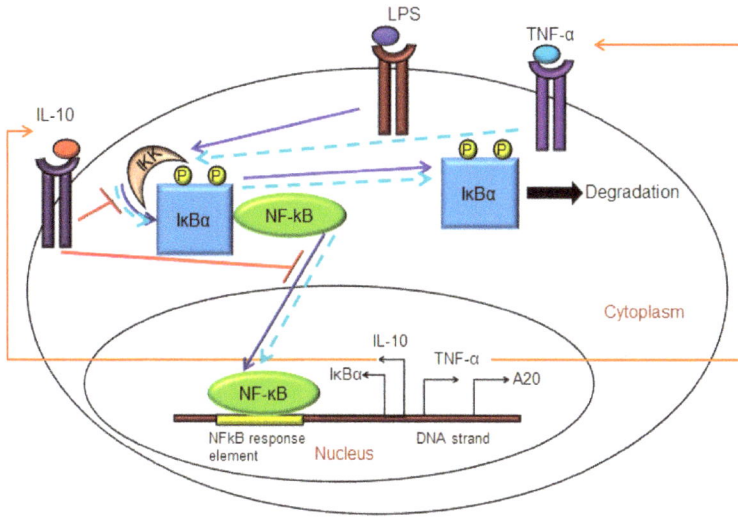

Figure 1. Schematic representation of NF-κB signal transduction pathway under LPS stimulation in macrophages. LPS binds to cell-surface TLR4, forms the LPS-TLR4 complex that initiates activation of IKK, subsequent rapid phosphorylation and dissociation of the IκBα-NFκB complex. Phosphorylated IκBα undergoes degradation, whereas free cytoplasmic NF-κB translocates into the nucleus, binds to DNA response elements and initiates the transcription of target genes TNF-α, IL-10, IκBα and A20. *De novo* synthesized TNF-α and IL-10 are secreted into the cell culture supernatant, where they bind to their respective cell surface receptors and initiate their positive (TNF-α) and negative (IL-10) feedback regulations on NF-κB. The LPS-induced NF-κB signaling pathway is indicated in solid blue arrows. TNF-α-induced positive feedback regulation of NF-κB is indicated in dashed cyan arrows, and IL-10-induced negative feedback regulation of NF-κB is indicated in solid red lines.

2. Materials and Methods

2.1. Model Formulation

The mathematical model presented in this paper is an integration of an inflammatory module and an anti-inflammatory module. The model is developed by representing biochemical reactions involved in the signal transduction pathway (Figure 2) as a set of non-linear ordinary differential equations (ODE) of the form:

$$\frac{dx}{dt} = f(x, u, p) \tag{1}$$

where x is a vector of states, u is a vector of inputs and p is a vector of parameters. The model comprises 29 differential equations (Table 1) and 37 parameters (Table 2). Each differential equation represents the rate of change of the concentration of a particular protein involved in the pathway.

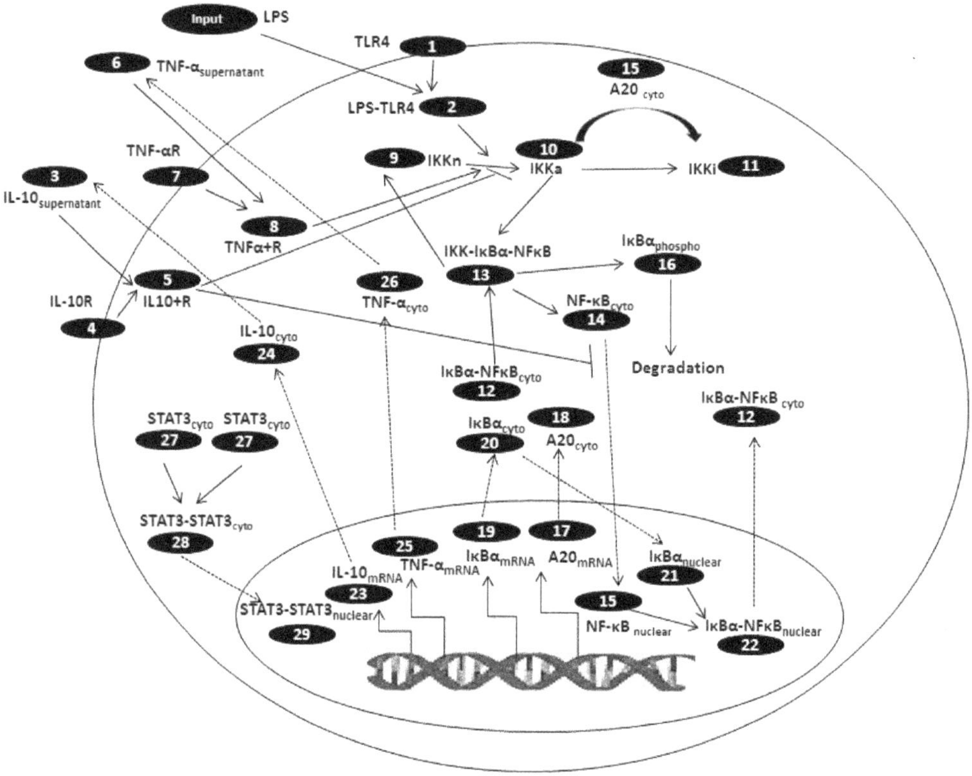

Figure 2. Implemented reaction network for the LPS-induced NF-κB signal transduction pathway with TNF-α (positive) and IL-10 (negative) feedback regulation.

The inflammatory (TNF-α) module is adapted from Huang *et al.* [25] and Lipniacki *et al.* [26]. While these models use TNF-α as the input, our model describes LPS (input)-induced signaling through TLR4 (LPS receptor), which leads to TNF-α production. Besides adding TLR4 to the model, the TNF-α receptor description is retained to represent the positive feedback of *de novo* synthesized TNF-α on NF-κB regulation. We have included a kinetic term for TNF-α mRNA transcription initiated by nuclear NF-κB and component balances for TNF-α in the cytoplasm and the supernatant. In addition, Lipniacki *et al.*, included TRADD, TRAF2, RIP-1, FADD, caspase-3 and caspase-8 proteins, which are left out of the model presented here, as we focused only on some of the key biochemical reactions involved in LPS-induced NF-κB activation, its effect on the production of TNF-α and IL-10 and, in turn, the role of these cytokines on the feedback regulation of NF-κB. The similarities between the model described in Lipniacki *et al.* [26], and our current ODE model lie in the formulation of the biochemical reactions involved in IKK activation, IκBα-NFκB phosphorylation, dissociation and nuclear transport of NF-κB, nuclear NF-κB-induced IκBα, A20 mRNA transcription, free NF-κB sequestration by *de novo* synthesized IκBα and IKK inactivation by A20. We added a balance for phosphorylated IκBα, as it is known to degrade after dissociation from the IκBα-NFκB complex.

Table 1. Differential equations representing biochemical reactions involved in LPS-induced NF-κB signal transduction pathway, as used in the ODE model.

1.	$\frac{d[\text{TLR4}]}{dt} = -kf_1 \times [\text{LPS}][\text{TLR4}] + kr_1 \times [\text{LPS} - \text{TLR4}]$
2.	$\frac{d[\text{LPS}-\text{TLR4}]}{dt} = kf_1 \times [\text{LPS}][\text{TLR4}] - kr_1 \times [\text{LPS} - \text{TLR4}]$
3.	$\frac{d[\text{IL}-10_{\text{sup}}]}{dt} = -kf_2 \times [\text{IL} - 10_{\text{sup}}][\text{IL} - 10\text{R}] + kr_2 \times [\text{IL}10 - \text{IL}10\text{R}] + ksec_{IL10} \times [\text{IL} - 10_{\text{cyto}}] \times \frac{0.36}{200} - kdeg_{IL-10sup} \times [\text{IL} - 10_{\text{sup}}]$
4.	$\frac{d[\text{IL}-10\text{R}]}{dt} = -kf_2 \times [\text{IL} - 10][\text{IL}10\text{R}] + kr_2 \times [\text{IL}10 - \text{IL}10\text{R}]$
5.	$\frac{d[\text{IL}10-\text{IL}10\text{R}]}{dt} = kf_2 \times [\text{IL} - 10_{\text{sup}}][\text{IL} - 10\text{R}] - kr_2 \times [\text{IL}10 - \text{IL}10\text{R}]$
6.	$\frac{d[\text{TNF}-\alpha_{\text{sup}}]}{dt} = -kf_3 \times [\text{TNF} - \alpha_{\text{sup}}][\text{TNF} - \alpha\text{R}] + kr_3 \times [\text{TNF}\alpha - \text{TNF}\alpha\text{R}] + ksec_{TNF\alpha} \times [\text{TNF} - \alpha_{\text{cyto}}] \times \frac{0.36}{200} - kdeg_{TNF\alpha sup} \times [\text{TNF} - \alpha_{\text{sup}}]$
7.	$\frac{d[\text{TNF}-\alpha\text{R}]}{dt} = -kf_3 \times [\text{TNF} - \alpha_{\text{sup}}][\text{TNF} - \alpha\text{R}] + kr_3 \times [\text{TNF}\alpha - \text{TNF}\alpha\text{R}]$
8.	$\frac{d[\text{TNF}\alpha-\text{TNF}\alpha\text{R}]}{dt} = kf_3 \times [\text{TNF} - \alpha_{\text{sup}}][\text{TNF} - \alpha\text{R}] - kr_3 \times [\text{TNF}\alpha - \text{TNF}\alpha\text{R}]$
9.	$\frac{d[\text{IKK}_n]}{dt} = -kfi \times k_{in} \times ([\text{LPS} - \text{TLR4}] + [\text{TNF}\alpha - \text{TNF}\alpha\text{R}]) \times [\text{IKK}_n] + ti_3 \times [\text{IKK}_a - \text{I}\kappa\text{B}\alpha\text{NF}\kappa\text{B}_{\text{cyto}}]$ where, $k_{in} = \max\left[(1 - \frac{[\text{IL}10-\text{IL}10\text{R}]}{[\text{IL}10-\text{IL}10\text{R}_{\text{max}}]}),0\right]$
10.	$\frac{d[\text{IKK}_a]}{dt} = kfi \times k_{in} \times ([\text{LPS} - \text{TLR4}] + [\text{TNF}\alpha - \text{TNF}\alpha\text{R}]) \times [\text{IKK}_n] - kk_3 \times k_{in} \times [\text{IKK}_a] \times [\text{I}\kappa\text{B}\alpha - \text{NF}\kappa\text{B}_{\text{cyto}}] - kk_1 \times [\text{IKK}_a] \times [\text{A20}_{\text{cyto}}]$
11.	$\frac{d[\text{IKK}_i]}{dt} = kk_1 \times [\text{IKK}_a] \times [\text{A20}_{\text{cyto}}]$
12.	$\frac{d[\text{I}\kappa\text{B}\alpha-\text{NF}\kappa\text{B}_{\text{cyto}}]}{dt} = kf_4 \times [\text{NF}\kappa\text{B}_{\text{cyto}}][\text{I}\kappa\text{B}\alpha_{\text{cyto}}] + eni \times [\text{I}\kappa\text{B}\alpha - \text{NF}\kappa\text{B}_{\text{nuclear}}] \times kv - kk_3 \times k_{in} \times [\text{IKK}_a] \times [\text{I}\kappa\text{B}\alpha - \text{NF}\kappa\text{B}_{\text{cyto}}]$
13.	$\frac{d[\text{IKK}_a-\text{I}\kappa\text{B}\alpha\text{NF}\kappa\text{B}_{\text{cyto}}]}{dt} = kk_3 \times k_{in} \times [\text{IKK}_a] \times [\text{I}\kappa\text{B}\alpha - \text{NF}\kappa\text{B}_{\text{cyto}}] - ti_3 \times [\text{IKK}_a - \text{I}\kappa\text{B}\alpha\text{NF}\kappa\text{B}_{\text{cyto}}]$
14.	$\frac{d[\text{NF}\kappa\text{B}_{\text{cyto}}]}{dt} = -kf_4 \times [\text{NF}\kappa\text{B}_{\text{cyto}}][\text{I}\kappa\text{B}\alpha_{\text{cyto}}] + ti_3 \times [\text{IKK}_a - \text{I}\kappa\text{B}\alpha\text{NF}\kappa\text{B}_{\text{cyto}}] - iln \times k_{in} \times [\text{NF}\kappa\text{B}_{\text{cyto}}]$
15.	$\frac{d[\text{NF}\kappa\text{B}_{\text{nuclear}}]}{dt} = iln \times k_{in} \times \frac{[\text{NF}\kappa\text{B}_{\text{cyto}}]}{kv} - kf_4 \times [\text{NF}\kappa\text{B}_{\text{nuclear}}][\text{I}\kappa\text{B}\alpha_{\text{nuclear}}]$
16.	$\frac{d[\text{I}\kappa\text{B}\alpha_{\text{phopsho}}]}{dt} = ti_3 \times [\text{IKK}_a - \text{I}\kappa\text{B}\alpha\text{NF}\kappa\text{B}_{\text{cyto}}] - kdeg_{I\kappa B\alpha} \times [\text{I}\kappa\text{B}\alpha_{\text{cyto}}]$
17.	$\frac{d[\text{A20}_{\text{mRNA}}]}{dt} = Sm \times p \times \frac{[\text{NF}\kappa\text{B}_{\text{nuclear}}]}{C+[\text{NF}\kappa\text{B}_{\text{nuclear}}]} - Dm \times [\text{A20}_{\text{mRNA}}]$
18.	$\frac{d[\text{A20}_{\text{cyto}}]}{dt} = a20_{trans} \times [\text{A20}_{\text{mRNA}}] - kdeg_{A20} \times [\text{A20}_{\text{cyto}}]$
19.	$\frac{d[\text{I}\kappa\text{B}\alpha_{\text{mRNA}}]}{dt} = Sm \times p \times \frac{[\text{NF}\kappa\text{B}_{\text{nuclear}}]}{C+[\text{NF}\kappa\text{B}_{\text{nuclear}}]} - Dm \times [\text{I}\kappa\text{B}\alpha_{\text{mRNA}}]$
20.	$\frac{d[\text{I}\kappa\text{B}\alpha_{\text{cyto}}]}{dt} = -kf_4 \times [\text{NF}\kappa\text{B}_{\text{cyto}}][\text{I}\kappa\text{B}\alpha_{\text{cyto}}] + i\kappa b\alpha_{trans} \times [\text{I}\kappa\text{B}\alpha_{\text{mRNA}}] - iki \times [\text{I}\kappa\text{B}\alpha_{\text{cyto}}] + eki \times [\text{I}\kappa\text{B}\alpha_{\text{nuclear}}] \times kv$
21.	$\frac{d[\text{I}\kappa\text{B}\alpha_{\text{nuclear}}]}{dt} = -kf_4 \times [\text{NF}\kappa\text{B}_{\text{nuclear}}][\text{I}\kappa\text{B}\alpha_{\text{nuclear}}] + iki \times \frac{[\text{I}\kappa\text{B}\alpha_{\text{cyto}}]}{kv} - eki \times [\text{I}k\text{B}\alpha_{\text{nuclear}}]$
22.	$\frac{d[\text{I}\kappa\text{B}\alpha-\text{NF}\kappa\text{B}_{\text{nuclear}}]}{dt} = kf_4 \times [\text{NF}\kappa\text{B}_{\text{nuclear}}][\text{I}\kappa\text{B}\alpha_{\text{nuclear}}] - eni \times [\text{I}\kappa\text{B}\alpha - \text{NF}\kappa\text{B}_{\text{nuclear}}]$
23.	$\frac{d[\text{IL}-10_{\text{mRNA}}]}{dt} = 0.4 \times Sm \times p \times \frac{[\text{NF}\kappa\text{B}_{\text{nuclear}}]}{C+[\text{NF}\kappa\text{B}_{\text{nuclear}}]} + 0.6 \times Sm\_il10 \times p \times \frac{[\text{STAT3}_{\text{nuclear}}]}{C\_\text{STAT3}+[\text{STAT3}_{\text{nuclear}}]} - Dm \times [\text{IL} - 10_{\text{mRNA}}]$
24.	$\frac{d[\text{IL}-10_{\text{cyto}}]}{dt} = il10_{trans} \times [\text{IL} - 10_{\text{mRNA}}] - ksec_{IL10} \times [\text{IL} - 10_{\text{cyto}}] - Dn \times [\text{IL} - 10_{\text{cyto}}]$
25.	$\frac{d[\text{TNF}-\alpha_{\text{mRNA}}]}{dt} = Sm \times p \times \frac{[\text{NF}\kappa\text{B}_{\text{nuclear}}]}{C+[\text{NF}\kappa\text{B}_{\text{nuclear}}]} - Dm \times [\text{TNF} - \alpha_{\text{mRNA}}]$
26.	$\frac{d[\text{TNF}-\alpha_{\text{cyto}}]}{dt} = tnf\alpha_{trans} \times [\text{TNF} - \alpha_{\text{mRNA}}] - ksec_{TNF\alpha} \times [\text{TNF} - \alpha_{\text{cyto}}] - Dn \times [\text{TNF} - \alpha_{\text{cyto}}]$
27.	$\frac{d[\text{STAT3}_{\text{cyto}}]}{dt} = -2 \times k_1 \times [\text{IL}10 - \text{IL}10\text{R}][\text{STAT3}_{\text{cyto}}]^2 + 2 \times k_2 \times [\text{STAT3} - \text{STAT3}_{\text{cyto}}]$
28.	$\frac{d[\text{STAT3}-\text{STAT3}_{\text{cyto}}]}{dt} = k_1 \times [\text{IL}10 - \text{IL}10\text{R}][\text{STAT3}_{\text{cyto}}]^2 - k_2 \times [\text{STAT3} - \text{STAT3}_{\text{cyto}}] - i_{stat3} \times [\text{STAT3} - \text{STAT3}_{\text{cyto}}] + eni \times [\text{STAT3} - \text{STAT3}_{\text{nuclear}}] \times kv$
29.	$\frac{d[\text{STAT3}-\text{STAT3}_{\text{nuclear}}]}{dt} = i_{stat3} \times \frac{[\text{STAT3}-\text{STAT3}_{\text{cyto}}]}{kv} - eni \times [\text{STAT3} - \text{STAT3}_{\text{nuclear}}]$

Table 2. List of parameters used in the ODE model.

Sr. No.	Parameter	Description	Value	Units	Comment
1.	kv	**Nuclear: Cytoplasmic (Volume)**	**1.17**	**NA**	**Estimated**
2.	kf_1	**LPS binding to receptor**	$\mathbf{2.64 \times 10^{-1}}$	$\mathbf{(\mu M^{-s})^{-1}}$	**Estimated**
3.	kr_1	Dissociation of LPS + receptor complex	1.25×10^{-3}	$(\mu M^{-s})^{-1}$	Huang *et al.* (2008) [25]
4.	kf_2	IL-10 binding to receptor	2.50×10^{-4}	$(\mu M^{-s})^{-1}$	Assumed
5.	kr_2	Dissociation of IL-10 + receptor complex	6.11×10^{-4}	$(\mu M^{-s})^{-1}$	Assumed
6.	kf_3	TNF-α binding to receptor	2.50×10^{-3}	$(\mu M^{-s})^{-1}$	Gray *et al.* [30]
7.	kr_3	Dissociation of TNF-α + receptor complex	1.25×10^{-3}	$(\mu M^{-s})^{-1}$	Rangamani *et al.* (2007) [27]
8.	kf_4	IκBa and NF-κB association	2.5×10^{-3}	$(\mu M^{-s})^{-1}$	Assumed
9.	kfi	**IKK activation**	$\mathbf{1.62 \times 10^{-3}}$	$\mathbf{s^{-1}}$	**Estimated**
10.	kk_1	Inactivation of IKK by A20	2.5×10^{-4}	$(\mu M^{-s})^{-1}$	Assumed
11.	kk_3	Association of IKK with IκBα-NFκB	1.0	$(\mu M^{-s})^{-1}$	Lipniacki *et al.* (2004) [26]
12.	ti_3	**Catalytic breakdown of IKK-IκBα-NFκB**	$\mathbf{1.72 \times 10^{-4}}$	$\mathbf{s^{-1}}$	**Estimated**
13.	iln	**NF-κB nuclear import**	$\mathbf{1.52 \times 10^{-3}}$	$\mathbf{s^{-1}}$	**Estimated**
14.	$a20_{trans}$	A20 translation	5.00×10^{-1}	s^{-1}	Lipniacki *et al.* (2004) [26]
15.	$kdeg_{A20}$	Degradation of A20 protein	3.00×10^{-4}	s^{-1}	Lipniacki *et al.* (2004) [26]
16.	$i\kappa ba_{trans}$	IκBα translation	5.00×10^{-1}	s^{-1}	Lipniacki *et al.* (2004) [26]
17.	$kdeg_{I\kappa Ba}$	Degradation of phosphorylated IκBα	1.28×10^{-4}	s^{-1}	Assumed half-life of 90 min
18.	$il10_{trans}$	IL-10 translation	5.00×10^{-1}	s^{-1}	Lipniacki *et al.* (2004) [26]
19.	$ksec_{IL10}$	Secretion of IL-10 from cytoplasm to supernatant	2.03×10^{-5}	s^{-1}	Assumed
20.	$kdeg_{IL10sup}$	Degradation of IL-10 in supernatant	7.40×10^{-5}	s^{-1}	Half-life of 2.6 h in supernatant. Fedorak *et al.* [31]
21.	$tnf\alpha_{trans}$	TNF-α translation	5.00×10^{-1}	s^{-1}	Lipniacki *et al.* (2004) [26]
22.	$ksec_{TNF\alpha}$	**Secretion of TNF-α from cytoplasm to supernatant**	$\mathbf{5.16 \times 10^{-5}}$	$\mathbf{s^{-1}}$	**Estimated**
23.	$kdeg_{TNFasup}$	**Degradation of TNF-α in supernatant**	$\mathbf{7.46 \times 10^{-5}}$	$\mathbf{s^{-1}}$	**Estimated**
24.	Dn	Degradation of intracellular cytokine	1.04×10^{-2}	s^{-1}	Huang *et al.* (2008) [25]
25.	iki	IκBα nuclear import	1.00×10^{-3}	s^{-1}	Lipniacki *et al.* (2004) [26]
26.	eki	IκBα nuclear export	5.00×10^{-4}	s^{-1}	Lipniacki *et al.* (2004) [26]
27.	eni	IκBα -NFκB nuclear export	1.00×10^{-2}	s^{-1}	Lipniacki *et al.* (2004) [26]
28.	k_1	STAT3 activation and dimerization	1.54×10^{-2}	$(\mu M^{-s})^{-1}$	Assumed
29.	k_2	Dissociation of STAT3 dimer	3.3×10^{-5}	s^{-1}	Assumed
30.	i_{stat3}	**STAT3 dimer nuclear import**	$\mathbf{3.56 \times 10^{-5}}$	$\mathbf{s^{-1}}$	**Estimated**
31.	Sm	Transcription due to NF-κB	1.00×10^{-1}	s^{-1}	Huang *et al.* (2008) [25]
32.	Sm_il10	IL-10 Translation due to STAT3	1.5	s^{-1}	Assumed
33.	p	Transcription parameter	5.00×10^{-3}	μM	Huang *et al.* (2008) [25]
34.	Dm	Degradation of mRNA	1.04×10^{-2}	s^{-1}	Huang *et al.* (2008) [25]
35.	C	Maximum NF-κB concentration in nucleus	1.08×10^{-1}	μM	Huang *et al.* (2008) [25]
36.	C_{STAT3}	Maximum STAT3 concentration in nucleus	5.00×10^{-2}	μM	Assumed
37.	$IL10\text{-}IL10R_{max}$	IL10-IL10R maximum concentration	2.56×10^{-6}	μM	Assumed

The anti-inflammatory (IL-10) module is adapted from the IL-6 and IL-10 model by Moya *et al.* [29]. Only the ODEs involved in IL-10 signaling through the IL-10 receptor (as mentioned in Moya *et al.* [29]) are included in the anti-inflammatory module of our current model to formulate the feedback effects of IL-10 on its own production (through positive feedback regulation of STAT3) and TNF-α production (through negative feedback regulation of NF-κB). Biochemical reactions, as described in Moya *et al.*, for STAT3 phosphorylation, dimerization and nuclear translocation to initiate transcription are retained in our current model. Transcription and translation of SOCS3 due to STAT3 and downstream biochemical reactions associated with SOCS3 are not included in the model presented here. A Michaelis–Menten-type kinetics for IL-10 transcription, initiated by the transcription factors, NF-κB and STAT3, and component balances for IL-10 in the cytoplasm and supernatant, have been included here.

Some values of the parameters (Table 2) and initial concentrations of proteins (Table 3) are adapted from the TNF-α signaling models by Huang *et al.* [25], Lipniacki *et al.* [26], Rangamani *et al.* [27], Hoffmann *et al.* [28] and the IL-6 and IL-10 model by Moya *et al.* [29]. The previously developed models consisted of 37 differential equations and 60 parameters for the TNF-α model by Huang *et al.* and 68 differential equations and 118 parameters for the IL-6 and IL-10 model by Moya *et al.* [29] Among the proteins included in these models, very few are quantifiable by experimental methods, making parameter estimation difficult. In our current integrated model, we have reduced the number of differential equations to 29 and the number of parameters to 37 by only focusing on the key proteins of the pathway. Using a smaller model increased parameter identifiability and simplified parameter estimation.

The ODE model is structurally divided into pro-inflammatory (TNF-α) and anti-inflammatory (IL-10) modules that are both initiated by LPS stimulation and NF-κB activation. Below is the description of the implemented reaction network as shown in Figure 2.

Pro-Inflammatory Module:

(1) Exogenous LPS binds to the cell surface receptor (TLR4);
(2) LPS-TLR4 complex initiates activation of IKK$_{neutral}$ to IKK$_{active}$;
(3) IKK$_{active}$ phosphorylates IκBα-NFκB and initiates dissociation of the inactive IκBα-NFκB complex into phosphorylated IκBα and NF-κB species;
(4) Free phosphorylated IκBα undergoes ubiquitination and degradation, whereas free cytoplasmic NF-κB translocates into nucleus;
(5) Nuclear NF-κB binds to response elements in the promoter regions of TNF-α, IκBα and A20 genes and leads to transcription and translation of the corresponding proteins and subsequent secretion of TNF-α into the supernatant. *de novo* intracellular IκBα sequesters both free cytoplasmic and nuclear NF-κB by binding to them, and A20 catalyzes the change of IKK$_{active}$ to the IKK$_{inactive}$ form;
(6) Secreted TNF-α in the cell culture supernatant binds to its cell surface receptor to form a complex that initiates similar pathways as LPS, resulting in the production of more TNF-α through a positive feedback regulation on NF-κB.

Table 3. State variables and their initial values as used in the ODE model.

Sr. No.	State variables	Initial values, μM
1.	TLR4	1.0×10^{-1}
2.	LPS-TLR4	0
3.	IL-10$_{\text{supernatant}}$	4.6×10^{-6}
4.	IL-10R	1.0×10^{-1}
5.	IL10-IL10R	0
6.	TNF-$\alpha_{\text{supernatant}}$	0
7.	TNF-αR	1.0×10^{-1}
8.	TNFα-TNFαR	0
9.	IKK$_{\text{neutral}}$	2.0×10^{-1}
10.	IKK$_{\text{active}}$	0
11.	IKK$_{\text{inactive}}$	0
12.	IκBα-NFκB$_{\text{cyto}}$	2.5×10^{-1}
13.	IKK- IκBαNFκB	0
14.	NFκB$_{\text{cyto}}$	3.0×10^{-3}
15.	NFκB$_{\text{nuclear}}$	0
16.	IκBα_{phopsho}	0
17.	A20$_{\text{mRNA}}$	0
18.	A20$_{\text{cyto}}$	4.8×10^{-3}
19.	IκBα_{mRNA}	0
20.	IκBα_{cyto}	2.5×10^{-3}
21.	IκBα_{nuclear}	0
22.	IκBα-NFκB$_{\text{nuclear}}$	0
23.	IL-10$_{\text{mRNA}}$	0
24.	IL-10$_{\text{cyto}}$	0
25.	TNF-α_{mRNA}	0
26.	TNF-α_{cyto}	0
27.	STAT3$_{\text{cyto}}$	5.92×10^{-1}
28.	STAT3-STAT3$_{\text{cyto}}$	0
29.	STAT3-STAT3$_{\text{nuclear}}$	0

Anti-Inflammatory Module:

(1) Nuclear NF-κB binds to the IL-10 gene promoter and initiates transcription of IL-10 mRNA and subsequent translation into IL-10 protein in the cytoplasm, which gets secreted into the supernatant;

(2) IL-10 secreted into the supernatant binds to its cell surface receptor, forming a ligand-receptor complex that inhibits activation of IKK$_{\text{neutral}}$ to IKK$_{\text{active}}$ and translocation of activated free cytoplasmic NF-κB into the nucleus, as well;

(3) The IL-10 + receptor complex activates a second transcription factor, presumably STAT3, which, in turn, regulates transcription of the IL-10 gene in a feed-forward manner.

Different LPS concentrations (0, 0.1, 1, 10 µg/mL) are used to stimulate the model. The IKK complex and NF-κB dimer (p50–p65) are considered as single proteins in the model. The signal transduction model comprises feedback regulatory loops involving TNF-α and IL-10. The positive feedback of TNF-α on its own production is represented by *de novo* TNF-α binding to its cell surface receptor, activating IKK$_{neutral}$ to IKK$_{active}$, leading to phosphorylation and dissociation of the IκBα-NFκB complex to release NF-κB, which translocates to the nucleus to initiate transcription of TNF-α. IL-10 has a negative feedback effect on TNF-α production by inhibiting NF-κB activation (phosphorylation and dissociation), and the extent of this inhibition is calculated on the basis of the ligand bound IL-10 receptor complex (IL10-IL10R) concentration, which is represented as k_{in} (Equation (2)):

$$k_{in} = \max\left[\left(1 - \frac{[IL10-IL10R]}{[IL10-IL10R_{max}]}\right),0\right] \tag{2}$$

The maximum attainable concentration of IL10-IL10R is denoted by IL10-IL10R$_{max}$ with an assumed value of 2.56×10^{-6}. k_{in} is multiplied by factors that are inhibited by IL-10, such as IKK activation and nuclear translocation of free cytoplasmic NF-κB [22,23] (Table 1). The higher the concentration of the IL10-IL10R complex, the lower will be the value of k_{in} and, hence, the lower will be the contribution of the terms mentioned above to the total outcome of NF-κB signaling, resulting in suppression of TNF-α production by IL-10. Positive feedback of IL-10 on its own production is represented by a set of differential equations that describe the IL-10 bound receptor complex phosphorylating transcription factor STAT3, which then dimerizes and translocates into the nucleus, binds to the promoter region of the IL-10 gene and initiates transcription of IL-10 mRNA and subsequent translation and secretion of IL-10 protein.

2.2. Parameter Selection and Estimation

The parameter estimation problem for a dynamic system described by ordinary differential equations (ODEs) can be mathematically formulated as follows:

$$\min_{p} \sum_{i} \sum_{k} w_{ik}(y_{ik} - \hat{y}_{ik})^2 \tag{3}$$

Subject to,

$$\dot{x}(t) = f(x, u, p), \, x(0) = x_0 \tag{4}$$

$$y = g(x) \tag{5}$$

$$x^{lb} \le x \le x^{ub} \tag{6}$$

$$p^{lb} \le p \le p^{ub} \tag{7}$$

where y_{ik} and $\hat{y}_{ik}$ are the simulated and measured output data of the *i*-th component at sampling time t_k, respectively (Equation (3)); p are the parameters to be estimated, which are selected by local sensitivity analysis; x are the state variables of the dynamic system with initial values x_0; and u are the inputs to the system (Equation (4)). In addition, the state variables x and parameters p are restricted within certain ranges, as shown in Equations (6) and (7), determined by the underlying biology and prior knowledge based on mathematical models developed by Lipniacki *et al.* [26], Huang *et al.* [25] and Rangamani *et al.* [27].

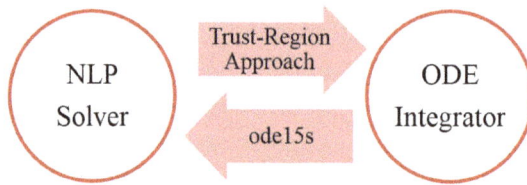

Figure 3. Algorithm for parameter estimation used to optimize model parameters. An optimization algorithm is applied in an outer loop, while the evaluation of the objective function and its gradients are performed by numerical integration of the ODEs in the inner loop. *fmincon* (MATLAB function) is used as the NLP (non-linear programming) solver and *ode15s* (MATLAB function) is used as the ODE integrator.

The simulated output vector y (Equation (5)), which is validated by experimental data, includes the intracellular ratio of phosphorylated NF-κB to total NF-κB (relative to the control) and the concentration of the cytokines, TNF-α and IL-10, in the cell culture supernatant. Since the experiments are conducted with four different levels of the input u (*i.e.*, different concentrations of LPS), four sets of measured outputs $\hat{y}_{ik}$ are obtained. Three sets of data obtained for 0, 0.1 and 1 μg/mL LPS stimulations are used for parameter estimation, and the fourth dataset for 10 μg/mL LPS stimulation is used for model validation.

First, the set of parameters that are to be estimated are selected by local sensitivity analysis and hierarchical clustering. Following this, the trust-region optimization technique is used to estimate the selected parameters. This technique is used in this work, as it is able to handle singular Hessian matrices, significant uncertainty in the parameters of the models and noisy data. In this technique, an optimization algorithm is applied in an outer loop, while the evaluation of the objective function and its gradients are performed by numerical integration of the ODEs in the inner loop [32,33] (shown in Figure 3). The trust-region method is guaranteed to converge to local optima with much weaker assumptions than line search methods. In this work, *fmincon* (MATLAB function) is used as the NLP (non-linear programming) solver and *ode15s* (MATLAB function) is used as the ODE integrator. It is worth noting that *ode15s* is specifically designed for stiff systems, such as the model discussed here, where both fast and slow dynamics exist, e.g., in our system, phosphorylation of NF-κB is a much faster process than *de novo* synthesis of TNF-α and IL-10.

2.3. Cell Culture and Experimental Set-Up

The murine macrophage cell line RAW264.7 (gift from Dr. Paul deFigueiredo, Texas A & M University) was routinely cultured in DMEM with 10% FBS. LPS (heat-killed *Salmonella enterica*) was purchased from Sigma Aldrich (St. Louis, MO, USA).

2.3.1. LPS Stimulation of Macrophages

RAW264.7 cells were seeded at a density of ~2.0×10^5 cells/well in a 96-well tissue culture plate and allowed to attach overnight. Cells were stimulated with different concentrations (0, 0.1, 1 and 10 μg/mL) of LPS diluted in growth medium. Whole cells were used to measure total and phosphorylated NF-κB after 5, 15, 30, 45, 60, 120 and 240 min post-LPS stimulation. Culture supernatants were collected after 2, 4, 8, 12, 16, 20 and 24 h post-LPS stimulation to measure secreted TNF-α and IL-10 concentrations.

2.3.2. Transcription Factor NF-κB Quantification by ELISA

The relative concentrations of total and phosphorylated NF-κB in LPS-stimulated RAW264.7 macrophage cells were determined using a commercially-available enzyme-linked-immunosorbent assay (ELISA) kit (R&D Systems, Minneapolis, MN, USA) according to the manufacturer's suggested protocol.

2.3.3. Cytokines TNF-α and IL-10 Quantification by ELISA

The concentrations of *de novo* synthesized TNF-α and IL-10 in LPS-stimulated RAW264.7 macrophage culture supernatant were determined by commercially available enzyme-linked immunosorbent assay (ELISA) kits (Thermo Scientific, Rockford, IL, USA), using the manufacturer's suggested protocol.

3. Results and Discussion

Based on published reports of LPS stimulation resulting in TNF-α [34] and IL-10 secretion [35] in RAW264.7 murine macrophages, as well as the established suppression of TNF-α by IL-10 in RAW264.7 cells [34], we developed an integrated ODE model to represent the production of TNF-α and IL-10 in RAW264.7 cells upon LPS stimulation and their regulatory feedback loops. Eight parameters of the model are selected for estimation using local sensitivity analysis and hierarchical clustering (shown in Figure 4) [36,37]. The y-axis represents the parameter distance ranging from zero to one (the larger the distance, the smaller the similarity between the parameters). The red line presents the cutoff value, which groups the entire set of parameters into eight pairwise indistinguishable clusters. The selected parameters, which have the largest sensitivity magnitude in each cluster, are highlighted in red. The values of the selected parameters are estimated using the trust-region optimization technique, as described in Materials and Methods section, and their estimated values are listed (in bold) in Table 2. One advantage of this approach is that the selected parameters used for estimation result in a more robust dynamic model with an accurate prediction capability [37]. Model simulations after parameter estimation predict rapid phosphorylation of NF-κB upon exposure to LPS, as shown in Figure 5A. The maximum fold change in the ratio of phosphorylated NF-κB to total NF-κB in LPS-treated cells relative to control increased with increasing LPS concentration and varied from ~2.0 at the highest LPS concentration to being essentially unchanged at the lowest concentration (Figure 5A).

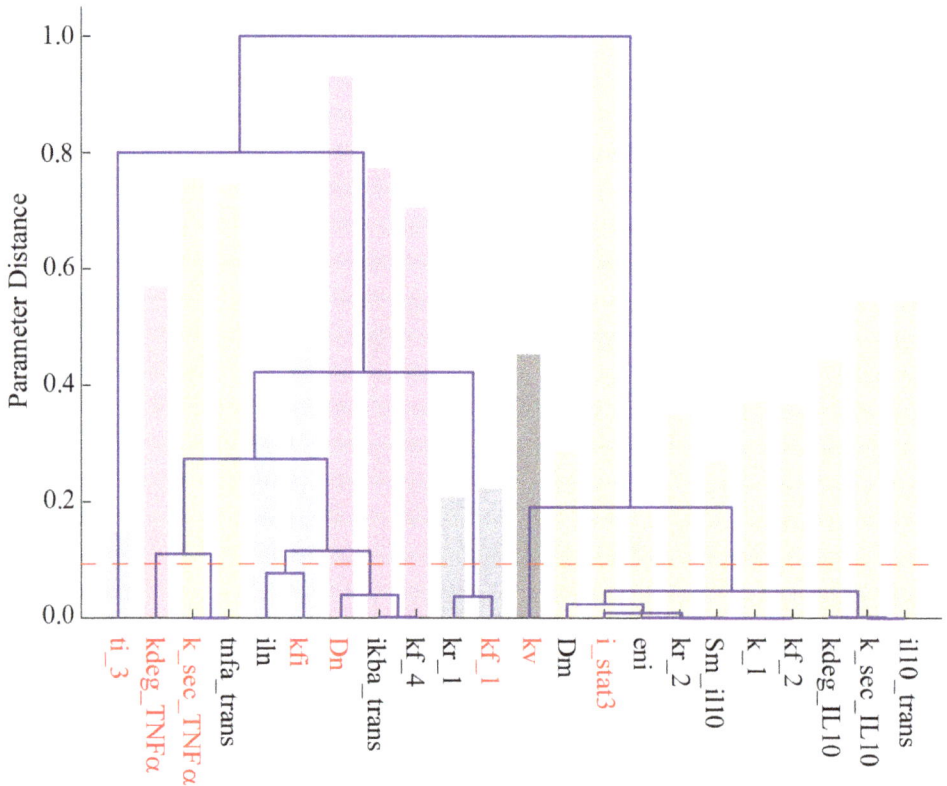

Figure 4. Representation of local sensitivity analysis results, used for selecting parameters that are to be estimated. The *y*-axis represents parameter distance ranging from zero to one. The red line represents the cutoff value, which groups the entire set of parameters into eight pairwise indistinguishable clusters. The selected parameters from each of the eight clusters are highlighted in red. The normalized sensitivity magnitudes of the parameters are reflected in the histograms.

Simulation of *de novo* synthesized TNF-α profile upon LPS stimulation shows that the TNF-α concentration reaches a maximum of ~1500 pg/mL at 4 h and starts declining thereafter. Even though LPS is continuously present, TNF-α is undetectable at 24 h for all LPS concentrations (Figure 5B). The maximum TNF-α concentration at 4 h increases with increasing concentrations of LPS. According to the model predictions, the *de novo* synthesized IL-10 concentration increases beyond 2 h of LPS stimulation, as shown in Figure 5C, and the concentration of IL-10 produced increases with increasing LPS concentrations.

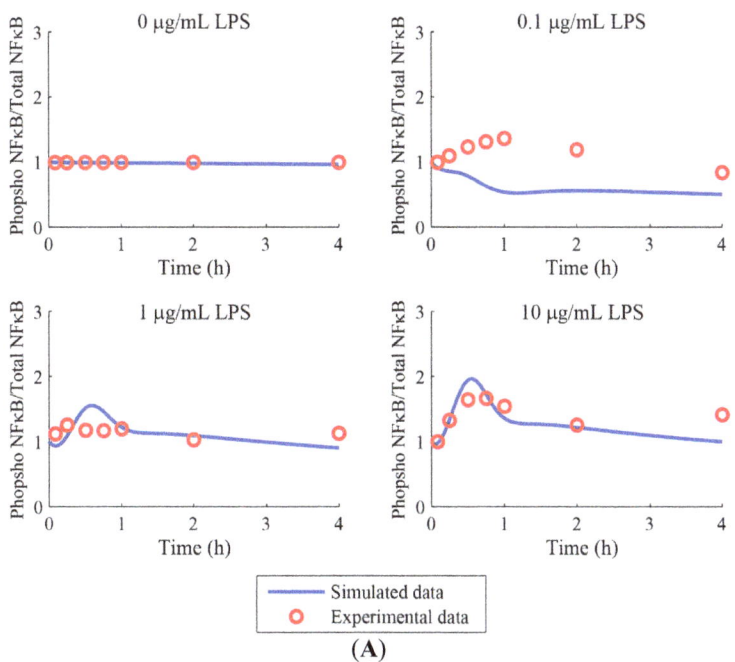

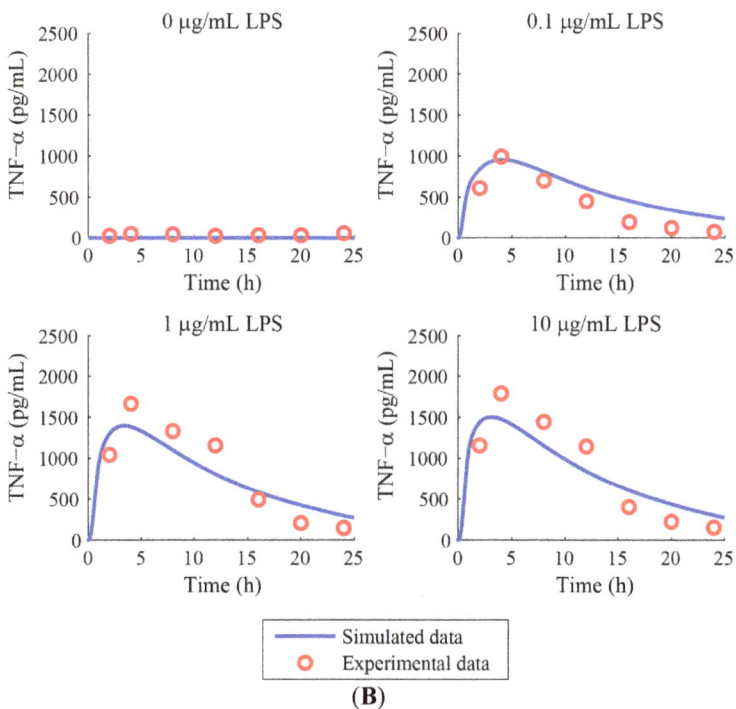

Figure 5. *Cont.*

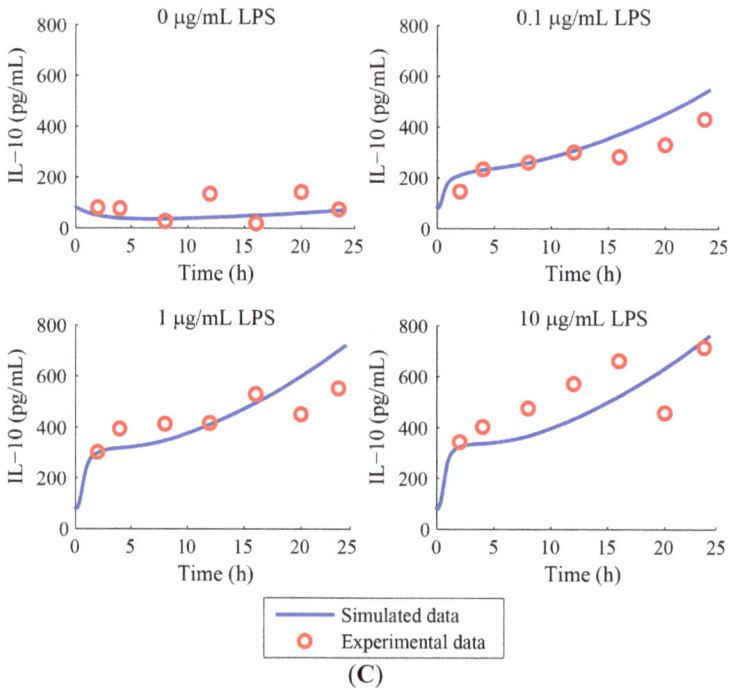

Figure 5. Comparison of model predictions and experimental data for LPS-stimulated RAW264.7 macrophages. (**A**) Phosphorylated NF-κB/Total NF-κB in LPS-treated macrophages relative to the control; (**B**) Profile of *de novo* synthesized TNF-α (pg/mL) upon LPS stimulation; (**C**) Profile of *de novo* synthesized IL-10 (pg/mL) upon LPS stimulation.

Model simulations are validated by experimental data obtained from LPS-stimulated RAW264.7 cells. The comparison between simulated and experimental data for the fold change in NF-κB (phosphorylated NF-κB/Total NF-κB, relative to control), *de novo* TNF-α and IL-10 concentration profiles after parameter estimation are shown in Figure 5A, 5B and 5C, respectively. In Figure 5A, the discrepancy between the model simulation and experimental data for 0.1 μg/mL LPS stimulation could arise from our assumption that the binding affinity between LPS and TLR4 is constant and concentrations of the LPS-TLR4 complex are linearly proportional to the concentrations of LPS tested. However, in reality, the ligand-receptor binding kinetics might follow a non-linear behavior, which is not accommodated in our computational model. The binding of LPS to TLR4 even at lower concentrations of LPS (e.g., 0.1 μg/mL) might result in higher concentrations of the LPS-TLR4 complex, resulting in more downstream phosphorylation of NF-κB (as indicated by the higher phosphorylated NF-κB/total NF-κB ratio for the experimental data in Figure 5A) than the model is able to predict. However, the dynamic model where the parameters have been estimated exhibits a reasonably good fit for phosphorylated NF-κB/total NF-κB profiles (relative to control) at 0 μg/mL, 1 μg/mL and 10 μg/mL LPS stimulations. Furthermore, the model exhibits reasonable agreement between simulated and experimental data for both training and validation datasets for the TNF-α and IL-10 dynamic concentration profiles.

The model suggests that the initial increases in TNF-α and IL-10 are due to NF-κB activation (*i.e.*, phosphorylation and dissociation of NF-κB from IκBα-NF-κB complex) in the cytoplasm and subsequent gene expression in the nucleus; and the decrease in TNF-α concentration after 4 h is due to the negative feedback of IL-10 on NF-κB activity (by inhibiting both IKK activation and nuclear translocation of phosphorylated NF-κB). Interestingly, the levels of IL-10 continue to increase, even when the levels of activated NF-κB are no longer increasing. It is possible that the initial burst of IL-10 produced through NF-κB activation can activate other transcription factors, which leads to an increase in IL-10 levels. An example of this could be the transcription factor STAT3, which has been shown to be activated by IL-10 [38]. Endogenous IL-10 in LPS stimulation of macrophages (RAW264.7) [39] forms the IL-10-IL10R complex that initiates phosphorylation of cytosolic STAT3, followed by its dimerization and translocation into the nucleus. The STAT3 dimer binds to the DNA response element and triggers transcription of IL-10 (Figure S1). Thus, increasing LPS concentration could lead to increasing concentrations of IL-10 due to the positive feedback of IL-10 on its own production.

It can be seen that the model predictions and experimental data are in reasonable agreement, which demonstrates that biochemical reactions, which form the structure of the model, are physiologically relevant and can depict the interplay between pro-inflammatory and anti-inflammatory immune responses to maintain equilibrium (homeostasis). Furthermore, cross-talk between positive and negative feedback regulatory loops, incorporated in the model, is integral towards mathematically representing biochemical and gene regulatory networks, as mentioned by Tian *et al.* [40]. Our model also shows that the anti-inflammatory functions in the RAW264.7 macrophage cell line is initially triggered by pro-inflammatory stimulation. This model structure can be extended to study other cell types with a modification in parameter values to fit model predictions to experimental datasets for specific cell types.

The integrated mathematical model of pro- and anti-inflammatory host immune response discussed in this paper is a step towards assimilating our knowledge and developing a quantitative understanding of the signal transduction pathways involved in maintaining immune homeostasis, disruption of which can lead to inflammatory disorders. This mathematical model can be further used to study intra- and inter-kingdom signaling, *i.e.*, the effect of bacterial metabolites (e.g., indole) synthesized by micro flora present in the host gastro-intestinal tract on host immune response [41]. This model can be used as the basic structure to incorporate additional transcription factors, which will be needed to study the signaling of indole and their interactions with NF-κB.

Acknowledgments

The authors acknowledge partial financial support from the National Science Foundation (Chemical, Bioengineering, Environmental and Transport Systems; CBET#084653 to A.J., CBET#0941313 to J.H. and A.J.).

Author Contributions

S.M., J.H. and A.J. conceived research; S.M., R.C.A. and A.J. designed the experiments; S.M. developed the ODE model and performed the experiments; W.D. performed parameter estimation; S.M., W.D., J.H., and A.J. analyzed the data; S.M., W.D., J.H. and A.J. wrote the paper.

Conflicts of Interest

The authors declare no conflict of interest.

References

1. Nathan, C. Points of control in inflammation. *Nature* **2002**, *420*, 846–852.
2. Hansson, G.K.; Libby, P.; Schönbeck, U.; Yan, Z.Q. Innate and adaptive immunity in the pathogenesis of atherosclerosis. *Circ. Res.* **2002**, *91*, 281–291.
3. Andersen-Nissen, E.; Smith, K.D.; Bonneau, R.; Strong, R.K.; Aderem, A. A conserved surface on Toll-like receptor 5 recognizes bacterial flagellin. *J. Exp. Med.* **2007**, *204*, 393–403.
4. Medzhitov, R.; Janeway, C. Innate immune recognition: mechanisms and pathways. *Immunol. Rev.* **2000**, *173*, 89–97.
5. Mantovani, A.; Cassatella, M.A.; Costantini, C.; Jaillon, S. Neutrophils in the activation and regulation of innate and adaptive immunity. *Nat. Rev. Immunol.* **2011**, *11*, 519–531.
6. Lodes, M.J.; Cong, Y.; Elson, C.O.; Mohamath, R.; Landers, C.J.; Targan, S.R.; Fort, M.; Hershberg, R.M. Bacterial flagellin is a dominant antigen in Crohn disease. *J. Clin. Invest.* **2004**, *113*, 1296–1306.
7. Montuschi, P. Pharmacotheraphy of patients with mild persistent asthma: strategies and unresolved issues. *Front. Pharmacol.* **2011**, *2*, doi:10.3389/fphar.2011.00035.
8. Trivedi, P.J.; Adams, D.H. Mucosal immunity in liver autoimmunity: A comprehensive review. *J. Autoimmun.* **2013**, *46*, 97–111.
9. Ghosh, G.; Wang, V.Y.F.; Huang, D.B.; Fusco, A. NF-κB regulation: Lessons from structures. *Immunol. Rev.* **2012**, *246*, 36–58.
10. Baeuerle, P.A.; Baltimore, D. A 65-kappaD subunit of active NF-kappaB is required for inhibition of NF-kappaB by I kappaB. *Genes Dev.* **1989**, *3*, 1689–1698.
11. Hayden, M.S.; Ghosh, S. Shared principles in NF-κB signaling. *Cell* **2008**, *132*, 344–362.
12. Bogdan, C.; Vodovotz, Y.; Nathan, C. Macrophage deactivation by interleukin 10. *J. Exp. Med.* **1991**, *174*, 1549–1555.
13. Collart, M.A.; Baeuerle, P.; Vassalli, P. Regulation of tumor necrosis factor alpha transcription in macrophages: Involvement of four kappa B-like motifs and of constitutive and inducible forms of NF-kappa B. *Mol. Cell. Biol.* **1990**, *10*, 1498–1506.
14. Saraiva, M.; Christensen, J.R.; Tsytsykova, A.V.; Goldfeld, A.E.; Ley, S.C.; Kioussis, D.; O'Garra, A. Identification of a macrophage-specific chromatin signature in the IL-10 locus. *J. Immunol.* **2005**, *175*, 1041–1046.
15. Scott, M.L.; Fujita, T.; Liou, H.C.; Nolan, G.P.; Baltimore, D. The p65 subunit of NF-kappa B regulates I kappa B by two distinct mechanisms. *Genes Dev.* **1993**, *7*, 1266–1276.

16. Zhang, S.Q.; Kovalenko, A.; Cantarella, G.; Wallach, D. Recruitment of the IKK signalosome to the p55 TNF receptor: RIP and A20 bind to NEMO (IKKγ) upon receptor stimulation. *Immunity* **2000**, *12*, 301–311.

17. Benkhart, E.M.; Siedlar, M.; Wedel, A.; Werner, T.; Ziegler-Heitbrock, H.W.L. Role of Stat3 in lipopolysaccharide-induced IL-10 gene expression. *J. Immunol.* **2000**, *165*, 1612–1617.

18. Lang, R.; Patel, D.; Morris, J.J.; Rutschman, R.L.; Murray, P.J. Shaping gene expression in activated and resting primary macrophages by IL-10. *J. Immunol.* **2002**, *169*, 2253–2263.

19. Niemand, C.; Nimmesgern, A.; Haan, S.; Fischer, P.; Schaper, F.; Rossaint, R.; Heinrich, P.C.; Müller-Newen, G. Activation of STAT3 by IL-6 and IL-10 in primary human macrophages is differentially modulated by suppressor of cytokine signaling 3. *J. Immunol.* **2003**, *170*, 3263–3272.

20. Lee, K.C.; Chang, H.H.; Chung, Y.H.; Lee, T.Y. Andrographolide acts as an anti-inflammatory agent in LPS-stimulated RAW264.7 macrophages by inhibiting STAT3-mediated suppression of the NF-κB pathway. *J. Ethnopharmacol.* **2011**, *135*, 678–684.

21. Bhattacharyya, S.; Sen, P.; Wallet, M.; Long, B.; Baldwin, A.S.; Tisch, R. Immunoregulation of dendritic cells by IL-10 is mediated through suppression of the PI3K/Akt pathway and of IκB kinase activity. *Blood* **2004**, *104*, 1100–1109.

22. Driessler, F.; Venstrom, K.; Sabat, R.; Asadullah, K.; Schottelius, A.J. Molecular mechanisms of interleukin-10-mediated inhibition of NF-κB activity: A role for p50. *Clin. Exp. Immunol.* **2004**, *135*, 64–73.

23. Schottelius, A.J.; Mayo, M.W.; Sartor, R.B.; Baldwin, A.S. Interleukin-10 signaling blocks inhibitor of κB kinase activity and nuclear factor κB DNA binding. *J. Biol. Chem.* **1999**, *274*, 31868–31874.

24. Singh, A.; Jayaraman, A.; Hahn, J. Modeling regulatory mechanisms in IL-6 signal transduction in hepatocytes. *Biotechnol. Bioeng.* **2006**, *95*, 850–862.

25. Huang, Z.; Senocak, F.; Jayaraman, A.; Hahn, J. Integrated modeling and experimental approach for determining transcription factor profiles from fluorescent reporter data. *BMC Syst. Biol.* **2008**, *2*, doi:10.1186/1752-0509-2-64.

26. Lipniacki, T.; Paszek, P.; Brasier, A.R.; Luxon, B.; Kimmel, M. Mathematical model of NF-κB regulatory module. *J. Theor. Biol.* **2004**, *228*, 195–215.

27. Rangamani, P.; Sirovich, L. Survival and apoptotic pathways initiated by TNF-α: Modeling and predictions. *Biotechnol. Bioeng.* **2007**, *97*, 1216–1229.

28. Hoffmann, A.; Levchenko, A.; Scott, M.L.; Baltimore, D. The IκB-NF-κB signaling module: Temporal control and selective gene activation. *Science* **2002**, *298*, 1241–1245.

29. Moya, C.; Huang, Z.; Cheng, P.; Jayaraman, A.; Hahn, J. Investigation of IL-6 and IL-10 signalling via mathematical modelling. *IET Syst. Biol.* **2011**, *5*, 15–26.

30. Gray, P.W.; Barrett, K.; Chantry, D.; Turner, M.; Feldmann, M. Cloning of human tumor necrosis factor (TNF) receptor cDNA and expression of recombinant soluble TNF-binding protein. *Proc. Natl. Acad. Sci. USA* **1990**, *87*, 7380–7384.

31. Fedorak, R.N.; Gangl, A.; Elson, C.O.; Rutgeerts, P.; Schreiber, S.; Wild, G.; Hanauer, S.B.; Kilian, A.; Cohard, M.; LeBeaut, A.; *et al.* Recombinant human interleukin 10 in the treatment of patients with mild to moderately active Crohn's disease. *Gastroenterology* **2000**, *119*, 1473–1482.

32. Tjoa, I.B.; Biegler, L.T. Simultaneous solution and optimization strategies for parameter estimation of differential-algebraic equation systems. *Ind. Eng. Chem. Res.* **1991**, *30*, 376–385.

33. Kravaris, C.; Hahn, J.; Chu, Y. Advances and selected recent developments in state and parameter estimation. *Comput. Chem. Eng.* **2013**, *51*, 111–123.

34. Riley, J.K.; Takeda, K.; Akira, S.; Schreiber, R.D. Interleukin-10 receptor signaling through the JAK-STAT pathway: Requirement for two distinct receptor-derived signals for anti-inflammatory action. *J. Biol. Chem.* **1999**, *274*, 16513–16521.

35. Rahim, S.S.; Khan, N.; Boddupalli, C.S.; Hasnain, S.E.; Mukhopadhyay, S. Interleukin-10 (IL-10) mediated suppression of IL-12 production in RAW 264.7 cells also involves c-rel transcription factor. *Immunology* **2005**, *114*, 313–321.

36. Chu, Y.; Hahn, J. Parameter set selection via clustering of parameters into pairwise indistinguishable groups of parameters. *Ind. Eng. Chem. Res.* **2008**, *48*, 6000–6009.

37. Dai, W.; Bansal, L.; Hahn, J.; Word, D. Parameter set selection for dynamic systems under uncertainty via dynamic optimization and hierarchical clustering. *AIChE J.* **2014**, *60*, 181–192.

38. Murray, P.J. The primary mechanism of the IL-10-regulated anti-inflammatory response is to selectively inhibit transcription. *Proc. Natl. Acad. Sci. USA* **2005**, *102*, 8686–8691.

39. Carl, V.S.; Gautam, J.K.; Comeau, L.D.; Smith, M.F. Role of endogenous IL-10 in LPS-induced STAT3 activation and IL-1 receptor antagonist gene expression. *J. Leukoc. Biol.* **2004**, *76*, 735–742.

40. Tian, X.J.; Zhang, X.P.; Liu, F.; Wang, W. Interlinking positive and negative feedback loops creates a tunable motif in gene regulatory networks. *Phys. Rev. E* **2009**, *80*, doi:10.1103/PhysRevE.1180.011926.

41. Bansal, T.; Alaniz, R.C.; Wood, T.K.; Jayaraman, A. The bacterial signal indole increases epithelial-cell tight-junction resistance and attenuates indicators of inflammation. *Proc. Natl. Acad. Sci. USA* **2010**, *107*, 228–233.

Modeling the Dynamics of Acute Phase Protein Expression in Human Hepatoma Cells Stimulated by IL-6

Zhaobin Xu, Jens O. M. Karlsson and Zuyi Huang

Abstract: Interleukin-6 (IL-6) is a systemic inflammatory mediator that triggers the human body's acute phase response to trauma or inflammation. Although mathematical models for IL-6 signaling pathways have previously been developed, reactions that describe the expression of acute phase proteins were not included. To address this deficiency, a recent model of IL-6 signaling was extended to predict the dynamics of acute phase protein expression in IL-6-stimulated HepG2 cells (a human hepatoma cell line). This included reactions that describe the regulation of haptoglobin, fibrinogen, and albumin secretion by nuclear transcription factors STAT3 dimer and C/EBPβ. This new extended model was validated against two different sets of experimental data. Using the validated model, a sensitivity analysis was performed to identify seven potential drug targets to regulate the secretion of haptoglobin, fibrinogen, and albumin. The drug-target binding kinetics for these seven targets was then integrated with the IL-6 kinetic model to rank them based upon the influence of their pairing with drugs on acute phase protein dynamics. It was found that gp80, JAK, and gp130 were the three most promising drug targets and that it was possible to reduce the therapeutic dosage by combining drugs aimed at the top three targets in a cocktail. These findings suggest hypotheses for further experimental investigation.

Reprinted from *Processes*. Cite as: Xu, Z.; Karlsson, J.O.M.; Huang, Z. Modeling the Dynamics of Acute Phase Protein Expression in Human Hepatoma Cells Stimulated by IL-6. *Processes* **2015**, *3*, 50-70.

1. Introduction

Interleukin-6 (IL-6) has been identified as one of the major systemic mediators that orchestrate acute phase response (APR) in the human body, as evidenced by the fact that IL-6 can stimulate the synthesis of most acute phase proteins in liver cells [1]. The mediation of APR by IL-6 begins with the release of IL-6 by leukocytes at the injury site. IL-6 then translocates to the liver, via the blood stream, where it stimulates hepatocytes and activates a cascade of intracellular signal transduction pathways. This leads to the activation of transcription factors, such as nuclear STAT3 dimer and C/EBPβ. These transcription factors, in turn, regulate the expression of acute phase proteins, such as haptoglobin, fibrinogen and albumin. The IL-6 signal transduction pathway has been extensively studied and the two major signaling pathways have been determined to be the JAK-STAT pathway and the MAPK-C/EBPβ pathway [2]. Nuclear STAT3 dimer and C/EBPβ have been identified as the transcription factors involved in these two signaling pathways, respectively [3,4]. Mathematical models have been developed for the JAK-STAT pathway [5] and the MAPK pathway [6–8]. Singh *et al.* (2006) presented the first comprehensive mathematical model for the IL-6 signal transduction pathway by integrating models for both the JAK-STAT and the MAPK pathways [9]. Moya *et al.* (2011) further extended the model of the MAPK pathway to predict the activation dynamics of transcription factor C/EBPβ [10]. However, acute phase proteins, which represent the end products of the IL-6 signal transduction pathway and participate in the human body's response to trauma or inflammation, were not included in these existing mathematical models. Ryll *et al.* (2011) presented the first attempt to incorporate expression of acute phase proteins (e.g., C-reactive protein, α2-macroglobulin, and fibrinogen) in a model of the IL-6 signaling network [11]. However, this was a qualitative logic model in which the detailed intermediate reactions for acute phase protein activation were neglected. To address this deficiency, the model presented by Moya *et al.* (2011) [10] will be extended in this work to predict the dynamics of acute phase protein expression in HepG2 cells. Haptoglobin, fibrinogen, and albumin were selected as the representative acute phase proteins in this work for model development. These were chosen because they represent both positive (*i.e.*, haptoglobin and fibrinogen) and negative (*i.e.*, albumin) acute phase proteins and quantitative experimental data for the expression dynamics of these three proteins in HepG2 cells stimulated by IL-6 were available in the literature [12].

Mathematical models have previously been developed, and used, to identify potential drug targets to treat human diseases. For example, Araujo *et al.* (2005) presented a modeling approach to study the effect of multiple drugs on EGFR signaling [13]. Yang *et al.* (2008) further extended Araujo's approach to determine multiple-target optimal intervention in the arachidonic acid metabolic network [14]. Such model-based approaches require that equations describing the signaling kinetics be integrated with models of target-drug binding kinetics. Since there is a lack of models of acute phase protein expression and limited information about relevant target-drug binding kinetics in the IL-6 pathway, model-based approaches have not previously been applied to the identification of drug targets for regulating the dynamics of acute phase proteins stimulated by IL-6. In order to address this, our extended IL-6 model will be used in a sensitivity analysis to identify potential drug targets for regulating the expression dynamics of albumin, haptoglobin, and fibrinogen. Targeting JAK kinases could be useful in the treatment of a variety of diseases, including rheumatoid arthritis (RA) [15], myeloproliferative disorders [16] and cancers [17]. The approved pan-JAK inhibitor, tofacitinib, has undergone extensive evaluation for RA and has demonstrated efficacy in various clinical trials, likely due to its suppression of the IL-6 and cytokine pathways [18]. Along with tofacitinib, a series of JAK inhibitors have been proposed recently using a novel fused triazolo-pyrrolopyridine scaffold [18]. Among them is imidazo-pyrrolopyridine. Since limited information is available for drugs that bind to targets other than JAK kinases in IL-6 signaling, this work focuses on JAK inhibitors only. Imidazo-pyrrolopyridine is used as the model drug to illustrate the approach developed to integrate the extended IL-6 model with target-drug binding kinetics for studying the effectiveness of single/multiple drug treatment in regulating the secretion of acute phase response proteins.

This paper is structured as follows: A recent model of IL-6 signaling [10] is extended, in Section 2, to predict the dynamics of acute phase protein expression. Experimental data presented in [12] are used to estimate the unknown model parameters, and the extended model is then validated against independent experimental data reported by [12] and [1]. Based upon the model developed in Section 2, sensitivity analysis is conducted, in Section 3, to identify reactions that play an important role in the regulation of the dynamics of haptoglobin, fibrinogen, and albumin. Molecular components involved in these reactions are regarded as potential drug targets. A model-based approach for virtual drug target screening is presented in Section 4 to evaluate the influence of the interaction between potential drugs and targets on the secretion rate of acute phase proteins. Discussion and concluding remarks related to the obtained results are given in Sections 5 and 6 respectively.

2. Model Development for the Kinetics of Acute Phase Proteins in IL-6 Stimulated Hepatocytes

2.1. IL-6 Signal Transduction Model

The starting point for the model used in this work is the IL-6 signal transduction model developed by one of the co-authors (ZH) as described in a previous study [10]. This model can be represented by a set of nonlinear ordinary differential equations (Equation (1)):

$$\frac{d\mathbf{x}}{dt} = f(\mathbf{x}, \mathbf{p}, u) \tag{1}$$

where $\mathbf{x}$ is a vector of the state variables of the model, $\mathbf{p}$ is a vector of the parameters, and u is the input to the system. The model consists of 68 ordinary differential equations representing the mass balances of the individual proteins and protein complexes, 117 parameters describing reaction constants, and one input given by the extracellular IL-6 concentration.

A simplified diagram of the proteins involved in the model is shown in Figure 1. Extracellular cytokine IL-6 initiates the APR by attaching to its receptor at the cell membrane and forming (IL6-gp80-gp130-JAK)2 complex. This phosphorylated dimer serves as the starting point for both the JAK-STAT and the MAPK pathways. In the JAK-STAT signaling, the phosphorylated dimer recruits the transcription factor STAT3, which is also tyrosine phosphorylated. The phosphorylated STAT3 dissociates from the receptor complex (IL6-gp80-gp130-JAK)2 and undergoes dimerization. The dimerized STAT3 complex translocates to the nucleus and functions as a transcription factor for the expression of SOCS3, which in turn binds to the receptor gp130 and blocks the activation of JAK, thus inhibiting both STAT3 activation and MAPK activation. In the MAPK signaling, phosphorylated gp130 recruits SHP2 which subsequently undergoes phosphorylation. The phosphorylated SHP2 interacts with Grb2 and SOS. The binding of Grb2 and SOS to the receptor complex leads to the activation of RAS, which further leads to the activation of the MAPK cascade up to transcription factor C/EBPβ.

In this work, the model reported by [10] is extended to predict the expression dynamics of haptoglobin, fibrinogen, and albumin in HepG2 cells stimulated by IL-6. Reactions for the synthesis of haptoglobin, fibrinogen, and albumin are first added into Moya's model. The unknown parameters in the extended model are estimated from the experimental data measured in HepG2 cells under constant exposure to 2 nM IL-6 for seven days [12]. The extended model is then validated by the two data sets: the first one is for HepG2 cells under a pulse-chase stimulation of IL-6 [12], while the second one is for steady state values of dose-dependent secretion rates of fibrinogen and albumin in IL-6-stimulated HepG2 cultures reported by [1].

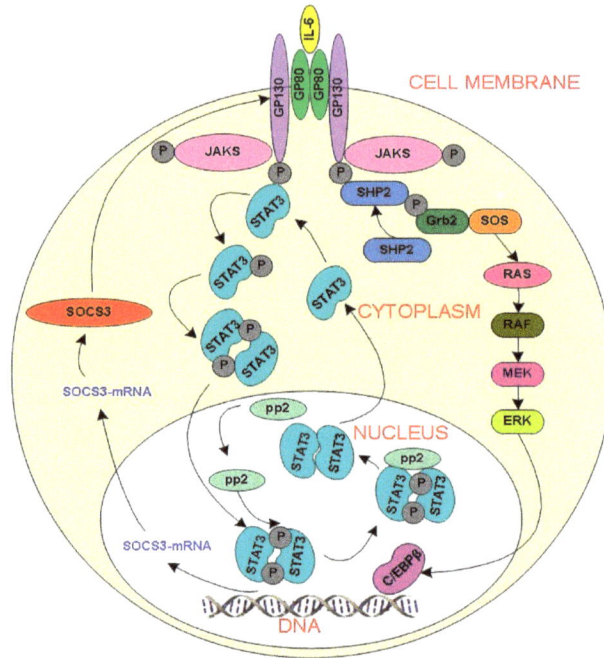

Figure 1. Implemented reaction network for Interleukin-6 (IL-6) induced signal transduction in hepatocytes. Adapted with permission from [10]. Copyright 2011, IET.

2.2. Extended Model of Acute Phase Protein Expression Dynamics

The model reported by [10] does not include reactions for the synthesis of haptoglobin, fibrinogen, and albumin. In this subsection, reactions describing the transcription of mRNA encoding haptoglobin, fibrinogen, and albumin, as well as reactions responsible for the translation and secretion of these three acute phase proteins, were added into the existing model of IL-6 signal transduction. Nuclear STAT3 dimer and C/EBPβ, whose activation levels can be regulated by IL-6 signaling, are the two major transcription factors for initiating the expression of acute phase proteins in the liver. In particular, C/EBPβ, nuclear STAT3 dimer, and C/EBPα are the transcription factors regulating the transcription of haptoglobin [19], fibrinogen [20], and albumin [19,21], respectively. The transcription factor C/EBPβ affects the expression dynamics of albumin by inhibiting the expression of C/EBPα [21], and thus C/EBPβ indirectly down-regulates the expression of albumin. Based on these observations, reactions shown in Equations (2) through (9) were added into the IL-6 model reported by [10] such that the secretion rates of extracellular haptoglobin, fibrinogen, and albumin (denoted as Ex-haptoglobin, Ex-fibrinogen, and Ex-albumin, respectively) can be predicted from the extended model.

$$C/EBP\beta \xrightarrow{V_{m\_h}, K_{m\_h}} mRNA - haptoglobin \tag{2}$$

$$mRNA - haptoglobin \xrightarrow{k_{t\_h}} Ex - haptoglobin \tag{3}$$

$$STAT3N^* - STAT3N^* \xrightarrow{V_{m\_f}, K_{m\_f}} mRNA - fibrinogen \tag{4}$$

$$mRNA - fibrinogen \xrightarrow{k_{t\_f}} Ex - fibrinogen \tag{5}$$

$$C/EBP\beta \xrightarrow{k_{i\_a}} C/EBP\alpha \tag{6}$$

$$C/EBP\alpha \xrightarrow{k_{d\_a}} degradation \tag{7}$$

$$C/EBP\alpha \xrightarrow{V_{m\_a}, K_{m\_a}} mRNA - albumin \tag{8}$$

$$mRNA - albumin \xrightarrow{k_{t\_a}} Ex - albumin \tag{9}$$

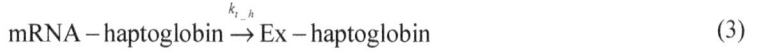

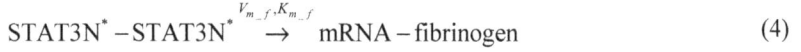

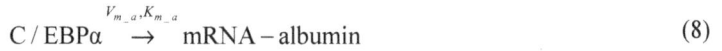

In Equations (2)–(9), the species mRNA-haptoglobin, mRNA-fibrinogen and mRNA-albumin refer to the mRNA encoding for the corresponding protein. Equation (2) describes the transcription process of haptoglobin, which is regulated by C/EBPβ [19] (the corresponding Michaelis-Menten coefficients are V_{m_h} and K_{m_h}). Equation (3) lumps the translation of the haptoglobin mRNA with the secretion process of synthesized haptoglobin. Lumping was performed to reduce the number of unknown parameters in the model, and is justifiable since one of the two sequential processes is rate-limiting. Thus, a mass action kinetic model with an effective rate constant k_{t_h} was used to describe the lumped translation and secretion processes for haptoglobin expression. Similarly, Equations (4) and (5) describe the transcription, translation, and secretion processes of fibrinogen. Compared to these two positive acute phase proteins, the regulation of the expression of albumin, given by Equations (6) to (9), is more complicated. C/EBPβ inhibits the activation of C/EBPα as given in Equation (6). This down-regulates the expression of albumin, because C/EBPα is required to initiate the transcription of albumin (as shown in Equation (8)). Equation (7) describes the degradation process of C/EBPα in HepG2 cells due to the cell growth, while Equation (9) represents the lumped translation and secretion of albumin. A one-day delay has been added in the albumin transcription process as suggested by the data presented in [12]. Figure 2 shows the schematic diagram of the extended IL-6 signal transduction pathway.

The ordinary differential equations describing the rates of newly added components involved in Equations (2) through (9) are developed based on mass balance, mass action kinetics and Michaelis-Menten kinetics. The resulting seven equations, Equations (A1)–(A7) (given in the Appendix), were integrated into our existing IL-6 model [10] to yield an extended IL-6 model with 75 ordinary differential equations and 128 parameters, to predict the secretion rates of extracellular haptoglobin, fibrinogen, and albumin.

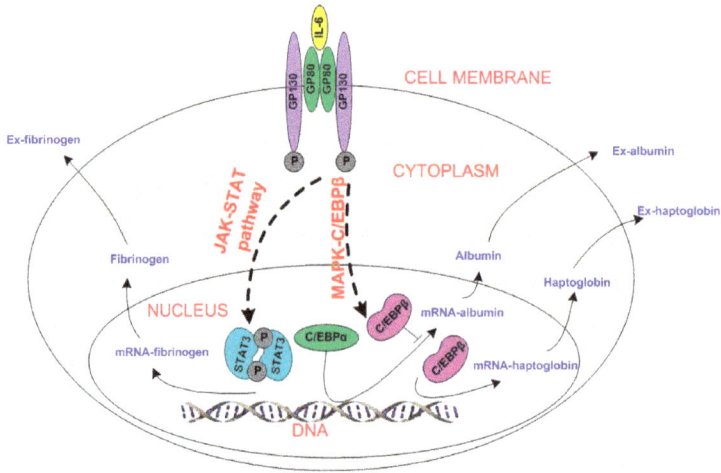

Figure 2. Extension of an existing Interleukin-6 (IL-6) signal transduction model [10] to include reactions describing the expression dynamics of haptoglobin, fibrinogen, and albumin. The two dashed lines represent the JAK-STAT and MAPK-C/EBPβ pathways (see Figure 1). Adapted with permission from [10]. Copyright 2011, IET.

2.3. Estimation of Unknown Parameters in the Extended Model

There are 11 unknown parameters in Equations (2) through (9). Fisher information matrixes (FIM) were used to perform identifiability analysis. All these parameters are identifiable. An experimental data set for HepG2 cultures under a seven-day exposure to 2 nM IL-6 [12] was used to estimate these unknown parameters. Parameter estimation was conducted with standard nonlinear least squares optimization routines, such as lsqnonlin, available from MATLAB (The Math Works: Natick, MA, USA). The estimated values of the 11 unknown parameters, as well as their bounds with a 95% confidence level, were listed in the Appendix. A comparison between the model output and the experimental data was shown in Figure 3. In addition to performing comparison by visual inspection, the relative errors (*Err*) were computed according to the Equation (10);

$$Err = \frac{\left\| \mathbf{Y} - \hat{\mathbf{Y}} \right\|}{\left\| \hat{\mathbf{Y}} \right\|} \times 100\% \tag{10}$$

where $\hat{\mathbf{Y}}$ is a vector of the experimentally measured outputs at different points in time, *i.e.*, $\left[\hat{y}(t_1) \quad \hat{y}(t_2) \quad \cdots \quad \hat{y}(t_n) \right]$, and $\mathbf{Y}$ is a vector of the outputs calculated by the model at corresponding time points, *i.e.*, $\left[y(t_1) \quad y(t_2) \quad \cdots \quad y(t_n) \right]$. The norm is the Euclidean (ℓ_2) norm. The values of *Err* for the secretion rates of haptoglobin, fibrinogen, and albumin were found to be 14.41%, 12.16%, and 7.17%, respectively. All of relative errors were below 15%, which indicates that the prediction was reasonably good in comparison to the magnitude of the error bars in the experimental data.

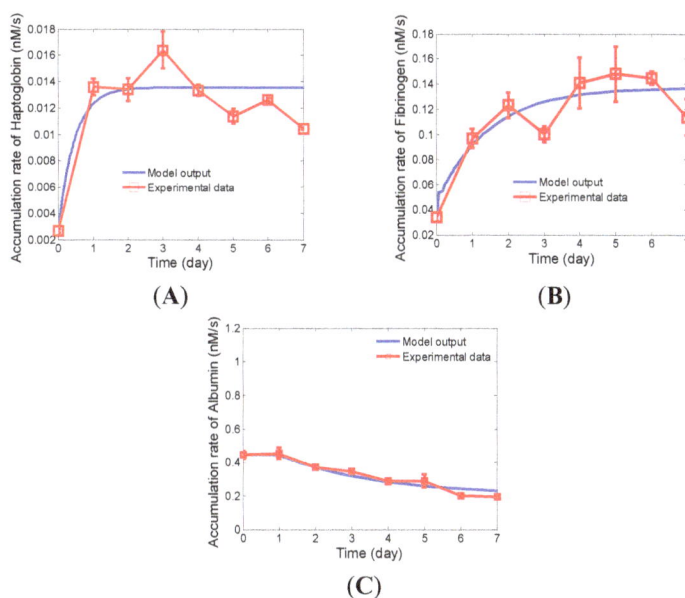

Figure 3. Comparison of model-predicted secretion rates of haptoglobin, fibrinogen, and albumin to experimental data obtained from the HepG2 cultures stimulated daily by 2 nM Interleukin-6 (IL-6): **(A)** the secretion rate of haptoglobin; **(B)** the secretion rate of fibrinogen; **(C)** the secretion rate of albumin.

2.4. Validation of the Developed Model

Validation of the developed model is a crucial step before it is used to generate potential biological hypotheses that are used for biological experiment design. In this subsection, two independent experimental data sets were used to verify the developed model.

2.4.1. Relaxation Kinetics of Albumin, Fibrinogen, and Haptoglobin in HepG2 Cultures under a Pulse-Chase Stimulation of IL-6

The developed model was used to predict the dynamics of the three acute phase proteins for a pulse-chase condition where HepG2 cultures were subjected to IL-6 stimulation for three days and thereafter maintained without IL-6 stimulation [12]. Figure 4 showed the comparison between the model prediction and the experimental data. The values of *Err* for the prediction of secretion rates of haptoglobin, fibrinogen, and albumin were 14.78%, 31.09%, and 5.15%, respectively. It can be seen from Figure 4A that the model predicted the relaxation kinetics of haptoglobin well. A large value of *Err* existed in the prediction of relaxation kinetics of fibrinogen. This is mainly due to the mismatch between the model prediction and experimental data in Day two described in Figure 4B, for which the error bar, *i.e.*, the variability of experimental measurements, was quite large. Corresponding explanation for this large error bar has been given in the original report by [12]. Other than Day two, the model prediction for fibrinogen secretion rate matched the experimental data well. The predicted albumin secretion kinetics agreed well with the experimental data, as shown in Figure 4C, although it was necessary to incorporate a 24 h time-delay in the albumin expression model in order to accurately describe the observed lag between changes in IL-6 concentration and the resulting albumin secretion rate.

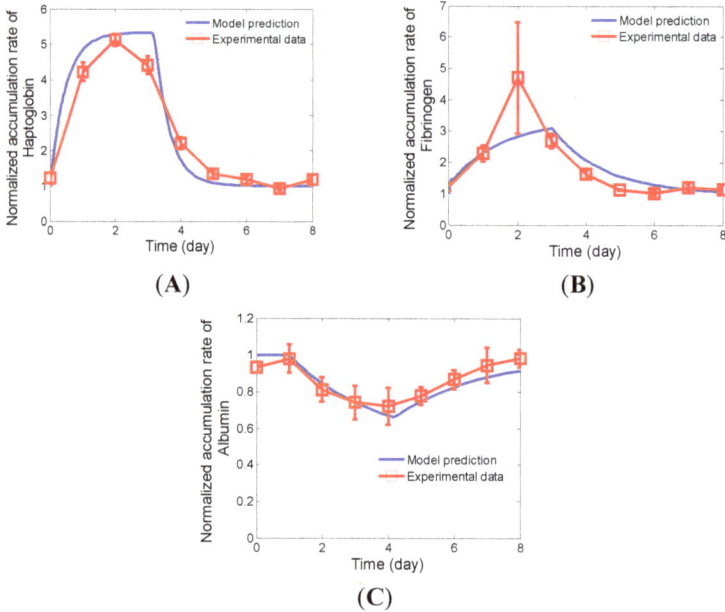

Figure 4. Comparison of model prediction to experimental secretion rates of haptoglobin, fibrinogen, and albumin for HepG2 cells under a pulse-chase stimulation of 2 nM Interleukin-6 (IL-6): (**A**) secretion rate of haptoglobin; (**B**) secretion rate of fibrinogen; (**C**) secretion rate of albumin.

2.4.2. Steady State Values of Dose-Dependent Secretion Rates of Fibrinogen and Albumin in IL-6 Stimulated HepG2 Cultures

In this subsection, the data published in [1], which includes the steady state secretion rates of fibrinogen and albumin obtained in human primary hepatocytes under the exposure to IL-6 of various concentrations, was used to further validate the extended model. Accordingly, different concentrations of IL-6 were used in this work as the inputs for the developed model, and the steady state values of the secretion rates of fibrinogen, albumin and haptoglobin were recorded, normalized by the secretion rates of these three proteins for the control condition, and plotted in Figure 5. Figure 5A showed that the steady state value of the secretion rate of fibrinogen increased with increasing IL-6 concentrations and leveled off afterwards, following a sigmoidal profile with a transition width corresponding to two orders of magnitude variation in the IL-6 concentration. In addition, the maximum steady-state secretion rate was approximately four-fold higher than for the control condition. Both of these predictions (magnitude and width of profile) were consistent with the corresponding characteristics of the dose-dependence profile reported by [1] for fibrinogen secretion. Figure 5B showed that the predicted steady-state albumin secretion rate decreased approximately two-fold as the IL-6 concentration increased. The experimental dose response data, measured by [1] for albumin secretion, exhibited an inhibition of similar magnitude (~2.5-fold). However, the width of the transition in the predicted dose response for albumin was narrower than the transition width reported by [1] for this negative acute phase protein. As seen in Figure 5C, the steady-state haptoglobin secretion rates for various IL-6 concentrations followed a similar trend as shown in Figure 5A for steady-state fibrinogen secretion rates, but with a narrower transition width and a higher enhancement in the maximum steady-state secretion rate (around 5 times of the secretion rate for the control condition). Although Heinrich et $al.$ [1] did not report the steady-state secretion rates of haptoglobin as a function of IL-6 concentration, the result shown in Figure 5C was generally consistent with the response expected for a positive acute phase. One interesting observation that was made in Figure 5 was that the ED_{50} value for the fibrinogen dose-dependence profile (ED_{50} = 0.74 nM) was approximately double the corresponding value for the albumin dose response curve (ED_{50} = 0.32 nM). This result was consistent with the relationship between the ED_{50} values that was observed by [1] for fibrinogen and albumin secretion. Given that measurements by [1] represent an independent data set that was not used for parameter estimation, the agreement between model predictions and the published dose response curves provided additional validation of the model.

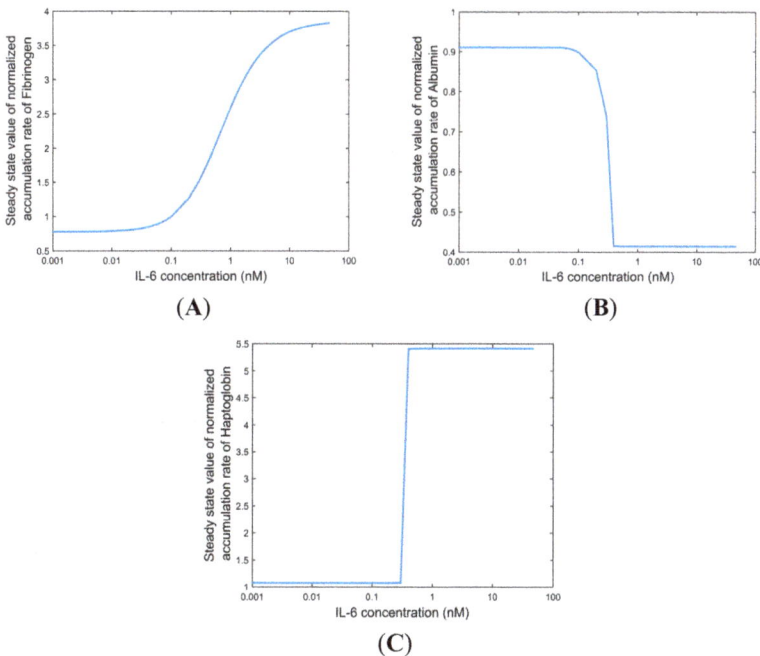

Figure 5. Predicted steady state values of secretion rates of fibrinogen, albumin, and haptoglobin under various stimulation doses of Interleukin-6 (IL-6): (**A**) Steady state values of secretion rates of fibrinogen; (**B**) Steady state values of secretion rates of albumin; (**C**) Steady state values of secretion rates of haptoglobin.

3. Investigation of Influence from the Reactions in IL-6 Signaling on the Expression Dynamics of Haptoglobin, Fibrinogen, and Albumin

In this section, sensitivity analysis was conducted to identify the reactions in the IL-6 signaling pathways that play an important role in the regulation of the expression dynamics of haptoglobin, fibrinogen, and albumin. Molecular components involved in these reactions were regarded as potential drug targets that will be further screened in Section 4.

The sensitivity analysis followed the approach of [22], in which parameters were varied by one order of magnitude above and below their nominal values. A sensitivity metric, $s_{i,j}$, was then quantified by Equation (11), in which the partial derivative of the output y_j with respect to parameter p_i (*i.e.*, a reaction rate constant) was normalized by the nominal values of p_i and y_j (*i.e.*, p_i^0 and y_j^0 respectively);

$$s_{i,j} = \frac{p_i^0 \partial y_j}{y_j^0 \partial p_i}\Big|_{\mathbf{P_0}} \tag{11}$$

where the vector $\mathbf{P_0}$ is a vector of nominal values of all parameters in the model. In this work, the output of the system, *i.e.*, y_j in Equation (11), was set to the seven-day mean value of the secretion rate of haptoglobin, fibrinogen, and albumin, respectively for j equal to 1, 2, and 3.

To identify the most important parameters for modeling the expression dynamics of each acute phase protein, the absolute value of $s_{i,j}$ was ranked in a decreasing order (listed in Table 1, only the top 20 parameters are listed).

As shown in Table 1, the parameters with the highest sensitivity measure values were primarily associated with those newly added reaction equations given by Equations (2) through (9) that described transcription, translation and secretion of haptoglobin, fibrinogen, and albumin, which was expected. The purpose for conducting sensitivity analysis, however, was to determine the effect of reactions from the original IL-6 signaling pathway on regulation of the expression of acute phase proteins. This information might reveal new mechanisms that can be used to manipulate the expression dynamics of acute phase proteins. In order to identify the reactions in IL-6 signal transduction that play an important role in regulation of the secretion rates of haptoglobin, fibrinogen, and albumin, the parameters shown in Table 1 were overlaid onto IL-6 signaling reaction networks in Figure 6. As seen from Table 1, most parameters that had an important impact on the secretion rate of haptoglobin also played an important role in regulating the secretion rate of albumin. This can be explained by the fact that the expression of both haptoglobin and albumin was regulated by MAPK-C/EBPβ signaling pathway. Reactions that were essential for the regulation of the expression of both haptoglobin and albumin constituted one group in Figure 6 (marked in blue pentagons), while the key reactions for regulating the expression of fibrinogen were associated with JAK-STAT pathway (marked in red ellipses in Figure 6). In addition, three reactions were identified from Table 1 for their important role in the regulation of expression dynamics of all three acute phase proteins. They were associated with the binding of IL-6 to its receptor and the formation of the receptor complex. These reactions initiated both the JAK-STAT and MAPK-C/EBPβ pathways. They were marked in purple squares in Figure 6.

Table 1. Sensitivity analysis results.

Rank, i	Impact on the Secretion Rate of Haptoglobin ($j = 1$)		Impact on the Secretion Rate of Fibrinogen ($j = 2$)		Impact on the Secretion Rate of Albumin ($j = 3$)	
	Parameter, p_i	Sensitivity, $|s_{i,j}|$	Parameter, p_i	Sensitivity, $|s_{i,j}|$	Parameter, p_i	Sensitivity, $|s_{i,j}|$
1	V_{m_h}	7.8168	V_{m_f}	7.0475	V_{m_a}	3.7746
2	$k_{t\text{-}h}$	4.0417	$K_{m\text{-}f}$	3.7901	k_{i_a}	1.0222
3	K_{m_h}	3.0482	k_{f7}	2.0167	K_{m_a}	1.0222
4	k_{f51}	0.5061	V_{m_24}	1.8237	k_{d_a}	1.0221
5	k_{f55}	0.3774	k_{26}	1.8237	k_{t_a}	0.5915
6	k_{58}	0.3091	k_{d31}	1.8100	k_{f51}	0.1788
7	k_{f71}	0.0075	k_{f27}	1.8055	k_{f55}	0.1070
8	k_{r71}	0.0075	k_{d30}	1.8046	k_{58}	0.0764
9	K_{m69}	0.0045	K_{m_24}	1.7836	k_{r71}	0.0114
10	V_{m69}	0.0044	$k_{t\text{-}f}$	1.7688	k_{f71}	0.0114
11	k_{70}	0.0043	k_{r27}	1.7573	K_{m69}	0.0068
12	k_{f1}	0.0019	k_8	0.9580	V_{m69}	0.0068
13	k_{54}	0.0011	k_{r7}	0.9403	k_{70}	0.0066
14	k_{f46}	0.0011	k_6	0.8621	k_{f1}	0.0015
15	k_6	0.0010	k_{f28}	0.7990	k_{54}	0.0011
16	k_{f3}	0.0008	k_{f1}	0.7675	k_{f66}	0.0010
17	k_{r1}	0.0008	k_{f3}	0.7628	k_{f46}	0.0009
18	k_{r3}	0.0007	k_{r1}	0.7621	k_{r1}	0.0009
19	k_{r5}	0.0006	k_{r3}	0.7619	k_{r3}	0.0009
20	k_{f5}	0.0006	k_{21}	0.7480	k_{f63}	0.0008

Figure 6 showed that reactions which influenced the regulation of fibrinogen secretion rate were mainly involved in the binding of extracellular IL-6 to its receptor on the cell membrane, the activation of STAT3C, and the expression of SOCS3. On the other hand, reactions that were related to the activation and deactivation regulation of Raf*, the activation of MEK, ERK-PP, and nuclear C/EBPβ were important to the expression of both haptoglobin and albumin.

The information in Figure 6 was used to select putative drug targets for further evaluation. In this work, good drug targets were regarded as those participating in reactions that affect the regulation of all three acute phase proteins. In addition, they should be in monomer form, as the monomer was the basic unit to form a complex [23]. Components with non-zero initial concentrations were also given priority, as they represented existing targets that the drug can bind to. Based upon these criteria, seven drug targets were selected from those important reactions as shown in Figure 6 (specifically; gp80, JAK, gp130, STAT3C, Raf, MEK, and C/EBPβ$_i$). Components gp80, JAK, gp130 were selected as they construct the receptor complex $(IL6 - gp80 - gp130 - JAK)_2^*$ which was involved in several important reactions. The other four drug targets were either directly involved or phosphorylated in those important reactions. These seven drug targets were further evaluated in Section 4 based upon the efficacy of the interaction with their drug counterparts on regulating the dynamics of acute phase proteins.

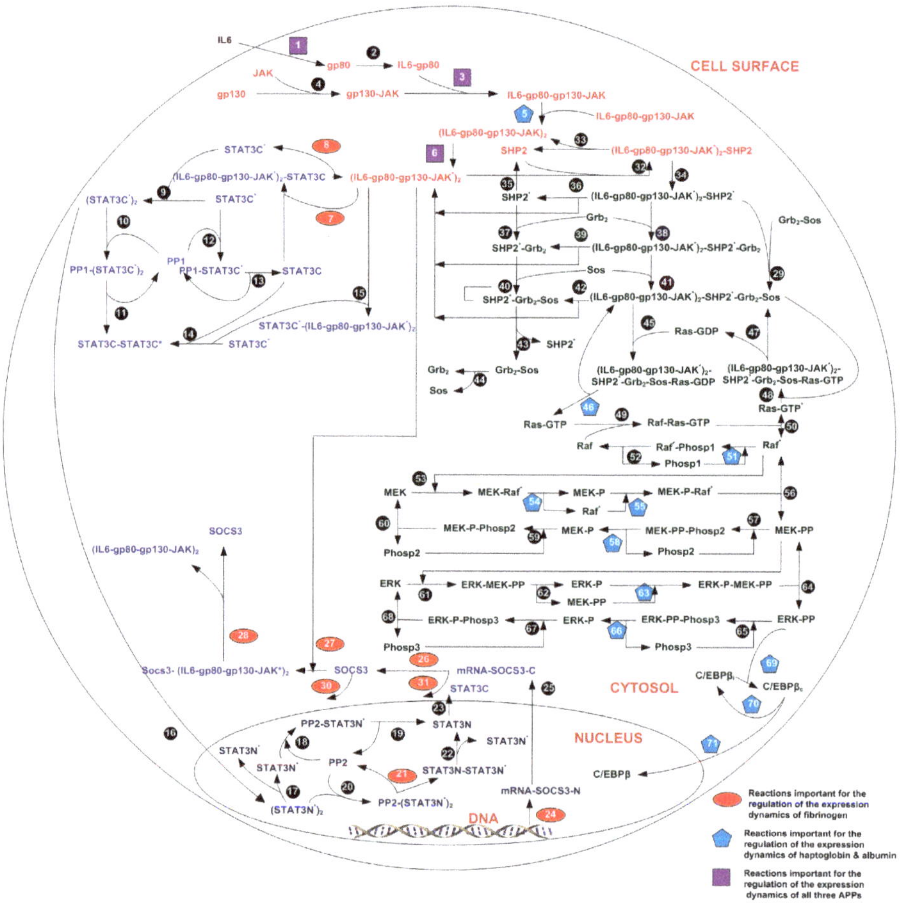

Figure 6. Identification of the reactions from the Interleukin-6 (IL-6) signal transduction pathway that have the largest impact on acute phase protein expression, based on results of sensitivity analysis (Table 1). The figure is adapted from [10], and the numerical labels correspond to the reaction numbering used in the model by [10]. Adapted with permission from [10]. Copyright 2011, IET.

4. Virtual Screening of Drug Targets and Drugs for Acute Phase Response

A good drug target should have high feasibility of binding a drug, that is, it should require low binding energy. On the other hand, the binding of a drug to a good target should cause a large influence on the dynamics of acute phase proteins. A model-based platform was developed in this section to incorporate the drug and target interaction in the extended IL-6 model to screen the seven drug targets selected in Section 3. The influence from drug binding kinetics was then investigated on the basis of the developed platform, which was followed by evaluating the treatment strategy of multiple drugs against multiple targets.

4.1. A Model-Based Platform to Study the Influence from the Drug (Imidazo-Pyrrolopyridine) on Acute Phase Protein Secretion

In this work, Equation (12) was used to describe the interaction of drug and its corresponding receptor. The competitive inhibition kinetics instead of more complicated inhibition kinetics (e.g., non-competitive binding) was preferred here to elucidate the developed approach, which can be revised and extended for other inhibition kinetics.

$$\text{Target} + \text{Drug} \underset{k_f/K_i}{\overset{k_f}{\longleftrightarrow}} \text{Complex} \tag{12}$$

In Equation (12) K_i was the equilibrium constant, k_f was the forward rate constant, and k_f/K_i was the backward rate constant. The value of K_i for a drug-target pair can be obtained from experiment or computational interpretation [24]. In order to quantify the influence from the drug on the signaling pathway and thus the system output, differential equations for the drug and the drug-target complex were added into the ODE model. The differential equation for the receptor was modified accordingly. Among the seven drug targets identified from Section 3, JAK had been extensively studied for its interaction with existing drugs. Therefore, JAK and its drug counterpart imidazo-pyrrolopyridine were used as the example to illustrate our approach in this section. The corresponding value of K_i was determined to be 2.5 nM^{-1} from experiment [18]. Since no information was found for k_f in the literature, a value of 0.01 $\text{nM}^{-1} \cdot \text{s}^{-1}$ was assigned to k_f to study the dynamics of the three acute phase proteins upon the treatment with various doses of imidazo-pyrrolopyridine in Figure 7. The value of 0.01 $\text{nM}^{-1} \cdot \text{s}^{-M}$ was selected for k_f here because a larger value didn't further change the expression dynamics of the three acute phase proteins in the simulation.

Since the EC50 of imidazo-pyrrolopyridine was 180 nM [18], the concentration of imidazo-pyrrolopyridine was increased from 0 to 240 nM in intervals of 60 nM, as shown in Figure 7. It can be seen that imidazo-pyrrolopyridine was predicted to greatly inhibit the secretion of fibrinogen, slightly inhibited the production of haptoglobin, and slightly promoted the secretion of albumin. The overall effect of this drug was to attenuate acute phase response, which was characterized by the decrease in the concentration of positive acute phase proteins (*i.e.*, fibrinogen and haptoglobin) but increase in the level of negative acute phase proteins (*i.e.*, albumin). In addition, the dose of imidazo-pyrrolopyridine had a significant influence on the production of fibrinogen especially, as a low dose (even lower than 60 nM) was able to repress most of the fibrinogen secretion. On the other hand, the production of hatoglobin and albumin did not change as much upon increasing the dose of imidazo-pyrrolopyridine, until a relatively high drug dose was applied.

The parameter k_f reflected the speed of the drug binding reaction. Figure 8 showed the kinetics of fibrinogen expression for four values of k_f, as no experimental data were found for the k_f value. Since imidazo-pyrrolopyridine had a large influence on the secretion of fibrinogen, only the result for fibrinogen was shown here to save space. It seems that a small increase in k_f value from zero was able to suppress the secretion of fibrinogen significantly. When k_f increased to 0.01 $\text{nM}^{-1} \cdot \text{s}^{-1}$, the binding of imidazo-pyrrolopyridine to JAK reached its saturated speed.

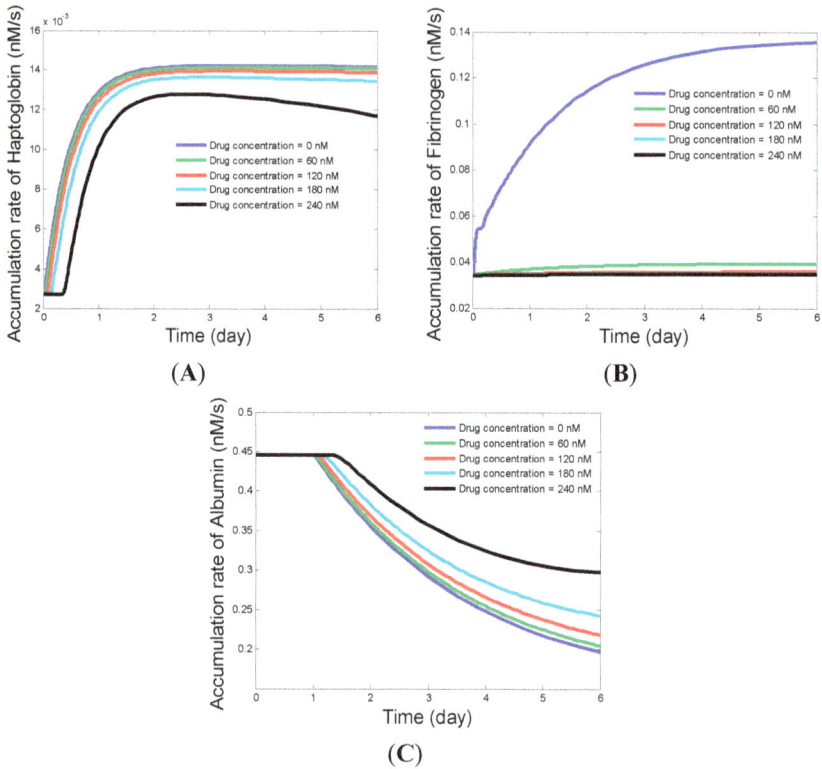

Figure 7. Dose effect of imidazo-pyrrolopyridine targeting at JAK on the production of three acute phase proteins: (A) haptoglobin; (B) fibrinogen; (C) albumin.

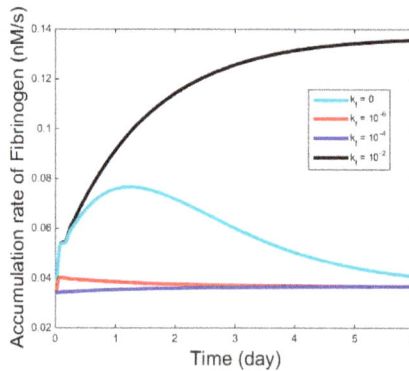

Figure 8. Fibrinogen expression kinetics predicted from the model with different k_f values.

4.2. Ranking Drug Targets Based upon the Influence from Their Interaction with the Drug on the Dynamics of Acute Phase Proteins

In this section, we further ranked the seven potential drug targets identified from Section 3 on the basis of the interaction between each drug and its target. For the same drug dose, the best drug

& target pair should return the highest effectiveness in regulating the dynamics of acute phase proteins. It was fortunate that the binding kinetics for the pair of imidazo-pyrrolopyridine and JAK was found in the literature (*i.e.*, the K_i value shown in Section 4.1). However, no binding kinetic data were found for the other six drug targets identified in Section 3, although data might exist in commercial database from pharmaceutical companies (which was not accessible by public). Since the value of K_i for the other six drug targets was not available in literature, it was assumed in this section that these targets were bound by the drug with the same kinetics as the one for imidazo-pyrrolopyridine and JAK. Based upon this, simulations were performed to evaluate the effectiveness of each drug-target pair on regulating the secretion rates of the three acute phase proteins. The effectiveness was quantified by the maximum percentage change in the secretion rate of each acute phase protein upon the binding of the drug to each target (Figure 9). The values of K_i and k_f were kept the same as those used in Section 4.1. The drug dose was set to 60 nM because the simulation result in Figure 7 implies that was a high enough concentration to suppress the fibrinogen secretion. The binding of the drug to each of gp80, JAK, and gp130 reduced the secretion rates of fibrinogen (by 74.4%, 71.0%, and 71.8%) and haptoglobin (by 44.5%, 4.2%, and 5.3%) but enhanced the production rate of albumin (by 22.9%, 1.7%, and 2.2%). The interaction from these drug-target pairs generally inhibited the acute phase response, especially in suppressing the secretion of fibrinogen. This can be explained by the fact that these three receptors played an important role in initiating both JAK-STAT and MAPK-C/EBPβ pathways. The binding of a drug to STATC inhibited the expression of fibrinogen (by 2.6%), slightly enhanced the section of haptoglobin (by 0.01%), and barely reduced the expression of albumin (by 0.006%). This made sense, as inhibition of STAT3C prevented the activation of nuclear STAT3 dimer and thus down-regulated the expression of fibrinogen. The deactivation of JAK-STAT pathway released some $(IL6 - gp80 - gp130 - JAK)_2^*$ complex to MAPK-C/EBPβ pathway for enhancing the production of haptoglobin. Therefore, the drug-STAT3C interaction only partially suppressed the acute phase response. The binding of the drug to Raf and C/EBPβi enhanced the secretion rate of albumin (by 0.08% and 54.2%), but reduced the secretion rate of haptoglobin (by 0.5% and 72.6%). Since Raf and C/EBPβi were the upstream components for the activation of nuclear C/EBPβ, inhibition of these two components by drugs down-regulated haptoglobin expression and restored albumin activation. The drugs binding to either Raf or C/EBPβi only partially inhibited the acute phase response, as the secretion of fibrinogen was enhanced by 0.2% and 0.001% upon the drug binding. The drug-MEK pair showed similar effect on acute phase response as the drug-Raf or drug-C/EBPβi pair, however, the effect from the drug-MEK pair was very limited (less than 0.001%). One potential reason for this was that the initial concentration of MEK (*i.e.*, 41,667 nM) overwhelmed the drug dose (*i.e.*, 60 nM) in this simulation.

Based upon the simulation result shown in Figure 9, components gp80, gp130, and JAK were ranked as the top three drug targets because: (1) they had a noticeable effect on the secretion rates of all three acute phase proteins; and (2) they counteracted the acute phase response, while the others only partially do so. While it was assumed that the same K_i was used for all the drugs in this study for screening drug targets, this assumption can be relaxed if kinetic data are available in the future.

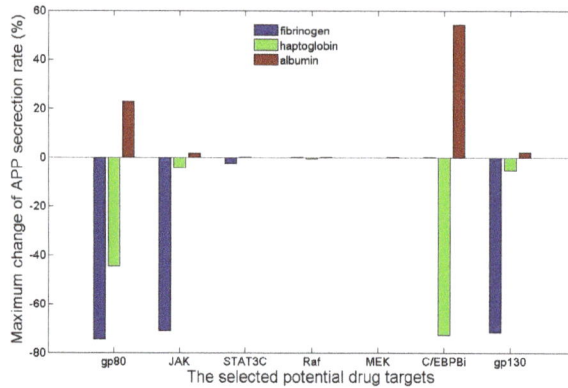

Figure 9. Maximum change of the secretion rates of the three acute phase proteins upon the binding of a drug with the concentration of 60 nM to each of the selected seven drug targets.

4.3. Influence of Multiple Drug Treatment on Acute Phase Protein Secretion

Only single drug-target pair was studied in Sections 4.1 and 4.2. We further applied the developed platform to quantify the effectiveness of a cocktail of drugs in regulating the dynamics of acute phase proteins. Since gp80, JAK, and gp130 were identified as good drug targets for regulating acute phase proteins, they constructed the drug target pool in this study. Similar to Section 4.2, a drug with similar binding kinetics was assumed for each of these three targets, and the corresponding drug and target binding reactions (e.g., Equation (12)) were added into the extended IL-6 ODE model. The production rate of fibrinogen for the cell treated with single drug that binds to gp80, JAK, and gp130 respectively was compared to the fibrinogen production rate in the cell treated with the three drugs that aim at gp80, JAK, and gp130 respectively in Figure 10. A low dose (*i.e.*, 10 nM), was used here for each drug in all these scenarios to prove that the same effectiveness could be obtained using low doses of drugs if multiple drugs were applied to multiple drug targets. As shown in Figure 10, treating the cell with multiple drugs against multiple targets was able to enhance the inhibition of fibrinogen accumulation (see the black curve in Figure 10). Compared to the effectiveness shown in Figure 7B where 60 nM imidazo-pyrrolopyridine was applied to its target (*i.e.*, JAK), the cocktail with three drugs returned a better performance in suppressing the secretion of fibrinogen even with a lower dose (*i.e.*, 10 nM). In case that 60 nM imidazo-pyrrolopyridine might cause side effect, it was possible to get the same inhibition effectiveness by reducing its dose by adding the drugs aimed at gp80 and gp130. The developed platform can thus be used as a tool to optimize the drug cocktail if the kinetic data and side effect information are available for drug candidates, and to provide a new strategy in future drug design for acute phase response.

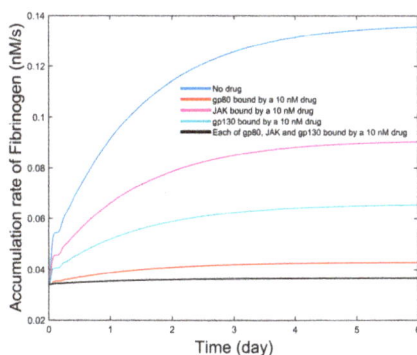

Figure 10. The accumulation rates of fibrinogen in the cell with only one of gp80, JAK, and gp130 bound by a single drug (10 nM) and in the cell with all gp80, JAK, gp130 bound by their drugs (each drug is of a 10 nM dose).

5. Discussion

This work presents the first comprehensive mathematical model for quantifying the kinetics of acute phase protein expression in the acute phase response mediated by IL-6. In addition to IL-6, it is possible for acute phase protein expression to be regulated by other cytokines, such as IL-1, TNF-α, IL-11, and OSM (oncostatin M) [25]. Although adding the signaling pathways associated with these cytokines can provide a more comprehensive model of the acute phase response, it would be a challenging task currently due to the following two reasons: (1) not all the reactions and pathways that connect these cytokines to acute phase proteins are known; (2) limited quantitative data are available for acute phase protein expression in cells stimulated by these cytokines. Therefore, we have focused on the IL-6 signaling pathway in this work. The extended IL-6 model will serve as a good starting point for incorporating other signaling pathways in the future, if quantitative data become available for acute phase protein expression following stimulation by other cytokines.

The presented IL-6 model was validated by two independent datasets that included measurements of three representative acute phase proteins for various stimulation patterns of IL-6. In addition, the dynamics of these three acute phase proteins were predicted for hepatic cells treated by single or multiple drugs. Although extensive literature review has been conducted, limited quantitative data were found to validate the predicted effects of drug treatment. Experimental research is therefore needed to further quantify the dynamics of acute phase proteins in hepatocyte cells treated with drugs targeting gp80, gp130, and JAK. Despite of this, the model is able to provide a general direction for drug target selection. For example, the model can rank potential drug targets based upon the predicted effect of the binding of a drug to each of these targets (as shown in Figure 9). With the same binding kinetics assumed for all target-drug pairs, the simulation result could reveal which drug targets, upon the competitive inhibition, have a relatively large influence on the kinetics of acute phase proteins. Since no binding affinity information is available and thus incorporated in the model, the ranking of drug targets may not completely accurate. However, the model dose provides a platform to qualitatively compare those potential drug targets. In addition, the model can be used to predict the dose response for each drug target (as shown in Figure 7 for JAK), which provides information on the sensitivity of a drug target to the drug dose. This provides additional information for drug target screening and drug dosage selection. Furthermore, the model is able to predict the effect of the drug on the kinetics of the molecules other than the three acute phase proteins. This may be helpful to quantify the side effect of the drug.

JAK, gp80, gp130 were predicted to be the three optimal drug targets for regulating the dynamics of the acute phase proteins (*i.e.*, fibrinogen, haptoglobin, and albumin) investigated in this work. Since these three macromolecules have distinct binding sites, they can be considered as separate drug targets. Seven potential drug targets were ranked according to the influence of their interaction with drugs on the dynamics of all three acute phase proteins. In the extended IL-6 model, fibrinogen is mainly regulated by the JAK-STAT pathway while albumin and haptoglobin are associated with the MAPK pathway. Because JAK, gp80, and gp130 are involved in reactions for activating both the JAK-STAT and MAPK pathways, they were found to be better intervention points than the other four drug targets. If we aim to regulate only one or two acute phase proteins, other drug target may be better than JAK, gp80, and gp130. For example, C/EBPβi is a better drug target than JAK, gp80, and gp130 for regulating the dynamics of albumin (see Figure 9).

6. Conclusions

This work developed the first comprehensive IL-6 model that can predict the expression dynamics of haptoglobin, fibrinogen, and albumin in HepG2 cultures stimulated by IL-6. The developed model was validated by two different sets of experimental data, and the relative errors of the model predictions for most cases were at, or below, 15%. Based on the developed model, sensitivity analysis was conducted to identify potential drug targets for regulating acute phase protein dynamics, which included gp80, JAK, gp130, STAT3C, Raf, MEK, and C/EBPβi. Imidazo-pyrrolopyridine targeted at JAK was used as an example drug to illustrate an approach in which the drug-target interaction is integrated with kinetic models to study the drug dose response. The simulation result showed that imidazo-pyrrolopyridine inhibited the acute phase response, especially the secretion of fibrinogen. The developed approach was used to further rank seven drug targets, with the assumption that each of them was targeted by a drug with similar binding kinetics. This assumption can be removed in the future when drug binding kinetic data are available for all drug targets. Upon binding to the drug, the targets gp80, JAK, and gp130 were found to have the largest effect on regulating the secretion of fibrinogen and on attenuating acute phase response. The developed platform was then applied to investigate the effectiveness of the drugs that bind to these three most effective targets on the regulation of fibrinogen. The simulation results show that the multiple-drug treatment approach can reduce the drug dosage to obtain the same treatment effectiveness when compared to single drug treatment approaches.

Acknowledgments

Z.H. was supported by Villanova University (grant SRF-RSG 217030-7252; VCASE seed grants 420492 and 420265). We appreciate Noelle Comolli's help in polishing the manuscript.

Author Contributions

Z.X. and Z.H. developed the ODE model, performed parameter estimation and sensitivity analysis. Z.X. proposed the virtual drug-screening framework. J.O.M.K. contributed to model development and sensitivity analysis, as well as interpretation of experimental data. All authors participated in the analysis of results and the preparation of the manuscript.

Appendix

Equations added to IL-6 signal transduction for predicting dynamics of extracellular haptoglobin, fibrinogen, and albumin are shown in Equations (A1)–(A7).

$$\frac{dC_{\text{mRNA–haptoglobin}}}{dt} = \frac{V_{m\_h}C_{\text{C/EBP}\beta}}{K_{m\_h}+C_{\text{C/EBP}\beta}} - k_{t\_h}C_{\text{mRNA–haptoglobin}} \tag{A1}$$

$$\frac{dC_{\text{ex-haptoglobin}}}{dt} = r_{h0} + k_{t\_h}C_{\text{mRNA–haptoglobin}} \tag{A2}$$

$$\frac{dC_{\text{mRNA–fibrinogen}}}{dt} = \frac{V_{m\_f}C_{\text{STAT3N}^*-\text{STAT3N}^*}}{K_{m\_f}+C_{\text{STAT3N}^*-\text{STAT3N}^*}} - k_{t\_f}C_{\text{mRNA–fibrinogen}} \tag{A3}$$

$$\frac{dC_{\text{ex-fibrinogen}}}{dt} = r_{f0} + k_{t\_f}C_{\text{mRNA–fibrinogen}} \tag{A4}$$

$$\frac{dC_{\text{C/EBP}\alpha}}{dt} = -k_{i\_a}C_{\text{C/EBP}\beta} - k_{d\_a}C_{\text{C/EBP}\alpha} \tag{A5}$$

$$\frac{dC_{\text{mRNA–albumin}}}{dt} = \frac{V_{m\_a}C_{\text{C/EBP}\alpha}(t-3600\times24)}{K_{m\_a}+C_{\text{C/EBP}\alpha}(t-3600\times24)} - k_{t\_a}C_{\text{mRNA–albumin}} \tag{A6}$$

$$\frac{dC_{\text{ex-albumin}}}{dt} = r_{a0} + k_{t\_a}C_{\text{mRNA–albumin}} \tag{A7}$$

where all variables are deviation variables that represent the concentration deviations from their nominal values once HepG2 cells are stimulated by IL-6. The initial values of all these variables are thus 0 nM. Constants r_{h0}, r_{f0}, and r_{a0} are the initial secretion rates of haptoglobin, fibrinogen, and albumin, respectively. Their values are determined to be 0.0027, 0.0341, and 0.4463 nM/s, from the experimental data for non-stimulated HepG2 cells. These reactions are integrated into the IL-6 signaling model presented in Moya *et al.* [10], to predict the expression dynamics of haptoglobin, fibrinogen, and albumin. The values of parameters from Equations (A1) to (A7) are listed in the following table.

Table A1. Values of the parameters from Equations (A1) to (A7).

Name	Value	Unit
V_{m_h}	0.06457	nM/s
K_{m_h}	99.7421	nM
k_{t_h}	2.5389×10^{-5}	1/s
V_{m_f}	1.1841	nM/s
K_{m_f}	58.1310	nM
k_{t_f}	7.8158×10^{-6}	1/s
k_{i_a}	1.0861×10^{-3}	1/s
k_{d_a}	0.06866	1/s
V_{m_a}	0.1470	nM/s
K_{m_a}	0.5118	nM
k_{t_a}	4.2195×10^{-6}	1/s

Conflicts of Interest

The authors declare no conflict of interest.

References

1. Heinrich, P.C.; Castell, J.V.; Andus, T. Interleukin-6 and the acute phase response. *Biochem. J.* **1990**, *2653*, 621–636.
2. Heinrich, P.C.; Behrmann, I.; Haan, S.; Hermanns, H.M.; Müller-Newen, G.; Schaper, F. Principles of interleukin (IL)-6-type cytokine signalling and its regulation. *Biochem. J.* **2003**, *374*, 1–20.
3. Akira, S. IL-6-regulated transcription factors. *Int. J. Biochem. Cell Biol.* **1997**, *29*, 1401–1418.
4. Schindler, C.; Darnell, J.E., Jr. Transcriptional responses to polypeptide ligands: The JAK-STAT pathway. *Annu. Rev. Biochem.* **1995**, *64*, 621–651.
5. Yamada, S.; Shiono, S.; Joo, A.; Yoshimura, A. Control mechanism of JAK/STAT signal transduction pathway. *FEBS Lett.* **2003**, *534*, 190–196.
6. Kholodenko, B.N.; Demin, O.V.; Moehren, G.; Hoek, J.B. Quantification of short term signaling by the epidermal growth factor receptor. *J. Biol. Chem.* **1999**, *274*, 30169–30181.
7. Brightman, F.A.; Fell, D.A. Differential feedback regulation of the MAPK cascade underlies the quantitative differences in EGF and NGF signalling in PC12 cells. *FEBS Lett.* **2000**, *482*, 169–174.
8. Schoeberl, B.; Eichler-Jonsson, C.; Gilles, E.D.; Müller, G. Computational modeling of the dynamics of the MAP kinase cascade activated by surface and internalized EGF receptors. *Nat. Biotechnol.* **2002**, *20*, 370–375.
9. Singh, A.; Jayaraman, A.; Hahn, J. Modeling regulatory mechanisms in IL-6 signal transduction in hepatocytes. *Biotechnol. Bioeng.* **2006**, *95*, 850–862.

10. Moya, C.; Huang, Z.; Cheng, P.; Jayaraman, A.; Hahn, J. Investigation of IL-6 and IL-10 signalling via mathematical modelling. *IET Syst. Biol.* **2011**, *5*, 15–26.

11. Ryll, A.; Samaga, R.; Schaper, F.; Alexopoulos, L.G.; Klamt, S. Large-scale network models of IL-1 and IL-6 signalling and their hepatocellular specification. *Mol. Biosyst.* **2011**, *7*, 3253–3270.

12. Karlsson, J.O.M.; Yarmush, M.L.; Toner, M. Interaction between heat shock and interleukin 6 stimulation in the acute-phase response of human hepatoma (HepG2) cells. *Hepatology* **1998**, *28*, 994–1004.

13. Araujo, R.P.; Petricoin, E.F.; Liotta, L.A. A mathematical model of combination therapy using the EGFR signaling network. *Biosystems* **2005**, *80*, 57–69.

14. Yang, K.; Bai, H.; Ouyang, Q.; Lai, L.; Tang, C. Finding multiple target optimal intervention in disease-related molecular network. *Mol. Syst. Biol.* **2008**, *4*, 228.

15. Stump, K.L.; Lu, L.D.; Dobrzanski, P.; Serdikoff, C.; Gingrich, D.E.; Dugan, B.J.; Angeles, T.S.; Albom, M.S.; Ator, M.A.; Dorsey, B.D.; *et al.* A highly selective, orally active inhibitor of Janus kinase 2, CEP-33779, ablates disease in two mouse models of rheumatoid arthritis. *Arthritis Res. Ther.* **2011**, *13*, R68.

16. Quintás-Cardama, A.; Kantarjian, H.; Cortes, J.; Verstovsek, S. Janus kinase inhibitors for the treatment of myeloproliferative neoplasias and beyond. *Nat. Rev. Drug Discov.* **2011**, *10*, 318.

17. Kremer, J.M.; Bloom, B.J.; Breedveld, F.C.; Coombs, J.H.; Fletcher, M.P.; Gruben, D.; Krishnaswami, S.; Burgos-Vargas, R.; Wilkinson, B.; Zerbini, C.A.; *et al.* The Safety and Efficacy of a JAK Inhibitor in Patients With Active Rheumatoid Arthritis Results of a Double-Blind, Placebo-Controlled Phase IIa Trial of Three Dosage Levels of CP-690,550 Versus Placebo. *Arthritis Rheum.* **2009**, *60*, 1895–1905.

18. Hurley, C.A.; Blair, W.S.; Bull, R.J.; Chang, C.; Crackett, P.H.; Deshmukh, G.; Dyke, H.J.; Fong, R.; Ghilardi, N.; Gibbons, P.; *et al.* Novel triazolo-pyrrolopyridines as inhibitors of Janus kinase 1. *Bioorg. Med. Chem. Lett.* **2013**, *23*, 3592–3598.

19. Alam, T.; An, M.R.; Papaconstantinou, J. Differential expression of three C/EBP isoforms in multiple tissues during the acute phase response. *J. Biol. Chem.* **1992**, *267*, 5021–5024.

20. Zhang, Z.; Fuller, G.M. Interleukin 1beta inhibits interleukin 6-mediated rat gamma fibrinogen gene expression. *Blood* **2000**, *96*, 3466–3472.

21. Ruminy, P.; Gangneux, C.; Claeyssens, S.; Scotte, M.; Daveau, M.; Salier, J.P. Gene transcription in hepatocytes during the acute phase of a systemic inflammation: From transcription factors to target genes. *Inflamm. Res.* **2001**, *50*, 383–390.

22. Huang, Z.Y.; Chu, Y.F.; Hahn, J. Model simplification procedure for signal transduction pathway models: An application to IL-6 signaling. *Chem. Eng. Sci.* **2010**, *65*, 1964–1975.

23. Marianayagam, N.J.; Sunde, M.; Matthews, J.M. The power of two: Protein dimerization in biology. *Trends Biochem. Sci.* **2004**, *29*, 618–625.

24. Swanson, J.M.; Henchman, R.H.; McCammon, J.A. Revisiting free energy calculations: A theoretical connection to MM/PBSA and direct calculation of the association free energy. *Biophys. J.* **2004**, *86*, 67–74.

25. Moshage, H. Cytokines and the hepatic acute phase response. *J. Pathol.* **1997**, *181*, 257–266.

A Computational Study of the Effects of Syk Activity on B Cell Receptor Signaling Dynamics

Reginald L. McGee, Mariya O. Krisenko, Robert L. Geahlen, Ann E. Rundell and Gregery T. Buzzard

Abstract: The kinase Syk is intricately involved in early signaling events in B cells and is required for proper response when antigens bind to B cell receptors (BCRs). Experiments using an analog-sensitive version of Syk (Syk-AQL) have better elucidated its role, but have not completely characterized its behavior. We present a computational model for BCR signaling, using dynamical systems, which incorporates both wild-type Syk and Syk-AQL. Following the use of sensitivity analysis to identify significant reaction parameters, we screen for parameter vectors that produced graded responses to BCR stimulation as is observed experimentally. We demonstrate qualitative agreement between the model and dose response data for both mutant and wild-type kinases. Analysis of our model suggests that the level of NF-κB activation, which is reduced in Syk-AQL cells relative to wild-type, is more sensitive to small reductions in kinase activity than Erkp activation, which is essentially unchanged. Since this profile of high Erkp and reduced NF-κB is consistent with anergy, this implies that anergy is particularly sensitive to small changes in catalytic activity. Also, under a range of forward and reverse ligand binding rates, our model of Erkp and NF-κB activation displays a dependence on a power law affinity: the ratio of the forward rate to a non-unit power of the reverse rate. This dependence implies that B cells may respond to certain details of binding and unbinding rates for ligands rather than simple affinity alone.

Reprinted from *Processes*. Cite as: Reginald L. McGee, Mariya O. Krisenko, Robert L. Geahlen, Ann E. Rundell and Gregery T. Buzzard A Computational Study of the Effects of Syk Activity on B Cell Receptor Signaling Dynamics. *Processes* **2015**, *3*, 75–97.

1. Introduction

Signaling through the B cell receptor (BCR) involves an intricate network of molecular reactions necessary for B cells to generate an immune response. The signaling network involves a variety of proteins including kinases and phosphatases and is particularly dependent on the protein-tyrosine kinase (PTK) Syk. To better understand the network, it is imperative to examine the roles of key signaling components like Syk and their most influential interactions. We will employ a computational approach to quantify the impact of Syk and other key enzymes and factors such as the effect of the amount of antigen on the B cell response.

Catalytically active Syk has been shown experimentally to play a central role in BCR signaling, but questions regarding its behavior still exist and the time frames in which critical interactions must occur have yet to be completely characterized. Experimentally, a mutated version of Syk called analog-sensitive Syk or Syk-AQL has been engineered to accept orthogonal inhibitors, *i.e.*, inhibitors that have been synthesized to render the mutant kinase inactive almost immediately [1]. Furthermore, by replacing wild-type Syk with Syk-AQL creates B cells whose signaling capacity can be reduced or interrupted completely by the addition of the orthogonal inhibitor, experimentalists could then control the time that Syk remains active following receptor engagement, which helped to confirm how BCR signaling is modulated by the actions of Syk. Recently, a Syk-deficient B cell line was generated in which Syk-AQL expression can be induced in response to the drug tetracycline. Thus, in addition to being able to turn Syk off when desired, its expression level before activation can be adjusted if needed.

Computational modeling allows us to gain insight into Syk's impact that was not previously possible with experimentation alone. We have developed a model built on a T cell receptor (TCR) signaling model originally created by Zheng *et al.* [2] and later expanded and used by Perley *et al.* [3] for cellular level control. Perley's success in using Zheng's model for prediction and open-loop control made it an ideal candidate to adapt for our B cell study. The signaling of B cell and of T cell can be divided into early interactions, which occur proximal to the membrane, and downstream interactions, which occur in the cytoplasm and ultimately lead to the nucleus. The dynamics of the downstream signaling are nearly identical between the cells, and thus this part of the Zheng model remained largely unchanged. The signaling dynamics of T cell and of B cell differ the most in their early signaling, which is where most model revisions were required.

In the past decade there have been a number of computational models, both stochastic and deterministic varieties, focusing on various aspects of B cell signaling, but none have considered impairment to Syk and the resulting effect on cell response. Stochastic simulations have been used by Tsourkas *et al.* [4] and Mukherjee *et al.* [5] while considering spatial dynamics of BCR signaling. The impact of affinity discrimination was considered by Tsourkas in their study, while Mukherjee investigated the roles of Syk and Lyn in immunoreceptor tyrosine-based activation motif (ITAM) phosphorylation. A deterministic model by Chaudhri *et al.* [6] considers a scope similar to Zheng's T cell model, with the model covering both membrane proximal, early signaling events and downstream signaling events. This model pays particular interest to the role of phosphatases in the signal transduction. In 2012, Barua *et al.* [7] developed a deterministic model of B cell early signaling in order to study the feedback loops involving Lyn and how varying stimulation to the BCR leads to a range of dynamics in Syk. Impressively, the model incorporates every phosphorylation event for all six signaling components considered.

Our model is novel in its incorporation of Syk-AQL dynamics and given its scope, the inclusion of both early and downstream signaling, this allows us to investigate the impact of Syk modulation on a large number of signaling components. Instead of considering all possible phosphorylations for our 32 signaling components, our model considers only the most critical events in order to represent relevant physiological behavior and minimize model complexity. Understanding the means by which cell responses are determined is also of particular interest and the model will allow us to investigate the impact of both the amount of antigen and the level of Syk activity on the response. In this initial study we are particularly interested in the regulation of Erk and NF-κB activity since both contribute to determining cell fate.

We will study how Erk and NF-κB phosphorylation change with modulation of receptor affinity. Using parameter values derived largely by fitting the Perley model [3] to T cell data as nominal points, we use B cell data from Healy *et al.* [8] to then determine points in parameter space that allow us to reproduce data from cellular assays. Then, using the difference of Erkp, the sum of singly and doubly phosphorylated Erk, and NF-κB as our metrics, we consider how cell response changes with receptor characteristics in both wild-type and mutant cells.

One interesting prediction of our model is that activation of Erkp and NF-κB depend on ligand binding rates in a way that is nearly independent of the reverse rate for low values of the reverse rate and in a way that depends on a fixed ratio of powers of forward and reverse binding rates for higher values of the reverse binding rate. This is illustrated and discussed in Section 4.3 and is one indication that affinity (the ratio of forward and reverse binding rates) alone is not sufficient to characterize the response of a B cell to a given antibody.

The cell line used and experimental procedures are described in the experimental section. In the model section, we describe model construction and explicitly show equations and a diagram for the signaling dynamics. In the methods section, we discuss the sensitivity analysis used and criteria used to screen the parameter space. The discussion section includes biological background for the model and findings of the sensitivity analysis, parameter screening, and contour analysis. We also present a comparison of our model output with a dataset from Chaudhri *et al.* [6], and finish with some discussion of future direction and limitations.

2. Model Development

2.1. Biological Background

Since our B cell model is derived from an existing T cell model, we note here some of the primary components of B cell signaling, with a focus on aspects that are unique to B cells. Conventional T cells bind peptide antigens presented by major histocompatibility molecules whereas B cells can bind multiple molecular species through polymorphic cell surface immunoglobulins that serve as antigen receptors. The B cells work collaboratively with T cells to respond to monomeric antigens or independently of T cells to respond to polymeric antigens that cluster the BCR.

Once an antigen is bound and the BCR is aggregated, the signaling mechanisms at the B cell membrane are activated and an intricate system of molecular interactions initiates [9]. There are many kinases involved in this process; two connected with events proximal to the receptor in B cells are the Src-family PTK Lyn and the PTK Syk [10]. Syk plays a central role in the overall response of a B cell [11]. Unlike signaling in T cells, which depends on the Src-family PTK Lck to phosphorylate the first tyrosine of the ITAM of the TCR, the first ITAM tyrosine in B cells can be phosphorylated by Syk when Lyn, which is homologous to Lck in T cells, is not present [12]. Furthermore, if Syk is not expressed, BCR signaling cannot proceed. Following the initiation of BCR signaling, the regulation of BCR, Syk, and Lyn activity is orchestrated by feedback loops involving the aforementioned PTKs and a collection of regulatory enzymes.

The regulatory enzymes considered are the tyrosine-phosphatase SHP1, the C-terminal Src kinase Csk and its binding protein Cbp, and the phosphatase CD45 [13,14]. In addition to regulating fully activated Lyn along with CD45, SHP1 also dephosphorylates the ITAMs of the BCR and tyrosines Y342 and Y346 of Syk, thus reducing their activity. Note that SHP1 does not complete these actions until it has been activated itself by a Lyn, which has been dephosphorylated at its inhibitory site by CD45. After binding with a phosphorylated Cbp, activated Csk counteracts the dephosphorylation of Lyn promoted by CD45. The phosphorylation of Cbp is also promoted by CD45 activity. Gaining a better grasp of the timing of the interplay in these feedback loops is an important task as it provides insight into exactly how the BCR and primary PTKs are regulated and thus illuminates the overall sequence of events for signal transduction.

Once active, Syk phosphorylates several substrates and the resulting signals propagate into several downstream pathways that lead to the activation of downstream targets such as Erk, NFAT, and NF-κB [9–11]. Following the translocation of these molecules into the nucleus, transcription begins and cell fate activation (proliferation), apoptosis (cell death), or anergy (chronic unresponsiveness) are determined. Interestingly, these cell responses have been found by Healy *et al.* [8] to correspond to characteristic combinations of the aforementioned targets. For example, anergic B cells exhibit signaling activity in the Erk and NFAT pathways, but not in the NF-κB pathway [8]. Again, Erk and NF-κB are of particular interest in this study due to their role in cell fate determination. The affinity of a receptor for an antibody $K_{\text{affinity}} = \frac{k_{\text{forward}}}{k_{\text{reverse}}}$ is a measure of how tightly a ligand binds to a receptor. However, for a given affinity, larger forward and reverse rates can allow the ligand to bind and unbind repeatedly and rapidly. Allowing this association and disassociation with the receptor to occur over prolonged periods of times and in the proper concentrations could be sufficient to simulate the low, chronic exposure of the BCR to antigen, which has been seen to induce anergy [15]. The causes for anergy have not been completely characterized, and we hope to use the model to gain a better understanding of the molecular triggers leading to anergy and the associated nonresponsiveness of B cells.

2.2. Model

The model we present was developed based on the deterministic model for the TCR MAPK pathway created by Zheng [2] and extended by Perley [3]. Due to similarities in the signaling network, much of the model structure for the medial and downstream pathways required little modification. In particular, the model structure and equations for the MAPK pathway, which contains Erk, and the NF-κB pathway, are analogous to those found in [16]. A diagram focusing on the signaling dynamics for our system is shown in Figure 1.

Figure 1. A depiction of the early, medial, and downstream signaling events induced by binding between B cell receptor and ligand, as described in the biological background section. Jagged arrows denote stimulations, curved arrows denote binding, straight arrows denote conversions, and color denotes species to appear repeatedly in the diagram. The plus and minus marks near the IκB-NF-κB disassociation reaction indicate which are positive feedbacks and which are negative feedbacks.

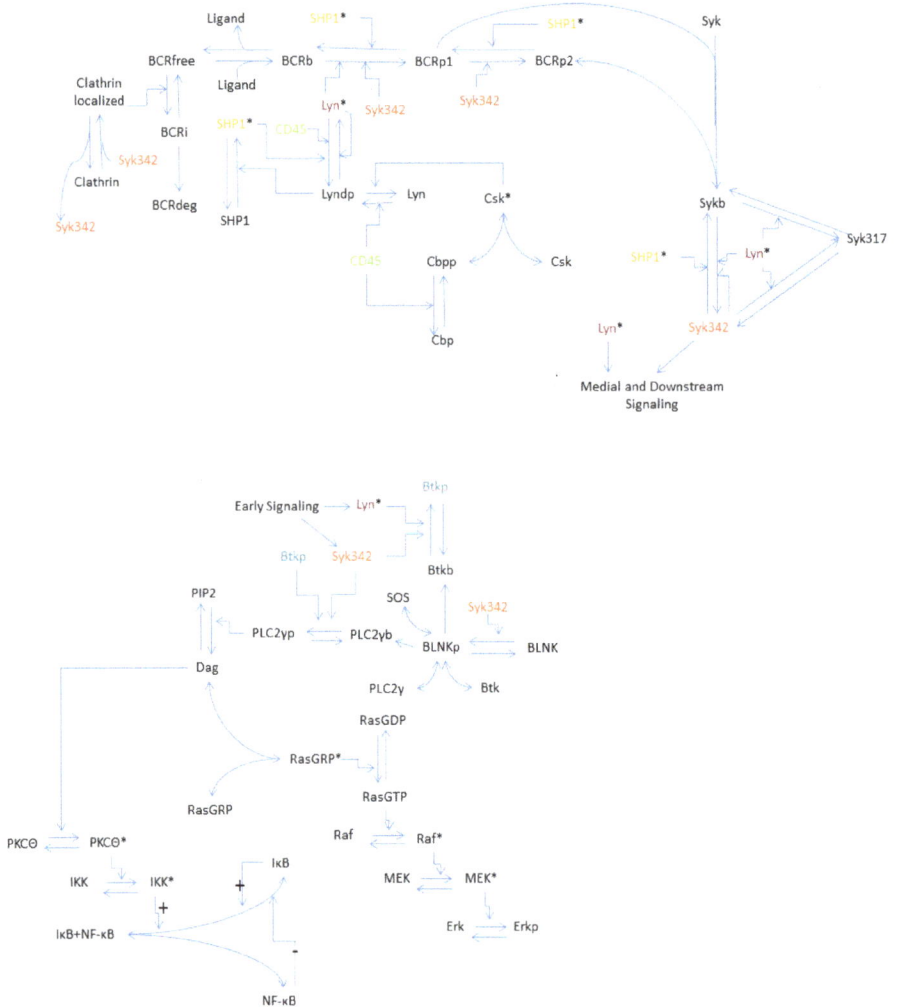

The model tracks the concentrations of 22 distinct species with the different forms of these species represented by individual variables. The model equations were formulated with mass action kinetics; conservation laws were used to reduce the number of variables in the system. The resulting model consists of 32 ordinary differential equations and has 114 parameters. Based on a sensitivity analysis described in the methods section, there are 12 kinetic parameters whose impact we will investigate in this study. Another important parameter is the ligand concentration, which is an external input to the model. We present equations for the revised early signaling dynamics only and refer to [3] for the remaining equations. We assume that there is no downstream activity prior to receptor stimulation; hence, our model is structured to be at zero at steady state for downstream variables like Erkp and NF-κB. Additionally, we do not have any feedback between our downstream and upstream components, so setting these variables to zero is more of a baseline value than an absolute value.

2.3. Model Equations

BCR Activation: The receptor dynamics considered here include engagement of the BCR, ITAM tyrosine phosphorylation, Syk binding to the BCR, and BCR internalization, recycling and degradation. The key variables include free BCR $x_{BCRfree}$, BCR bound by ligand x_{BCRb}, singly-phosphorylated BCR x_{BCRp1}, and doubly-phosphorylated BCR x_{BCRp2}. The model formulation reflects how the kinase Syk can bind to either form of the phosphorylated BCR. Due to the positive promotion of ITAM tyrosine phosphorylation by membrane proximal PTKs [11], we assume that if Syk binds to a singly-phosphorylated BCR, that receptor will become doubly-phosphorylated before the kinase can unbind. Thus there is no term for unbinding from Sykb to BCRp1 in any of the model equations. Receptor internalization x_{BCRi} is promoted by clathrin, which is localized $x_{Clathrin_{local}}$ to the membrane by Syk.

$$\frac{dx_{BCRfree}}{dt} = [\text{BCR recycling}] - [\text{BCR internalization}]...$$

$$+ [\text{Ligand disassociation}] - [\text{Ligand association}]$$

$$= r_{Recycling} \cdot x_{BCRi} - r_{Internalization} \cdot x_{Clathrin_{local}} x_{BCR_{Free}} \cdots$$

$$+ r_{disassociation} \cdot x_{BCRb} - r_{association} \cdot [\text{Ligand}] \cdot x_{BCR_{Free}}$$

$$\frac{dx_{BCRb}}{dt} = [\text{Ligand binding}] - [\text{Ligand disassociation}]...$$

$$+ [\text{ITAM1 dephosphorylation}] - [\text{ITAM1 phosphorylation}]$$

$$= r_{association} \cdot [\text{Ligand}] \cdot x_{BCR_{Free}} - r_{disassociation} \cdot x_{BCRb} \cdots$$

$$+ \left(r_{BCRp1dephosphorylation} \cdot x_{SHP1*} + r_{BCRp1dephosphorylation \, by \, phosphatases} \right) x_{BCRp1} \cdots$$

$$- \left(r_{BCRp1phosphorylation1} \cdot x_{Lyn*} + r_{BCRp1phosphorylation2} \cdot x_{Syk342} \right) x_{BCRb}$$

$$\frac{dx_{BCRp1}}{dt} = [\text{ITAM1 phosphorylation}] - [\text{ITAM1 dephosphorylation}]...$$

$$+ [\text{ITAM2 dephosphorylation}] - [\text{ITAM2 phosphorylation}] - [\text{Syk-BCR binding1}]$$

$$= (r_{BCRp1 phosphorylation1} \cdot x_{Lyn*} + r_{BCRp1 phosphorylation2} \cdot x_{Syk342}) x_{BCRb}...$$

$$- (r_{BCRp1 dephosphorylation} \cdot x_{SHP1*} + r_{BCRp1 dephosphorylation\ by\ phosphatases}) x_{BCRp1}...$$

$$+ (r_{BCRp2 dephosphorylation} \cdot x_{SHP1*} + r_{BCRp2 dephosphorylation\ by\ phosphatases}) \cdot x_{BCRp2}...$$

$$- r_{BCRp2\ phosphorylation} \cdot x_{Syk342} x_{BCRp1} - r_{Syk-BCR\ binding1} \cdot x_{Syk} x_{BCRp1}$$

$$\frac{dx_{BCRp2}}{dt} = [\text{ITAM2 phosphorylation}] - [\text{ITAM2 dephosphorylation}]$$

$$+ [\text{Syk-BCR unbinding}] - [\text{Syk-BCR binding2}]$$

$$= r_{BCRp2\ phosphorylation} \cdot x_{Syk342} x_{BCRp1}...$$

$$- (r_{BCRp2 dephosphorylation} \cdot x_{SHP1*} + r_{BCRp2 dephosphorylation\ by\ phosphatases}) \cdot x_{BCRp2}...$$

$$+ r_{Syk-BCR\ unbinding} \cdot x_{Sykb} - r_{Syk-BCR\ binding2} \cdot x_{Syk} x_{BCRp2}$$

$$\frac{dx_{BCRi}}{dt} = [\text{BCR internalization}] - [\text{BCR recycling}] - [\text{BCR degradation}]$$

$$= r_{Internalization} \cdot x_{Clathrin_{local}} x_{BCR_{Free}} - r_{Recycling} \cdot x_{BCRi} - r_{Degradation} \cdot x_{BCRi}$$

$$\frac{dx_{Clathrin_{local}}}{dt} = [\text{Clathrin localization via Syk342}] - [\text{Clathrin delocalization}]$$

$$= r_{Clathrin\ localization} \cdot x_{Syk342} x_{Clathrin} - r_{Clathrin\ delocalization} \cdot x_{Clathrin_{local}}$$

Syk Activation: We consider four forms of Syk, three of which have been modified through binding or phosphorylation. The variable x_{Syk} represents the amount of kinase that has not been activated and is unbound. The variable x_{Sykb} is the basally active form of the kinase that has been bound to the BCR. The catalytically active form of Syk that has been phosphorylated at tyrosine Y342 and Y346 is denoted by x_{Syk342} and is responsible for enhancing signaling propagation. If either active form of the kinase becomes phosphorylated at tyrosine Y317 it is rendered inactive. This inactive form is denoted by x_{Syk317}. The forms represented by x_{Syk342} and x_{Syk317} are still assumed to be bound to the BCR. As discussed below, each of these four forms of Syk can bind to an orthogonal inhibitor; this binding also renders Syk inactive.

$$\frac{dx_{Sykb}}{dt} = [\text{Syk-BCR binding}] - [\text{Syk-BCR unbinding}]...$$

$$+ [\text{Syk dephosphorylation at Y342}] - [\text{Syk phosphorylation at Y342}]...$$

$$+ [\text{Syk dephosphorylation at Y317}] - [\text{Syk phosphorylation at Y317}]$$

$$= (r_{Syk-BCR\ binding1} \cdot x_{BCRp1} + r_{Syk-BCR\ binding2} \cdot x_{BCRp2}) x_{Syk} - r_{Syk-BCR\ unbinding} \cdot x_{Sykb}...$$

$$+ (r_{Syk342\ dephosphorylation} \cdot x_{SHP1*} + r_{Syk342\ dephosphorylation\ by\ phosphatases}) x_{Syk342}...$$

$$- (r_{Syk342\ via\ Lyn*} \cdot x_{Lyn*} + r_{Syk342\ autophosphorylation} \cdot x_{Syk342}) x_{Sykb}...$$

$$+ r_{Syk317\ dephosphorylation} \cdot x_{Syk317} - r_{Syk317\ phosphorylation1} x_{Lyn*} x_{Sykb}$$

$$\frac{dx_{Syk342}}{dt} = [\text{Syk phosphorylation at Y342}] - [\text{Syk dephosphorylation at Y342}]...$$

$$+ [\text{Syk dephosphorylation at Y317}] - [\text{Syk phosphorylation at Y317}]$$

$$= (r_{Syk342 \ via \ Lyn*} \cdot x_{Lyn*} + r_{Syk342 \ autophosphorylation} \cdot x_{Syk342}) x_{Sykb}...$$

$$- (r_{Syk342 \ dephosphorylation} \cdot x_{SHP1*} + r_{Syk342 \ dephosphorylation \ by \ phosphatases}) x_{Syk342}...$$

$$+ r_{Syk317 \ dephosphorylation} \cdot x_{Syk317} - r_{Syk317 \ phosphorylation2} \cdot x_{Lyn*} x_{Syk342}$$

$$\frac{dx_{Syk317}}{dt} = [\text{Syk phosphorylation at Y317}] - [\text{Syk dephosphorylation at Y317}]$$

$$= (r_{Syk317 \ phosphorylation1} \cdot x_{Sykb} + r_{Syk317 \ phosphorylation2} \cdot x_{Syk342}) x_{Lyn*}...$$

$$- 2r_{Syk317 \ dephosphorylation} \cdot x_{Syk317}$$

Syk-AQL dynamics: As discussed in the Introduction, Syk-AQL allows for Syk activity to be modulated through the addition of the orthogonal inhibitor (OI). The binding of the OI to the mutant Syk-AQL is also modeled using mass action kinetics and can be seen in Figure 2.

Figure 2. Orthogonal Inhibitor (OI) binding kinetics. Sykj denotes any of the modeled forms of Syk.

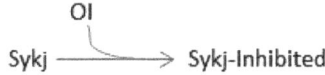

This binding results in the term $r_{Inhibitor \ association} \cdot [\text{OI}] \cdot x_{Syk_j}$ where j denotes any one of the forms of Syk modeled. These terms are subtracted from each of the equations for their respective forms of Syk. By conservation of mass, we have the following equation for orthogonally inhibited Syk:

$$\frac{dx_{Syk-inh}}{dt} = [\text{Inhibitor association - Syk bound}] + [\text{Inhibitor association - Syk Y342}]...$$

$$+ [\text{Inhibitor association - Syk Y317}] + [\text{Inhibitor association - free Syk}]$$

$$= r_{Inhibitor \ association} \cdot [\text{OI}] \cdot x_{Sykb} + r_{Inhibitor \ association} \cdot [\text{OI}] \cdot x_{Syk342}...$$

$$+ r_{Inhibitor \ association} \cdot [\text{OI}] \cdot x_{Syk317} + r_{Inhibitor \ association} \cdot [\text{OI}] \cdot x_{freeSyk}.$$

Lyn Activation: For the Src-family PTK Lyn (x_{Lyn}) to become fully activated (x_{Lyn*}), it must be dephosphorylated at Y508 (x_{Lyndp}) and then go through an autophosphorylation reaction. We consider both events with the following equations:

$$\frac{dx_{Lyndp}}{dt} = [\text{Lyn dephosphorylation}] - [\text{Lyn phosphorylation}]...$$

$$+ [\text{Lyn de-autophosphorylation}] - [\text{Lyn autophosphorylation}]$$

$$= r_{Lyn \ dephosphorylation} x_{CD45} x_{Lyn} - r_{Lyn \ phosphorylation} x_{Csk*} x_{Lyndp}...$$

$$+ r_{Lyn \ de-autophosphorylation} x_{Lyn*} - (r_{Lyn* \ phosphorylation} + r_{Lyn* \ autophosphorylation} x_{Lyn*}) x_{Lyndp}$$

$$\frac{dx_{Lyn*}}{dt} = [\text{Lyn autophosphorylation}] - [\text{Lyn de-autophosphorylation}]$$

$$= \left(r_{Lyn*\ phosphorylation} + r_{Lyn*\ autophosphorylation} x_{Lyn*}\right) x_{Lyndp} \cdots$$

$$- \left(r_{Lyn\ de-autophosphorylation1} \cdot x_{CD45} + r_{Lyn\ de-autophosphorylation2} \cdot x_{SHP1*}\right) x_{Lyn*} \cdots$$

$$- r_{Lyn\ de-autophosphorylation\ by\ phosphatases} \cdot x_{Lyn*}$$

Regulatory Enzyme Dynamics: Following the initiation of BCR signaling, the regulation of BCR, Syk, and Lyn activity is orchestrated by feedback loops involving the aforementioned PTKs and a collection of regulatory enzymes. The dynamic members of the regulatory subsystem are SHP1, Csk and Cbp, with their dynamics being driven by the amount of CD45. The activated forms of SHP1, Csk and Cbp are denoted by the variables x_{SHP1*}, x_{Csk*} and x_{Cbpp*}, respectively, and are modeled with the following equations:

$$\frac{dx_{SHP1*}}{dt} = [\text{SHP1 activation}] - [\text{SHP1 inactivation}]$$

$$= r_{SHP1\ activation} \cdot x_{Lyndp} x_{SHP1} - r_{SHP1\ inactivation} \cdot x_{SHP1*}$$

$$\frac{dx_{Csk*}}{dt} = [\text{Csk activation}] - [\text{Csk disassociation}]$$

$$= r_{Csk\ activation} \cdot x_{Cbpp} x_{Csk} - r_{Csk\ disassociation} \cdot x_{Csk*}$$

$$\frac{dx_{Cbpp*}}{dt} = [\text{Cbp phosphorylation}] - [\text{Cbp dephosphorylation}]$$

$$= r_{Cbp\ phosphorylation} \cdot x_{Lyndp} x_{Cbp} - r_{Cbp\ dephosphorylation} \cdot x_{CD45} x_{Cbp}$$

Medial Signaling Dynamics (BLNK, BTK, PLC2γ): The second messenger PLC2γ is critical for transducing a signal downstream following Syk activation. Before becoming fully activated, PLC2γ must bind to the linker protein BLNK and be phosphorylated by Syk and the Bruton's tyrosine kinase (BTK). Here BTK must also bind to BLNK, and it is phosphorylated by Syk and Lyn before it becomes fully activated. These events are modeled by the following equations:

$$\frac{dx_{BLNKp}}{dt} = [\text{BLNK phosphorylation}] - [\text{BLNK dephosphorylation}]$$

$$= r_{BLNK\ phosphorylation} \cdot x_{Syk342} x_{BLNK} \cdots$$

$$- \left(r_{BLNK\ dephosphorylation} \cdot x_{SHP1*} + r_{BLNK dephosphorylation\ by\ phosphotases}\right) x_{BLNKp}$$

$$\frac{dx_{BTKb}}{dt} = [\text{BLNK-BTK binding}] - [\text{BLNK-BTK unbinding}]$$

$$= r_{BLNK-BTK\ binding} \cdot x_{BTK} x_{BLNKp} - r_{BLNK-BTK\ unbinding} \cdot x_{BTKb}$$

$$\frac{dx_{BTKp}}{dt} = [\text{BTK phosphorylation by Syk342}] + [\text{BTK phosphorylation by Lyn*}]$$

$$- [\text{BTK dephosphorylation}]$$

$$= \left(r_{BTK\ phosphorylation\ by\ Syk342} \cdot x_{Syk342} + r_{BTK\ phosphorylation\ by\ Lyn*} \cdot x_{Lyn*}\right) x_{BTKb} \cdots$$

$$- r_{BTK\ dephosphorylation} \cdot x_{BTKp}$$

$$\frac{dx_{PLC2\gamma b}}{dt} = [\text{BLNK-PLC2}\gamma \text{ binding}] - [\text{BLNK-PLC2}\gamma \text{ unbinding}]$$

$$= r_{BLNK-PLC2\gamma \text{ binding}} \cdot x_{PLC2\gamma} x_{BLNKp} - r_{BLNK-PLC2\gamma \text{ unbinding}} \cdot x_{PLC2\gamma b}$$

$$\frac{dx_{PLC2\gamma p}}{dt} = [\text{PLC2}\gamma \text{ phosphorylation by Syk342}] + [\text{PLC2}\gamma \text{ phosphorylation by BTK*}]...$$

$$- [\text{PLC2}\gamma \text{ dephosphorylation}]$$

$$= (r_{PLC2\gamma \text{ phosphorylation by Syk342}} \cdot x_{Syk342} + r_{PLC2\gamma \text{ phosphorylation by BTK*}} \cdot x_{BTK*})x_{BTKb}...$$

$$- r_{PLC2\gamma \text{ dephosphorylation}} \cdot x_{PLC2\gamma p}$$

Described in [3,16] are the remaining equations not shown here, *i.e.*, equations for Erkp, IκB and NF-κB, which are referenced below.

3. Materials and Methods

In this section we describe the experimental methods used to obtain Erk phosphorylation data and the algorithmic methods used for sensitivity analysis and parameter screening.

3.1. Experimental Protocols

3.1.1. Cell Lines

Chicken DT40 B-cells lacking Syk were obtained from Dr. Tomohiro Kurosaki. Cells were cultured in RPMI 1640 media supplemented with 10% fetal calf serum, 1% chicken serum, 50 μM 2-mercaptoethanol, 1 mM sodium pyruvate, 100 IU/mL penicillin G, and 100 μg/mL streptomycin. Stable DT40 cell lines expressing analog sensitive Syk-AQL-EGFP (R428Q/M429L/M442A, referred to as Syk-AQL) were constructed using the Lenti-X Tet-On Advanced Inducible 105 Expression System (Clontech, Mountain View, CA, USA). To constitutively express the tetracycline-controlled transactivator, rtTA, in the Tet-On inducible system, the HEK293 cells were first infected with viral particles containing the pLVX Tet-On Advanced Regulator. Lentiviral particles were generated by co-transfecting HEK293T cells with 4 μg of pLVX-Tet-On, 4 μg of pHR'-CMV-ΔR8.20 vpr, and 2 μg of pHR'-CMV-VSVG using Lipofectamine 2000 (Invitrogen, Carlsbad, CA, USA). The supernatants containing viral particles were harvested 48 h post-transfection and used to infect Syk-deficient DT40 cells. Two days after infection, cells were selected with 500 μg/mL G418 and screened for rtTA expression. Cells constitutively expressing rtTA protein were infected with lentiviral particles packaged with pLVX-Tight-Puro-Syk-AQL-EGFP as described above. After 48 h, cells were selected with 1 μg/mL puromycin and these cells were treated with 1 μg/mL doxycycline to induce Syk-AQL expression followed by screening for expression by Western blotting.

3.1.2. Cellular Activation Assay

For the analysis of Erk phosphorylation, DT40 Syk-AQL-EGFP cells were treated with or without goat-anti-mouse IgM (10 µg/mL) for the indicated periods of time at 37 °C and then lysed in buffer containing 25% sucrose, 2.5% SDS, 25 mM Tris/2.5 mM EDTA, 2.5 mg pyronin Y, and 2% 2-mercaptoethanol. The DNA in lysates was sheared by passing through a 26 G × 1/2 in needle. Proteins in the lysate were separated by SDS-PAGE, transferred to polyvinylidene difluoride membrane, and analyzed by Western blotting with anti-pERK (Cell Signaling p44/p42 MAPK (T202/Y204) rabbit 4370S), and anti-Syk (Santa Cruz N-19 rabbit) antibodies. The results of these assays were used for parameter screening and will be referenced in the results section.

3.2. Sensitivity Analysis

Our first objective was to identify parameters that produce behavior that fits B cell data. This is important as our nominal parameters largely came from parameters estimated by fitting to T cell data. In order to obtain a computationally tractable search, we first conducted a sensitivity analysis to identify the parameters with the greatest influence on model output associated with our available data.

We were concerned with fitting the model output to Erkp and IκB data reported by Healy *et al.* in [8], Erkp data obtained as described in the previous section, and NF-κB data reported by Oh *et al.* in [1]. Recall that Erkp denotes the sum of singly and doubly phosphorylated Erk. The experimental conditions we sought to simulate were administrations of ligand at time $t = 0$ in doses ranging from 5.5 to 150 µg/mL. The data from Healy *et al.* was used for model fitting through parameter screening, and so the form of this data was taken into consideration during sensitivity analysis. The measurements taken by Healy *et al.* were reported relative to the basal or unstimulated phosphorylation of a species. Thus, to evaluate the fitness of a set of parameters, we ran the model to steady state and recorded the value of x_{out}^{Basal}, where $out = Erkp$ or $I\kappa B$ throughout all subsequent sections, before continuing the simulation with the addition of ligand.

Fitness was quantified using the objective

$$J_{out} = (y_{obs} - y_{simulated})/\sigma_{out} \tag{1}$$

Given that the basal and simulated values both depend on parameters, for our initial sensitivity analysis we chose to compute the sensitivity of

$$y_{sim} = \frac{x_{out}^{Stimulation}(t_{obs}, p_k)}{x_{out}^{Basal}(p_k)} \tag{2}$$

with respect to variations in parameters p_k. Note that p_k is the kth point in our parameter screening. To estimate the uncertainty in the data σ_{out} for Equation (1), we assumed a linear dependence of σ_{out} on y_{obs} and conducted a linear regression using the information in Figure 2C of [8]. We found that the error in the measurements could be reasonably approximated by

$$\sigma_{out} = 0.0127 + 0.3084 \cdot y_{obs} \tag{3}$$

where y_{obs} is an observed measurement. Given the total number of model parameters and the cost associated with varying them, we partitioned parameters into seven distinct groups and conducted a sensitivity analysis with respect to each group when determining which parameters to screen initially. These groups of parameters were detrained using natural divisions such as BCR dynamics, Syk dynamics, regulatory enzyme and Lyn dynamics, Erk pathway dynamics, *etc.* A study by Zheng [2] comparing local derivative-based sensitivity methods and global variance-based methods found that global parameter sensitivities were necessary to capture model behavior when considering a large parameter space, but that there were no significant difference between Sobol analysis and the other variance based methods considered. Given the relative independence of these groups, we calculated only primary Sobol sensitivities [17] to estimate the sensitivity of the outputs, normalized as in Equation (2), to each specified parameter. The sensitivity, $S_{p_k}(t) = \frac{Var_{p_k}(E_{p \neq p_k}[x_{out}^{Stimulation}|\, p_k])}{Var(x_{out}^{Stimulation})}$, for a given parameter was computed at integer values $t = 0, \ldots, 30$ using the method based on sparse-grid interpolation as described in [18]. This expression is designed to capture the relative sensitivity of the output as a function of one particular parameter p_k, averaged over the other parameters. That is, if we fix p_k, we can determine the average behavior as we vary the remaining parameters, and then determine how this average changes as a function of p_k. These calculations were carried out in log space in each parameter, with a range of one order of magnitude above and below the nominal value for each parameter.

To match the conditions in much of that in [8], we used 20 µg/mL for the stimulation amount at time $t = 0$. For each parameter considered, we calculated the median of the sensitivities for that parameter over the times considered. The value 0.15 was found to be a natural threshold for each group, and if the median was less than 0.15, we concluded the parameter was insensitive and excluded it from future parameter screening. This criterion left us with 12 parameters to consider for the parameter screening. Plots of these sensitivity values are included in Supplementary Materials.

3.3. Parameter Screening

Using Latin Hypercube Sampling (LHS), we screened parameter space for points that we consider acceptable if they produce simulations satisfying $|J_{out}| \leq \eta$ or equivalently

$$y_{obs} - \eta\sigma \leq y_{sim} \leq y_{obs} + \eta\sigma \tag{4}$$

The first screening used the following data from Healy *et al.*: an Erk measurement at time $t = 5$ minutes and dose responses for IκB all measured at time $t = 15$ minutes. The doses provided in the dose response experiment were 5.5, 16.5, 50, and 150 µg/mL.

The second screening was with respect to our data and also used the condition Equation (4) to determine acceptability. However, in the acceptability condition for this screening, we modified the calculation of y_{sim} in that we calculate phosphorylation relative to the ending value rather than the basal value. That is, the signal intensities for the Western blots from our data were normalized by their ending phosphorylation levels to avoid the magnification of errors that would result from a small initial value. Applying the same normalization procedure to simulated data gives the form

$$y_{sim} = \frac{x_{out}^{Stimulation}(t_{obs}, p_k)}{x_{out}^{Stimulation}(t_{final}, p_k)} \tag{5}$$

In this case the uncertainty in our experimental measurements was determined by calculating the standard deviation of the three replicates.

The data from Oh *et al.* [1] consisted of data for wild-type and mutant B cells, which featured Syk-AQL. The mutant data was reported relative to wild-type activity. The wild-type data was used to ensure that we achieved reasonable behavior in NF-κB after using data from Healy *et al.* to fit IκB, the model variable that directly preceded NF-κB. Using sensitive parameters relating to Syk dynamics as a guide to select a small subset of parameters to tune manually, the mutant data was used to determine a separate parameter set to reproduces this mutant (Syk-AQL) behavior.

3.4. Contour Analysis

In order to investigate the dependence of Erk and NF-κB activation on ligand–receptor binding rates, we simulated the model at a dose of 20 µg/mL anti-BCR over a product grid of forward and reverse binding rates. This was done for each of wild-type, mutant with no orthogonal inhibitor, and mutant with a dose of 1 µM orthogonal inhibitor. The parameter grid was constructed using evenly spaced points in log scale over ranges for forward and reverse binding rates found in literature [4,6].

In order to avoid numerical inaccuracies associated with overly stiff parameters, we halted any simulation that took longer than fifteen minutes during the contour analysis. For these grid points, we used the built-in MATLAB function *griddata* in order to interpolate the corresponding values.

Table 1. Sensitive parameters.

	Reactions	**Parameters**
Group 1	BCR dynamics	$rw0_{kf}$
Group 2	Syk activation	$rw7_{kr}, rw9_{kf}$
Group 3	Regulatory enzyme dynamics	N/A
Group 4	Medial signaling	$rw15_{kf}, rw16_{kf}, rw16_{kr}$
Group 5	Medial signaling	$r12s_{kf}, r13_{kf}, r13_{kr}$
Group 6	Erk pathway dynamics	$r18_{kf}, r19_{kf}$
Group 7	NF-κB pathway dynamics	$r38_{kf}$

The parameter screening, sensitivity analysis, and contour analysis scripts were implemented in MATLAB 2012a. The script was parallelized to run on a Sun Server X3-2 server with two Intel Xeon E5-2690 processors and 160 GB RAM. For the parameter screening and sensitivity analyses, parameter ranges were set to two orders of magnitude on either side of the nominal parameters for each group except the fourth group (see Section 4.2 and Table 1).

4. Results

4.1. Sensitive Parameters

Primary Sobol sensitivities [17] were calculated for each output and we analyzed the distribution of the sensitivities over time. The following parameters were considered during the parameter screening. Inclusion in the parameter screening meant that the parameter did not violate the criterion that median $S(t) < 0.15$. The box and whisker plots for parameter sensitivities leading to this criterion are included in the supplementary information Figure S1. We ultimately sought for the model to produce a graded response to increasing dosages of anti-BCR stimulation in Syk342 and then let that gradation propagate downstream; thus, the results of the sensitivity analysis match what one would expect as they correspond to key signaling reactions.

The parameters in group one correspond to BCR dynamics, and $rw0_{kf}$ specifically represents the forward rate of the ligand binding reaction to the BCR. Group two contains parameters related to Syk activation, and $rw7_{kr}$ is the reverse rate of the phosphorylation reaction for the Y342 tyrosine on Syk. Parameter $rw9_{kf}$ is the forward rate in the phosphorylation reaction for the Y317 tyrosine on Syk that has already been phosphorylated at Y342. Group three is comprised of parameters from the regulatory enzyme subsystem. A sensitivity analysis was not conducted with respect to this group due to issues with stiffness. We describe next steps to examine this stiffness and future plans to expand the regulatory enzyme subsystem to become fully dynamic in Section 5.

Groups four and five consisted of parameters relating to rates for reactions involving medial signaling components BLNK, PLCγ, Bruton's Tyrosine Kinase (BTK). For group four, parameter $rw15_{kf}$ is the rate at which BLNK is phosphorylated by Syk342, while $rw16_{kf}$ and $rw16_{kr}$ are the forward and backward rates for the binding of PLCγ to the linker protein BLNK. In group five, $r12s_{kf}$ is the forward rate at which Syk342 phosphorylates bound PLCγ. Parameters $r13_{kf}$ and $r13_{kr}$ represent the rate at which $PLC\gamma$ phosphorylates the phospholipid PIP$_2$ and the corresponding rate of dephosphorylation. Finally, group six is made up of medial signaling parameters for reactions involving the kinase PKC and also the downstream MAPK pathway leading to Erk. Parameter $r18_{kf}$ is the rate at which Erk is phosphorylated by MEK. Parameter $r19_{kf}$ is the rate at which the enzyme SOS binds to phosphorylated BLNK. Finally, Group seven consists of parameters for reactions related to the NFκB pathways. Here $r38_{kf}$ is the rate of phosphorylation of IκB by the kinase IKK.

4.2. Parameter Screening and Fitting

To find a set of parameters that qualitatively match a variety of data, we first screened with respect to the data from Healy *et al.* [8] and required $|J_{Erkp}| \leq 1$ and $|J_{I\kappa B(dose1)}| \leq 2$. We found seven parameter vectors that met the criteria among the 1800 candidates considered. Due to large per-simulation time requirements, large objective values for doses #3 and #4 of the dose response experiments, and tradeoffs between Erk costs and IκB costs, we determined we would need to manually tune parameters related to IκB to achieve reasonable fits at all four doses.

Figure 3. Simulations using p_{WT}^* compared with experimental data. On the left, a simulation for the Erkp time course (normalized by total Erk) with 20 µg/mL anti-BCR is shown with the mean from Healy *et al.* [8] at time $t = 5$ and one standard deviation interval of uncertainty. On the right, simulations using p_{WT}^* and 10 µg/mL anti-BCR (normalized by Erk at time $t = 15$) are compared with Erkp triplicate data from Section 3.1.2.

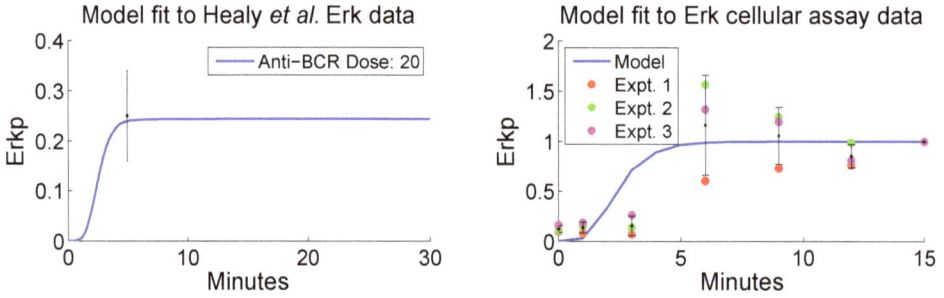

Figure 4. Simulations using p_{WT}^* compared with IκB data from Healy *et al.* [8]. Simulations for non-degraded IκB (normalized by total IκB) are shown (left to right, top to bottom) for 5.5, 16.5, 50 and 150 µg/mL anti-BCR, with all measurements taken at time $t = 15$ and one standard deviation interval of uncertainty.

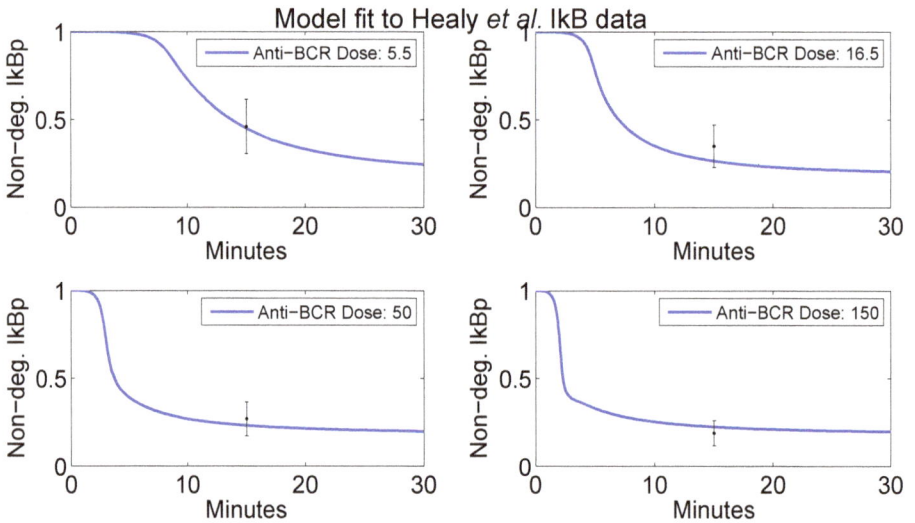

We next screened the seven accepted parameter vectors for fitness to the Erkp data obtained as in Section 3.1.2 and normalized as in Equation (5) with a threshold of $\eta = 1$. From this screening, we selected one parameter vector p based upon the smoothness of its Erkp time course, time to full Erkp activation, agreement with intermediate Erkp data points, and smoothness of non-degraded IκB time courses. Simulations using p are shown in Figure 3. For this Erkp data, we did not do any local optimization, but rather focused on the qualitative response. The right panel in Figure 3 indicates both the variability in experimental data and good qualitative agreement between these data and simulation.

From the vector p, we improved our fits for IκB by manually tuning parameters in the IκB pathway. The parameters that were adjusted were the rate of IκB production and the rate of NF-κB production. This final manual tuning led us to the parameter vector we call p^*_{WT}. The final fits for IκB can be seen in Figure 4. We were able to achieve qualitative agreement to the wild-type NF-κB data of Oh *et al.* [1] without any further changes to parameters, as seen in the left panel of Figure 5.

Figure 5. Anti-BCR dose response curves compared with experimental data from Oh *et al.* [1]. On the left, a simulation using p^*_{WT} (normalized by WT activity at the maximum dose) is shown to qualitatively agree with the wild-type NF-κB data ($\cdot$). On the right, a simulation with the parameter vector p^*_{Mutant} (also normalized by WT activity at the maximum dose) is shown with NF-κB data ($\cdot$) from B cells with Syk-AQL activity.

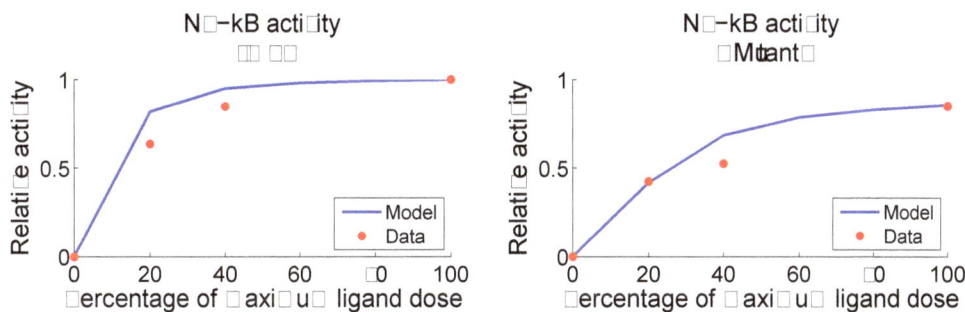

Since mutant Syk-AQL has experimentally different NF-κB response compared with WT, we manually tuned the sensitive parameters associated with Syk dynamics. We found that increasing the rate of Y317 phosphorylation $rw9_{kf}$ allowed us to fit two of the three nonzero data points. The agreement to the mutant data with this new parameter vector p^*_{Mutant} can be seen in the right panel of Figure 5. Intuitively, this corresponds to inhibiting a larger fraction of Syk, and thus there is less Syk available to propagate a signal. Interestingly, we could also achieve the same fits to mutant data by lowering the total amount of Syk in the cell. This was reminiscent of the effects of the drug tetracycline, which can regulate the amount of kinase prior to stimulation. Note that the measurements used from Oh *et al.* were reported relative to phosphorylation levels observed following an experiment where cells were stimulated using PMA and ionomyocin. We do not simulate the effects of ionomyocin in this work since calcium is not modeled, so our simulated activity in the right panel of Figure 5 is relative to the final phosphorylation level observed in simulated wild-type activity.

In Figure 6, we plot predicted dose response curves associated with the parameter vector p^*_{Mutant} as a function of ligand dose, one curve for each of several doses of orthogonal inhibitor (the OI doses are specified in μM in the legend). The simulation values are given at $t = 5$ minutes. To investigate the qualitative response, we express the ligand dose in each case as a percentage of saturating dose. As seen in Figure 6, our model exhibits a clear dose response to antigen. Additionally, it is clear in the figure that the orthogonal inhibitor limits the Erkp response; activity level is reduced as the amount of inhibitor increases, suggesting that active Syk is critical to propagate the signal and may be a limiting quantity.

Figure 6. Anti-BCR dose response curves for baseline Syk-AQL activity and inhibited activities. The curves show simulated relative activity for Erkp measured at $t = 5$ after applying ligand and orthogonal inhibitor (μM) simultaneously. All curves have been normalized by Erk activity at the maximum dose with no orthogonal inhibitor added. The color of the curve corresponds to the amount of orthogonal inhibitor specified in the legend.

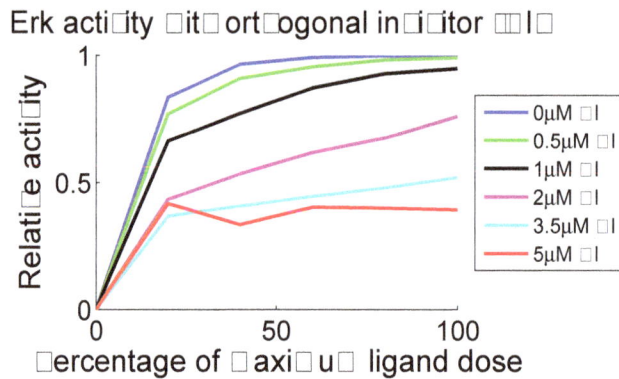

4.3. Contour Analysis

As shown by Healy *et al.* in [8], there is full signaling activity through the Erk pathway and limited activity in the NF-κB pathway during an anergic response. To investigate a variety of affinities that could induce anergy, we vary the forward and reverse kinetic rates for BCR binding and consider the cell activity as a function of the binding rates. We seek to find areas of the grid of binding rates that lead to high Erkp activity and low NF-κB activity. We have constructed contour plots for normalized Erkp activity minus normalized NF-κB activity for several scenarios: WT B cells, mutant B cells without OI added, mutant B cells with 1 μM of OI added. The contour plots allow us to ascertain relationships between the binding rates associated with the responses we found.

As seen in Figure 7, for each scenario the response at low values of the reverse binding rate is qualitatively different from the response at higher values of reverse binding rate. At low values, the response depends only on the forward binding rate, while at higher values the response depends more or less linearly in log space on both binding rates. The slope for this linear relationship is not 1, however, which would be the case if the response depended on the standard affinity, $K_a = \frac{rw0_{kf}}{rw0_{kr}}$. As seen in the contour plots, the response above the value $rw0_{kr} > -0.5$ is reasonably described as a function of $\log rw0_{kf} - \alpha \log rw0_{kr}$. This leads to a kind of power law affinity, $K_{a,\alpha} = \frac{rw0_{kf}}{(rw0_{kr})^\alpha}$, where the multiplier $\alpha = 3/4$ is the reciprocal of the slope of the linear relationship in the contour plot. The origin of the power law affinity will be investigated in future analysis of the dynamical system.

Figure 7. Contour plots for wild-type (WT), mutant without orthogonal inhibitor and mutant with 1 µM orthogonal inhibitor. The diagonal black line has a slope equal to $4/3$. Regions with high values correspond to large Erkp response and small NF-κB response (both responses normalized by their maximum WT activity), and hence possible regions of anergy. Both rates are shown in log scale.

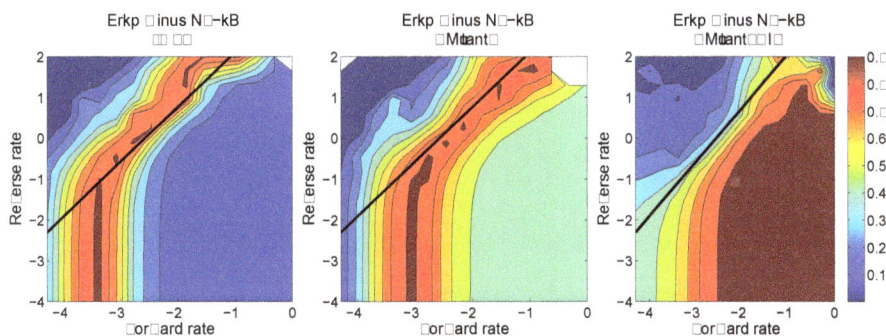

To illustrate these dependencies, we plot in Figure 8 the responses in the low reverse rate region against the forward rate $rw0_{kf}$ and the responses in the high reverse rate region against the power law affinity. As expected from the contour plots, the plots in Figure 8 show a clear dependence on forward rate alone in the region of low reverse rate and a reasonably clear dependence on the power law affinity in the region of high reverse rate.

There are higher plateaus of Erkp-NF-κB present in the mutant plots (middle and right) of Figure 8. Plots of each quantity separately (not shown) demonstrate that plateau levels of Erkp are relatively unchanged while NF-κB is suppressed in these mutants. These higher plateaus lead to the question of whether it is easier to induce and observe anergy in B cells with Syk-AQL than in WT. If so, this could have important implications for attempts to produce mice with these mutant B cells.

In order to further understand the effect of Syk-AQL and OI on Syk, we consider the allocation of Syk in each scenario. Using the power law affinity, we find that the variables x_{Sykb}, x_{Syk342}, and x_{317} all follow a sigmoidal course. Note that a percentage of total Syk is also allocated to other variables, such as free, unbound Syk, and to Syk bound to clathrin; since our focus is on the active forms of Syk, we omit these other forms. We find that Syk-AQL with no orthogonal inhibitor mimics fairly closely the response of wild-type, except that Syk342 is somewhat reduced. As expected, Syk-AQL with orthogonal inhibitor shows a marked decrease in these three forms of Syk, with the balance migrating to inhibited Syk.

Figure 8. Plots of normalized Erkp minus normalized NF-κB (each normalized by their maximum WT activity) over a product grid of forward and reverse binding rates as in the contour plots above, but separated into regions of high and low reverse rates. The first column is wild-type simulation, the second column is mutant simulation without orthogonal inhibitor, and the third column is mutant simulation with 1 μM orthogonal inhibitor. Rates are shown in log scale.

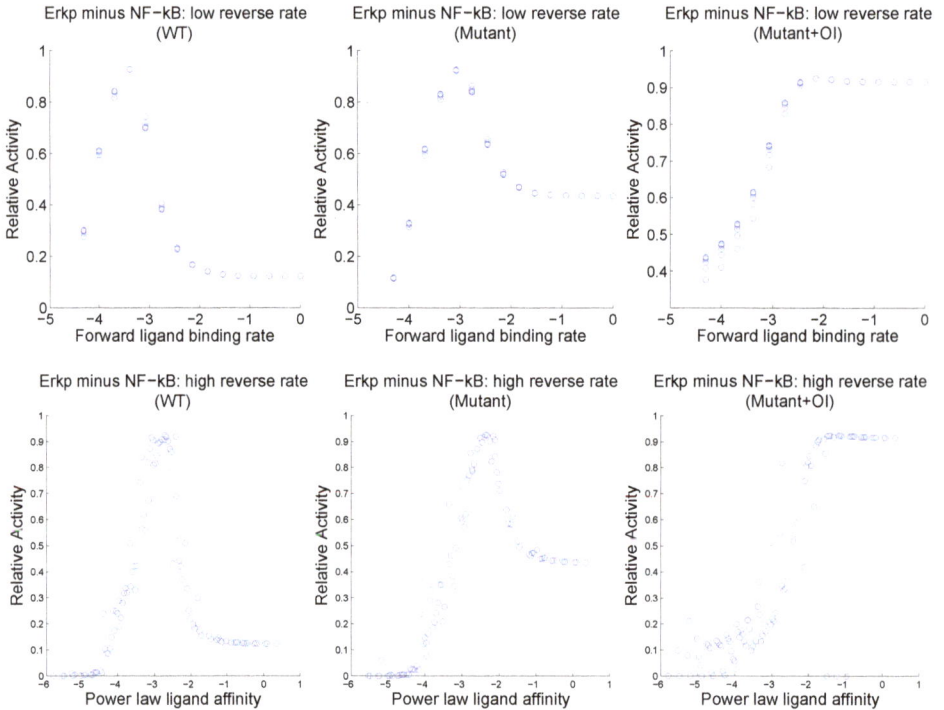

The analysis in this section has several possible biological implications. The moderate reduction in Syk-AQL activity compared with wild-type suggests that the level of NF-κB activation, which is reduced in Syk-AQL cells relative to wild-type, is more sensitive to small reductions in kinase activity than Erkp activation, which is essentially unchanged. Since this profile of high Erkp and reduced NF-κB is consistent with anergy, this implies that anergy is particularly sensitive to small changes in catalytic activity. A second possible implication derives from the observed dependence of Erkp and NF-κB on the power law affinity. This implies that B cells may respond to certain details of binding and unbinding rates for ligands rather than simple affinity alone. These observations provide a platform upon which to plan future experimental approaches and to predict experimental outcomes to further evaluate the role of Syk and changes in its catalytic activity in determining cell fate decisions following BCR engagement.

Figure 9. Plots for three forms of Syk in the model as a function of the power law affinity constant for wild-type and mutant behavior. We notice lower phosphorylation levels at both tyrosine Y317 and Y342 in the mutant cells. After the addition of 1 μM orthogonal inhibitor to the mutant cell there is the expected decrease in overall activity; the balance is accounted for by inactive forms of Syk.

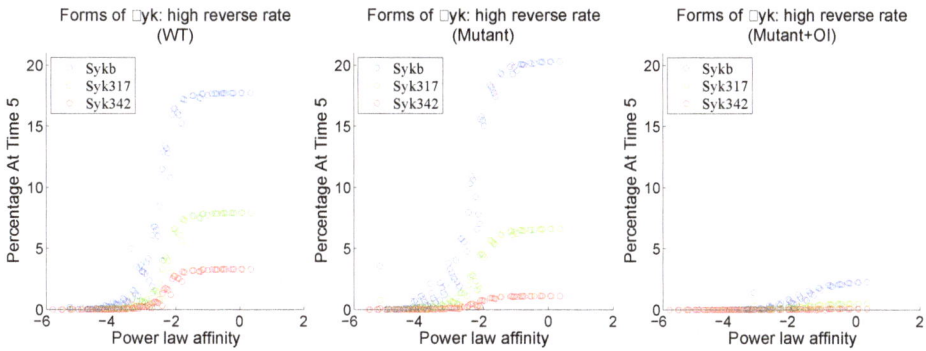

4.4. Independent Dataset Comparison

We compared the model to an independent dataset from Chaudhri *et al.* [6]. These data were not used in screening the parameters; the comparison is presented in Figure 10. The Chaudhri data include ligand concentrations that are much smaller than those available in our training data and indicate a relatively large activation even at very small ligand concentrations. Our model displays significantly smaller activity levels than those seen in the Chaudhri data at these low ligand concentrations. We believe that further parameter screening could produce better agreement to these data, but the underlying question is somewhat deeper in view of the phenomenon of anergy, in which B cells display reduced response to higher levels of ligand concentration. Experiments have shown that a low constant signal [15] can drive a B cell to become anergic and thus relatively unresponsive to the presence of antigen. Hence the question is not only what is the effective level of phosphorylation of Erk at low doses of ligand but also what is the effect of such low doses over extended periods of time. This is consistent with our model predictions of relatively high levels of Erkp activity and low levels of NF-κB activity in response to small amounts of active Syk. However, our model also suggests that the details of forward and reverse binding rates may also play a role in anergy.

Figure 10. Anti-BCR dose response curves resulting from p^*_{WT}; the figure shows ligand dose response for Erkp resulting from p^*_{WT} as compared with data from Chaudhri *et al.* [6]. As with the data, the simulation curve is normalized by the simulated value at the maximum dose 0.5 µg/mL.

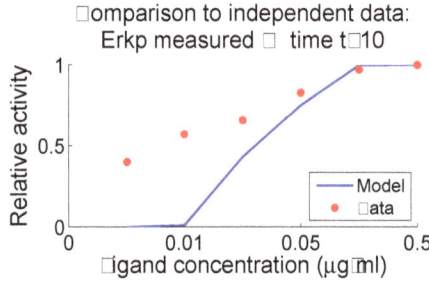

5. Conclusions and Future Directions

We have constructed a deterministic model of B cell signaling, with a focus on the role of Syk in modulating the activity of Erk and NF-κB. In particular, we include dynamics for the mutant kinase Syk-AQL, which experimentally displays dynamics that are qualitatively similar to wild-type dynamics in the absence of orthogonal inhibitor but can be modulated through the addition of orthogonal inhibitor. With the correct choice of parameters, our model reproduces data from recent cellular assays and qualitatively matches trends from datasets in the literature.

We sought to explore the kinetic rate constants associated with ligand binding that produced high relative activation of Erkp and low relative activation of NF-κB. These signaling conditions have been previously associated with anergy. We found that at different levels of $rw0_{kr}$ our responses actually depended on quantities other than the standard affinity constant. For low levels of $rw0_{kr}$, the model predicts that the response depends only on the forward rate of BCR binding $rw0_{kf}$. At higher levels of $rw0_{kr}$, the model predicts that the response depends on a power law form of the affinity constant, $K_{a,\alpha} = \frac{rw0_{kf}}{(rw0_{kr})^\alpha}$. These predictions were robust for WT and mutant simulations. Given the complexity of the dynamical system, a model reduction will likely be necessary in order to analytically investigate the origin of the power law affinity underlying the model response.

Insight into the model prediction that NF-κB is more sensitive than Erkp to changes in signaling activity is found when considering the relative amplification in each pathway. For both Erkp and NF-κB, we considered the relative change in response between wild-type and mutant with orthogonal inhibitor simulations. The relative changes were both with respect to the signaling component DAG, the last signaling component to influence both pathways. We calculated the difference between Erkp in wild-type and mutant+OI simulations and then divided by wild-type Erkp simulation to get the normalized change in Erkp. We made a similar calculation using DAG, normalized by wild-type DAG simulation, and then took the ratio of the normalized change in Erkp to the normalized change in DAG. This gives us a measure of the amplification of the DAG signal in the response of Erkp.

We likewise calculated the simulated amplification of DAG in the response of NF-κB. We found the amplification for Erkp to be $\approx$0.28 and the amplification for NF-κB to be $\approx$0.99. That is, the response of NF-κB to DAG is nearly 1:1, while the response of Erkp to DAG is reduced to roughly one-fourth of the incoming signal. These estimates agree with the findings in contour analysis that if there is a reduction in signaling activity to Syk, and thus DAG, then NF-κB will be more affected than Erkp. The mechanisms and parameters in these two pathways are structurally distinct: the Erkp pathway is based on mass-action kinetics, while the NF-κB pathway includes promotion of PKCΘ^* by DAG and a feedback loop involving NF-κB. Further experiments are needed to validate these predictions. One approach to this might be to use the DAG analog PMA as a means of effectively altering the level of DAG and investigate the resulting changes in Erkp and NF-κB experimentally.

Planned expansions to the model include stimulation by ionomycin, the addition of Ca^{2+} dynamics, and the addition of the NFAT pathway. We plan also to restructure the dynamics of CD45, which is constant in the current version of the model; this modification will impact the regulatory enzyme dynamics as they are driven by CD45 activity.

One of the difficulties with this model is the stiffness of the differential equations; for a large subset of parameter space the model takes one to tens of minutes for a simulation of 30 minutes. This stiffness limits our ability to explore the parameter space fully. Model stiffness prevented sensitivity analysis with respect to the parameters for regulatory enzyme dynamics, which made up group three of Table 1. Stiffness also presented issues during other sensitivity analysis trials and during the parameter screening and so we will seek to address this issue in future studies. We believe the improvements to CD45 dynamics will alleviate at least some of the issues with stiffness.

As seen in the right panel of Figure 6, there is a discrepancy between our model and the activity observed by Chaudhri *et al.* [6]. It is not clear whether this limitation can be resolved via the tuning of ligand binding parameters or if there are additional mechanisms needed to capture the response to lower levels of ligand.

In general, the ways in which the modulation of Syk changes the response of Erk, NFAT, and NF-κB is an important question of interest for our group. Our model is an early attempt to disentangle the behavior of Syk from these downstream responses. While there is much left to be improved in our model, we believe that it will be an important tool in our search to understand the mechanisms underlying the onset of anergy in B cells. Beyond that, we believe that our model may be used as in [3] as the basis for model-informed control strategies to achieve desired cellular responses.

Acknowledgments

G.T.B., R.L.M., and A.E.R. were supported in part by NSF Grant DMS-0900277. R.L.G. was supported in part by National Institutes of Health Grant AI098132. M.O.K. was supported by National Science Foundation Graduate Research Fellowship DGE-0833366.

Author Contributions

Reginald McGee developed the B cell model, coded it into MATLAB, ran the sensitivity analysis, fit the model to data, and wrote the bulk of the paper. Mariya Krisenko conducted the experiments for Erk phosphorylation that were used in model fitting and wrote the experimental methodology section. All authors collaborated to make any inclusions, removals, or revisions to the signaling network used for the model, with much of the details of signaling mechanisms in B cells coming from Robert Geahlen and Mariya Krisenko. Ann Rundell and Greg Buzzard provided technical details of model construction, sensitivity analysis and parameter fitting. Ann Rundell, Greg Buzzard, and Robert Geahlen edited the manuscript.

Conflicts of Interest

The authors declare no conflict of interest.

References

1. Oh, H.; Ozkirimli, E.; Shah, K.; Harrison, M.L.; Geahlen, R.L.   Generation of an Analog-sensitive Syk Tyrosine Kinase for the Study of Signaling Dynamics from the B Cell Antigen Receptor. *J. Biol. Chem.* **2007**, *282*, 33760–33768.
2. Zheng, Y.; Rundell, A.   Comparative study of parameter sensitivity analyses of the TCR-activated Erk-MAPK signalling pathway. *IEE Proc. Syst. Biol.* **2006**, *153*, 201–211.
3. Perley, J.P.; Mikolajczak, J.; Harrison, M.L.; Buzzard, G.T.; Rundell, A.E. Multiple Model-Informed Open-Loop Control of Uncertain Intracellular Signaling Dynamics. *PLoS Comput. Biol.* **2014**, *10*, 1296–1310.
4. Tsourkas, P.K.; Somkanya, C.D.; Yu-Yang, P.; Liu, W.; Pierce, S.K.; Raychaudhuri, S. Formation of BCR oligomers provides a mechanism for B cell affinity discrimination. *J. Theor. Biol.* **2012**, *307*, 174–182.
5. Mukherjee, S.; Zhu, J.; Zikherman, J.; Parameswaran, R.; Kadlecek, T.A.; Wang, Q.; Au-Yeung, B.; Ploegh, H.; Kuriyan, J.; Das, J.; *et al.* Monovalent and Multivalent Ligation of the B Cell Receptor Exhibit Differential Dependence upon Syk and Src Family Kinases. *Sci. Signal.* **2013**, *6*, ra1.
6. Chaudhri, V.K.; Kumar, D.; Misra, M.; Dua, R.; Rao, K.V.S. Integration of a Phosphatase Cascade with the Mitogen-activated Protein Kinase Pathway Provides for a Novel Signal Processing Function. *J. Biol. Chem.* **2010**, *285*, 1296–1310.
7. Barua, D.; Hlavacek, W.S.; Lipniacki, T. A Computational Model for Early Events in B Cell Antigen Receptor Signaling: Analysis of the Roles of Lyn and Fyn. *J. Immunol.* **2012**, *189*, 646–658.
8. Healy, J.I.; Dolmetsch, R.E.; Timmerman, L.A.; Cyster, J.G.; Thomas, M.L.; Crabtree, G.R.; Lewis, R.S.; Goodnow, C.C. Different Nuclear Signals Are Activated by the B Cell Receptor during Positive Versus Negative Signaling. *Immunity* **1997**, *6*, 419–428.

9. Skaggs, B.J.; Clark, M.R. Proximal B cell receptor signaling pathways. *Signal Transduct.* **2004**, *4*, 173–194.

10. Kurosaki, T.; Hikida, M. Tyrosine kinases and their substrates in B lymphocytes. *Immunol. Rev.* **2009**, *228*, 132–148.

11. Geahlen, R.L. Syk and pTyr'd: Signaling through the B cell antigen receptor. *Biochim. Biophys. Acta* **2009**, *1793*, 1115–1127.

12. Ma, H.; Yankee, T.M.; Hu, J.; Asai, D.J.; Harrison, M.L.; Geahlen, R.L. Visualization of Syk-antigen receptor interactions using green fluorescent protein: Differential roles for Syk and Lyn in the regulation of receptor capping and internalization. *J. Immunol.* **2001**, *166*, 1507–1516.

13. Veillette, A.; Latour, S.; Davidson, D. Negative regulation of immunoreceptor signaling. *Ann. Rev. Immunol.* **2002**, *20*, 669–707.

14. Reth, M.; Brummer, T. Feedback regulation of lymphocyte signalling. *Nat. Rev. Immunol.* **2004**, *4*, 269–278.

15. Andrews, S.F.; Wilson, P.C. The anergic B cell. *Blood* **2010**, *115*, 4976–4978.

16. Perley, J.P.; Mikolajczak, J.; Buzzard, G.T.; Harrison, M.L.; Rundell, A.E. Resolving Early Signaling Events in T-Cell Activation Leading to IL-2 and FOXP3 Transcription. *Processes* **2014**, *2*, 867–900.

17. Saltelli, A.; Chan, K.; Scott, E.M. *Sensitivity Analysis. Wiley Series in Probability and Statistics;* Wiley: Chichester, UK, 2000.

18. Buzzard, G. Global sensitivity analysis using sparse grid interpolation and polynomial chaos. *Reliab. Eng. Syst. Saf.* **2012**, *107*, 82–89.

Dynamic Modeling of the Human Coagulation Cascade Using Reduced Order Effective Kinetic Models

Adithya Sagar and Jeffrey D. Varner

Abstract: In this study, we present a novel modeling approach which combines ordinary differential equation (ODE) modeling with logical rules to simulate an archetype biochemical network, the human coagulation cascade. The model consisted of five differential equations augmented with several logical rules describing regulatory connections between model components, and unmodeled interactions in the network. This formulation was more than an order of magnitude smaller than current coagulation models, because many of the mechanistic details of coagulation were encoded as logical rules. We estimated an ensemble of likely model parameters ($N = 20$) from *in vitro* extrinsic coagulation data sets, with and without inhibitors, by minimizing the residual between model simulations and experimental measurements using particle swarm optimization (PSO). Each parameter set in our ensemble corresponded to a unique particle in the PSO. We then validated the model ensemble using thrombin data sets that were not used during training. The ensemble predicted thrombin trajectories for conditions not used for model training, including thrombin generation for normal and hemophilic coagulation in the presence of platelets (a significant unmodeled component). We then used flux analysis to understand how the network operated in a variety of conditions, and global sensitivity analysis to identify which parameters controlled the performance of the network. Taken together, the hybrid approach produced a surprisingly predictive model given its small size, suggesting the proposed framework could also be used to dynamically model other biochemical networks, including intracellular metabolic networks, gene expression programs or potentially even cell free metabolic systems.

Reprinted from *Processes*. Cite as: Sagar, A.; Varner, J.D. Dynamic Modeling of the Human Coagulation Cascade Using Reduced Order Effective Kinetic Models. *Processes* **2015**, *3*, 178–203.

1. Introduction

Developing mathematical models of biochemical networks is a significant facet of systems biology. Modeling approaches differ in their degree of detail, where the choice of approach is often determined by prior knowledge, or model requirements [1]. Ordinary differential equation (ODE) models are common tools for modeling biochemical systems because of their ability to capture dynamics and encode mechanism. However, ODE models typically come with difficult (or sometimes impossible) parameter identification problems. For example, Gadkar *et al.* showed that even with near-perfect information, it was often impossible to identify all the parameters in typical signal transduction models [2]. However, it is not clear whether we actually need precise estimates for all model parameters. Bailey suggested more than a decade ago, that achieving qualitative or even quantitative understanding of biological systems should not require complete structural and parametric knowledge [3]. Since Bailey's complex biology with no parameters hypothesis, Sethna showed that model performance is typically sensitive to only a few parameters, a characteristic seemingly universal to multi-parameter models referred to as *sloppiness* [4]. Thus, reasonable predictions may be possible, despite parametric uncertainty, if a few critical parameters are well-defined. For example, Tasseff *et al.*, showed in a model of Retinoic acid (RA) induced differentiation of HL-60 cells, that correct predictions were possible even when 75% of the parameters were known only to an order of magnitude [5]. Perhaps more importantly, ODE models require significant mechanistic information, thereby limiting their utility in poorly understood systems, or conversely explode in size when considering multiple pathways or subsystems. Toward this challenge, logical modeling is an emerging paradigm that encodes causal relationships between model components using quasi-mechanistic non-linear transfer functions [6]. Logical models are highly flexible, and despite their simplicity, they have captured rich behaviors in a variety of systems important to human health [7–9]. However, modeling complex dynamics with logical models is challenging. Thus, there is an unmet need for a third approach which combines ODEs and logical models, where ODEs could encode mechanistic information, while missing or incomplete mechanistic knowledge can be approximated using a logical approach.

In this study, we developed a hybrid approach which combined ODE modeling with logical rules to model a well studied biochemical network, the human coagulation system. Coagulation is an archetype proteolytic cascade involving both positive and negative feedback [10–12]. Coagulation is mediated by a family proteases in the circulation, called factors and a key group of blood cells, called platelets. The central process in coagulation is the conversion of prothrombin (fII), an inactive coagulation factor, to the master protease thrombin (FIIa). Thrombin generation involves three phases, initiation, amplification and termination [13,14]. Initiation requires a trigger event, for example vessel injury, which leads to the activation of factor VII (FVIIa). Two converging pathways, the extrinsic and intrinsic cascades, then process and amplify this initial coagulation signal. The extrinsic cascade is generally believed to be the main mechanism of thrombinogenesis in the blood [15–17]. Initially, thrombin is produced upon cleavage of prothrombin by fluid phase activated factor X (FXa), which itself has been activated by Tissue Factor/activated factor VII (TF/FVIIa) [10]. Picomolar amounts of thrombin then activate the cofactors factors V and VIII (fV and fVIII) and platelets, leading to the formation of the tenase and prothrombinase complexes on activated platelets. These complexes amplify the early coagulation signal by further activating FXa, and directly converting prothrombin to thrombin. There are several control points in the cascade that inhibit thrombin formation, and eventually terminate thrombin generation. Tissue Factor Pathway Inhibitor (TFPI) inhibits FXa formation catalyzed by TF/FVIIa, while antithrombin III (ATIII) neutralizes several of the proteases generated during coagulation, including thrombin. Thrombin itself also inadvertently plays a role in its own inhibition; thrombin, through interaction with thrombomodulin, protein C and endothelial cell protein C receptor (EPCR), converts protein C to activated protein C (APC) which attenuates the coagulation response by proteolytic cleavage of fV/FVa and fVIII/FVIIIa. Termination occurs after either prothrombin is consumed, or thrombin formation is neutralized by inhibitors such as APC or ATIII.

Previous coagulation models have typically been formulated as systems of nonlinear ordinary differential equations, using mass action or more complex kinetics, to describe the rates of biochemical conversions [18–22]. Mechanistic ODE coagulation models from our laboratory [23,24] were built upon the earlier studies of Jones and Mann [25], Hockin *et al.* [26], and later Butenas *et al.* [27] who developed and then subsequently refined highly mechanistic coagulation models. Other laboratories have also expanded upon Hockin *et al.* for example by exploring the intrinsic pathway, the role of stochastic fluctuations in coagulation [28], and the dynamics of thrombin mediated clot formation [29] and fibrinolysis [30]. Other aspects of coagulation have also been modeled, such as platelet biochemistry [31], multi-scale models of clot formation [32,33], and transport inside clots [34]. However, these previous studies were largely based upon extensive mechanistic knowledge. This is possible because blood, while enormously complex, can be systematically interrogated. Other systems, such as intracellular signaling networks, are much more difficult to experimentally interrogate. Towards this unmet need, we formulated a hybrid modeling approach which combines ODEs and logical rules to model biochemical processes for which a complete mechanistic understanding is missing. We tested this approach by modeling the human coagulation cascade. Others have also constructed reduced order human coagulation models. Recently, Papadopoulos and co-workers constructed a phenomenological mathematical model for thrombin generation [35]. Using four ordinary differential equations and six parameters, they derived an expression describing the temporal evolution of thrombin generation in a variety of cases. The reduced order Papadopoulos model showed good agreement with experimental data, and underscored that model reduction is possible even for complex positive feedback systems like coagulation. However, the Papadopoulos model was focused on thrombin generation, and had a lesser emphasis on the influence of physiological inhibitors such as ATIII or the protein C pathway. In this study, we focused on building a reduced order coagulation model that included the physiological inhibitors using a hybrid strategy. The hybrid model consisted of only five differential equations augmented with several logical rules. Thus, the model was more than an order of magnitude smaller than comparable purely ODE models in the literature. We estimated the model parameters from *in vitro* extrinsic coagulation data sets, in the presence of ATIII, with and without the protein C pathway. We then compared the model predictions with thrombin data sets, for both normal and hemophilic coagulation, that were not used for model training. Once validated, we performed flux and sensitivity analysis on the model to estimate which parameters were critical to model performance in several conditions. The reduced order hybrid approach produced a surprisingly predictive coagulation model, suggesting this framework could potentially be used to model other biochemical networks important to human health.

2. Results

2.1. Formulation of Reduced Order Coagulation Models

We developed a reduced order extrinsic coagulation model to test our hybrid modeling approach (Figure 1). The core of our model was based upon the earlier work of Ismagilov and coworkers [36–39], where we added initiation, factor dependence, and specific inhibition terms to the earlier simplified model. A trigger event initiates thrombin formation (FIIa) from prothrombin (fII) through a lumped initiation step. This step loosely represents the initial activation of thrombin by activated FXa. Once activated, thrombin catalyzes its own formation (amplification step), and inhibition via the conversion of protein C to activated protein C (APC). Antithrombin III (ATIII) inhibits amplification, while APC and tissue factor pathway inhibitor (TFPI) potentially inhibit both initiation and amplification. All initiation and inhibition processes, as well as the dependence of amplification upon other coagulation factors, was approximated using our rule-based approach (Figure 2). Individual regulatory contributions to the activity of pathway enzymes were integrated into control coefficients (v's) using an integration rule (min/max). These control coefficients then modified the rates of model processes at each time step. Hill-like transfer functions $0 \leq f(\mathcal{Z}) \leq 1$ quantified the contribution of components upon a target process. Components were either individual inhibitor or activator levels or some function of levels, e.g., the product of factor levels. In this study, $\mathcal{Z}$ corresponded to the abundance of individual inhibitors or activators, with the exception of the dependence of amplification upon specific coagulation factors (modeled as the product of factors). When a process was potentially sensitive to multiple inputs, logical integration rules were used to select which transfer functions influenced the process at any given time. In our proof of concept model, we used a winner takes all strategy; the maximum or minimum transfer function was selected at any given time step. However, other integration rules are certainly possible. Taken together, while the reduced order coagulation model encodes significant biological complexity, it is highly compact (consisting of only five differential equations). Thus, it will serve as an excellent proof of principle example to study the reduction of a highly complex human subsystem.

2.2. Identification of Model Parameters Using Particle Swarm Optimization

A critical challenge for any dynamic model is the estimation of kinetic parameters. We estimated kinetic and control parameters simultaneously from eight *in vitro* time-series coagulation data sets with and without the protein C pathway. The residual between model simulations and experimental measurements was minimized using particle swarm optimization (PSO). A population of particles ($N = 20$) was initialized with randomized kinetic and control parameters and allowed to search for parameter vectors that minimized the residual. However, not all parameters were varied simultaneously. We partitioned the parameter estimation problem into two subproblems based upon the biological organization of the training data; (i) estimation of parameters associated with thrombin formation in the absence of the protein C pathway and (ii) estimation of parameters associated with the protein C pathway. Only those parameters associated with each subproblem were varied during the optimization procedure for that subproblem, e.g., thrombin parameters were *not* varied during the protein C subproblem. The PSO procedure was run for 20 generations for each subproblem, where each generation was 1200 iterations. The best particle from each generation was used to generate the particle population for the next generation. We rotated the subproblems, starting with subproblem 1 in the first generation.

Figure 1. Schematic of the connectivity of the reduced order coagulation model. A trigger compound, e.g., TF/FVIIa initiates thrombin production (FIIa) from prothrombin (fII). Once activated, thrombin catalyzes its own activation (amplification step), as well as its own inhibition via the conversion of protein C to activated protein C (APC). APC and tissue factor pathway inhibitor (TFPI) inhibit initiation and amplification, while antithrombin III (ATIII) directly inhibits thrombin. All inhibition steps and trigger-induced initiation were modeled using a rule-based approach. Likewise, the dependence of amplification on other coagulation factors was also modeled using a rule-based approach. The abundance of the highlighted species (in the dashed boxes) was governed by an ordinary differential equation. All other species were assumed to be constant.

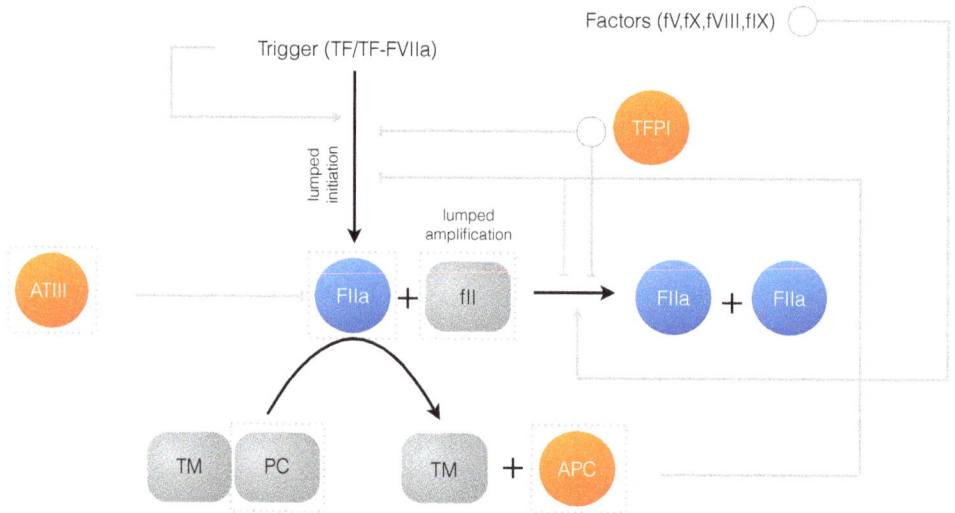

The experimental training data for parameter estimation was reproduced from the experiments of Butenas and co-workers [40]. In these experiments thrombin generation was initiated by FVIIa-TF using mean plasma concentrations of coagulation proteins and inhibitors. To prepare FVIIa-TF, TF (0.5 nmol/L) was relipidated into 400 μmol/L of phospholipid vesicles (PCPS) by incubation in 20 mmol/L HEPES, 150 mmol/L NaCl, and 2 mmol/L CaCl$_2$ pH 7.4 (HBS/Ca^{2+}) for 30 min at 37 °C. The relipidated TF was incubated with 10 pmol/L factor VIIa for 20 min to allow the formation of FVIIa-TF. Factors V, VIII and thrombomodulin (Tm) (when protein C activation is required) were added to FVIIa-TF complex. Thrombin generation was then initiated by adding equal volumes of this mixture with a mixture containing prothrombin, factor IX and factor X, TFPI, AT-III and protein C (added when required), protein S (added when required) and factor XI (added when required). In the training data, 5 pmol/L FVIIa-TF was used along with 200 μmol/L of phospholipid vesicles (PCPS) to initiate thrombin generation. All other the coagulation factors and inhibitors *i.e.*, factors X, IX,

V, and VIII, prothrombin, TFPI and AT-III, protein C and protein S (when applicable) were at their mean plasma concentrations.

Figure 2. Schematic of rule-based effective control laws. Traditional enzyme kinetic expressions, e.g., Michaelis-Menten or multiple saturation kinetics are multiplied by an enzyme activity control variable $0 \leq v_j \leq 1$. Control variables are functions of many possible regulatory factors encoded by arbitrary transfer functions of the form $0 \leq f_j(\mathcal{Z}) \leq 1$. At each simulation time step, the v_j variables are calculated by evaluating integration rules such as the max or min of the set of transfer functions $f_1, \ldots, f_n$ influencing the activity of enzyme E_j.

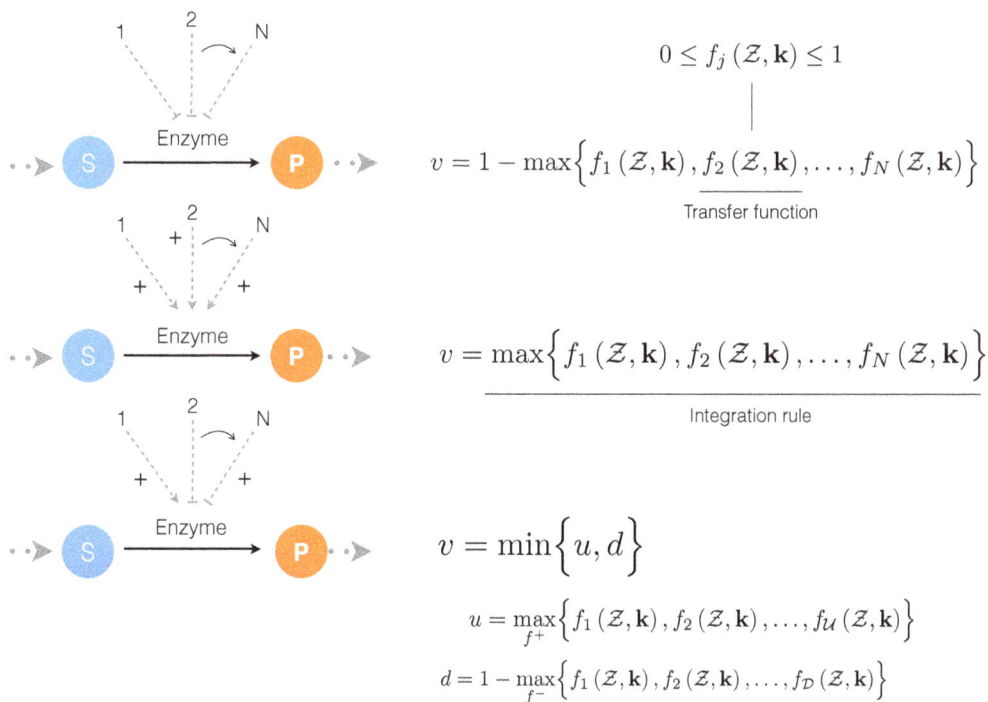

$$0 \leq f_j(\mathcal{Z}, \mathbf{k}) \leq 1$$

$$v = 1 - \max\left\{ f_1(\mathcal{Z}, \mathbf{k}), f_2(\mathcal{Z}, \mathbf{k}), \ldots, f_N(\mathcal{Z}, \mathbf{k}) \right\}$$

Transfer function

$$v = \max\left\{ f_1(\mathcal{Z}, \mathbf{k}), f_2(\mathcal{Z}, \mathbf{k}), \ldots, f_N(\mathcal{Z}, \mathbf{k}) \right\}$$

Integration rule

$$v = \min\left\{ u, d \right\}$$

$$u = \max_{f^+}\left\{ f_1(\mathcal{Z}, \mathbf{k}), f_2(\mathcal{Z}, \mathbf{k}), \ldots, f_\mathcal{U}(\mathcal{Z}, \mathbf{k}) \right\}$$

$$d = 1 - \max_{f^-}\left\{ f_1(\mathcal{Z}, \mathbf{k}), f_2(\mathcal{Z}, \mathbf{k}), \ldots, f_\mathcal{D}(\mathcal{Z}, \mathbf{k}) \right\}$$

Figure 3. Reduced order coagulation model training simulations. Reduced order coagulation model parameters were estimated using particle swarm optimization (PSO) without the protein C pathway as a function of prothrombin. Solid lines denote the simulated mean value of the thrombin profile for $N = 20$ independent particles, points denote experimental data. The shaded region denotes the 99% confidence estimate of the mean simulated thrombin value (uncertainty in the model simulation). (**A–C**) depict training data and results for 150%, 100% and 50% of physiological prothrombin levels in the absence of protein C pathway. The experimental training data was reproduced from the study of Butenas *et al.* [40]. Thrombin generation in these experiments was initiated using 5 pmol/L FVIIa-TF in the presence of 200 μmol/L of phospholipid vesicles (PCPS). As depicted in (**A–C**) the prothrombin levels were at 150%, 100% and 50% of their physiological concentration in the absence of protein C pathway. All factors and control proteins in these experiments were at their physiological concentration unless otherwise denoted.

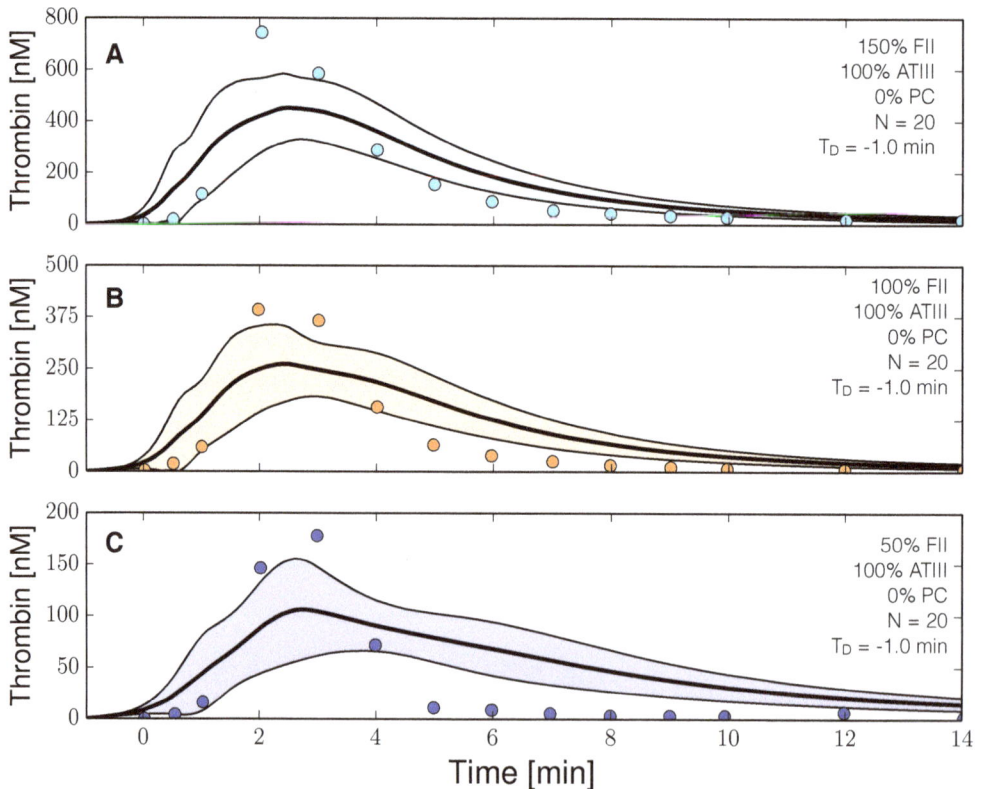

Figure 4. Reduced order coagulation model training simulations. Reduced order coagulation model parameters were estimated using particle swarm optimization (PSO) with the protein C pathway as a function of prothrombin. Only APC pathway parameters were allowed to vary in these simulations keeping the parameters estimated without protein C pathways constant. Solid lines denote the simulated mean value of the thrombin profile for $N = 20$ independent particles, points denote experimental data. The shaded region denotes the 99% confidence estimate of the mean simulated thrombin value (uncertainty in the model simulation). (A–C) depict training data and results for 150%, 100% and 50% of physiological prothrombin levels in the presence of the protein C pathway. The experimental training data was reproduced from the study of Butenas *et al.* [40,41]. Thrombin generation in these experiments was initiated using 5 pmol/L FVIIa-TF in the presence of 200 μmol/L of phospholipid vesicles (PCPS). As depicted in (A–C) the prothrombin levels were at 150%, 100% and 50% of their physiological concentration in the presence of protein C pathway. All factors and control proteins in these experiments were at their physiological concentration unless otherwise denoted.

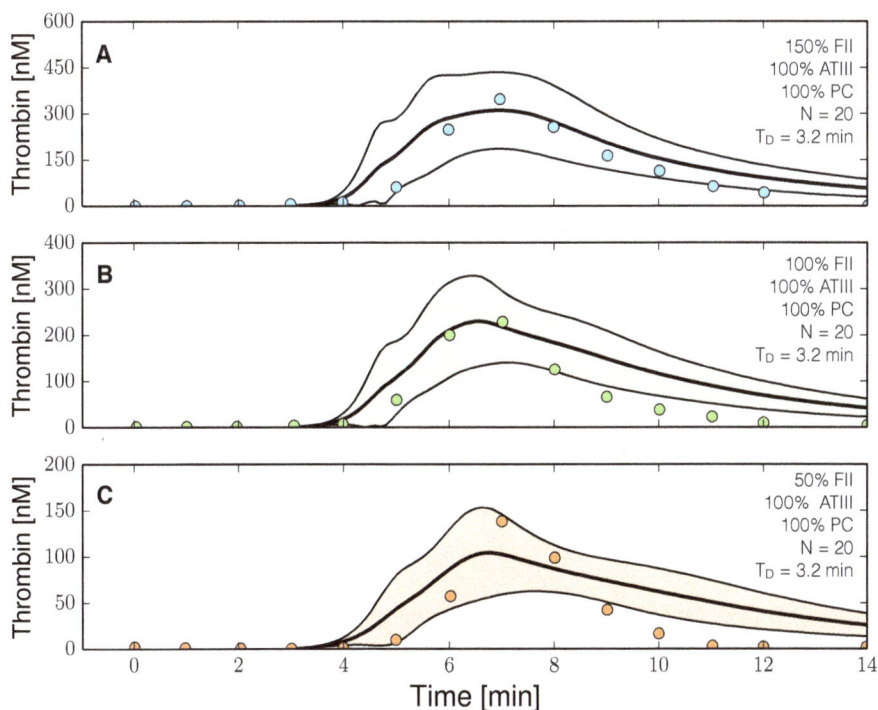

The reduced order coagulation model captured the role of initial prothrombin abundance, and the decay of the thrombin signal following from ATIII activity (Figure 3). However, we systematically under-predicted the thrombin peak and the strength of ATIII inhibition in this training data set. On the other hand, with fixed thrombin parameters, we captured peak thrombin values and the decay of the thrombin signal (at least for the 150% fII case) in the presence of both ATIII and the protein C pathway (Figure 4). Lastly, we were unable to capture global differences in initiation time *across* separate data sets with a single ensemble of model parameters. These differences likely resulted from normal experimental variability. For example, different thrombin generation experiments within the training data (at the same physiological factor levels) had significantly different initiation times (data not shown). However, the inability to globally capture initiation time also highlighted a potential shortcoming of the initiation module within the model. To capture the variability in initiation time *across* training data sets, we included a constant time-delay parameter (T_D) for each data group. The delay parameter was constant within a data set, but allowed to vary *across* training data sets. Introduction of the delay parameter allowed the model to simulate multiple training data sets using a single ensemble of model parameters. Taken together, the model identification results suggested that our hybrid approach could reproduce a panel of thrombin generation data sets in the neighborhood of physiological factor and inhibitor concentrations. However, it was unclear whether the reduced order model could predict new data, without updating the model parameters.

2.3. Validation of the Reduced Order Coagulation Model

We tested the predictive power of the reduced order coagulation model with validation data sets not used during model training. Two validation data sets were used, thrombin generation for various prothrombin and ATIII concentrations with the protein C pathway, and thrombin generation in normal *versus* hemophilic plasma in the presence of the protein C pathway. Lastly, we compared the qualitative output of the model to rFVIIa addition in the presence of hemophilia. The hemophilia case was an especially difficult test as it was taken from a different study which used a plasma-based *in vitro* assay involving platelets instead of phospholipid vesicles (PCPS). All kinetic and control parameters were fixed for the validation simulations. The only globally adjustable parameter T_D, was fixed within each validation data set but allowed to vary between data sets. The reduced order model predicted the thrombin generation profile for ratios of prothrombin and ATIII in the absence of the protein C pathway (Figure 5). Simulations near the physiological range (fII,ATIII) = (100%, 100%) or (125%, 75%) tracked the measured thrombin values (Figure 5B,C). On the other hand, predictions for factor levels outside of the physiological range (fII,ATIII) = (50%, 150%) or (150%, 50%), while qualitatively consistent with measured thrombin values, did show significant deviation from the measurements (Figure 5A,D). Likewise, simulations of thrombin generation in normal *versus* hemophilia (missing both fVIII and fIX) were consistent with measured thrombin values (Figure 6). We modeled the dependence of thrombin amplification on factor levels using a product rule ($\mathcal{Z} = fV \times fX \times fVIII \times fIX$), which was then was integrated using a min integration rule into the control variable governing amplification. Thus, in the absence of fVIII or fIX, the amplification control variable evaluated to zero, and the only thrombin produced was from initiation (Figure 6B). However, the decay of the thrombin signal was underpredicted in the normal case (Figure 6A), while the activated thrombin level was overpredicted in hemophilia simulations, although thrombin generation was far less than normal (Figure 6B). Taken together, the reduced order model performed well in the physiological range of factors, even with unmodeled components such as platelet activation in the hemophilia data set.

Figure 5. Reduced order coagulation model predictions *versus* experimental data for normal coagulation. The reduced order coagulation model parameter estimates were tested against data not used during model training. Simulations of different levels of prothrombin and ATIII were compared with experimental data in the absence of the protein C pathway. Solid lines denote the simulated mean value of the thrombin profile for $N = 20$ independent particles, points denote experimental data. The shaded region denotes the 99% confidence estimate of the mean simulated thrombin value (uncertainty in the model simulation). (**A–D**) prediction results for (FII,ATIII): (50%, 150%), (100%, 100%), (125%, 75%) and (150%,50%) of physiological prothrombin and ATIII levels in the absence of the protein C pathway. The experimental validation data was reproduced from the study of Butenas *et al.* [40]. Thrombin generation in these experiments was initiated using 5 pmol/L FVIIa-TF in the presence of 200 μmol/L of phospholipid vesicles (PCPS). As depicted in (**A–D**) the prothrombin and ATIII levels were at (50%, 150%), (100%, 100%), (125%, 75%) and (150%, 50%) of their physiological concentrations in the absence of protein C pathway. All factors and control proteins were at their physiological concentration unless otherswise denoted.

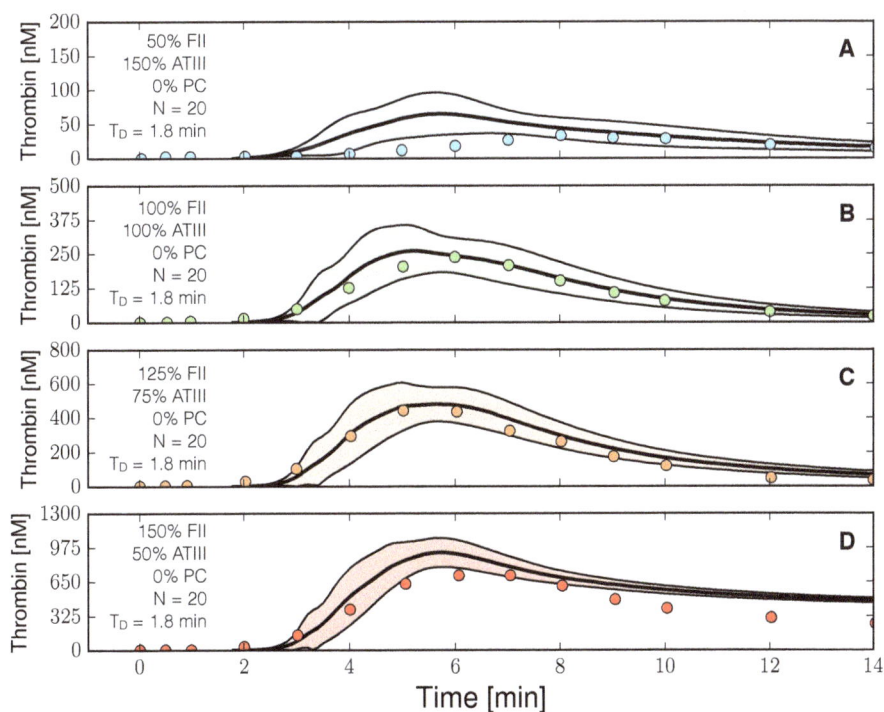

Figure 6. Reduced order coagulation model predictions *versus* experimental data with and without coagulation factors VIII (fVIII) and IX (FIX). The reduced order coagulation model parameter estimates were tested against data not used during model training. Simulations of normal thrombin formation with ATIII and the protein C pathway were compared with thrombin formation in the absence of fVIII and fIX. Solid lines denote the simulated mean value of the thrombin profile for N = 20 independent particles, points denote experimental data. The shaded region denotes the 99% confidence estimate of the mean simulated thrombin value (uncertainty in the model simulation). (**A,B**) prediction results for normal thrombin generation and thrombin generation in hemophilia. All factors and control proteins were at their physiological concentration unless others noted. Coagulation was initiated with 0.2 nmol/L FVIIa. The experimental validation data was reproduced from the study of Allen *et al.* [42].

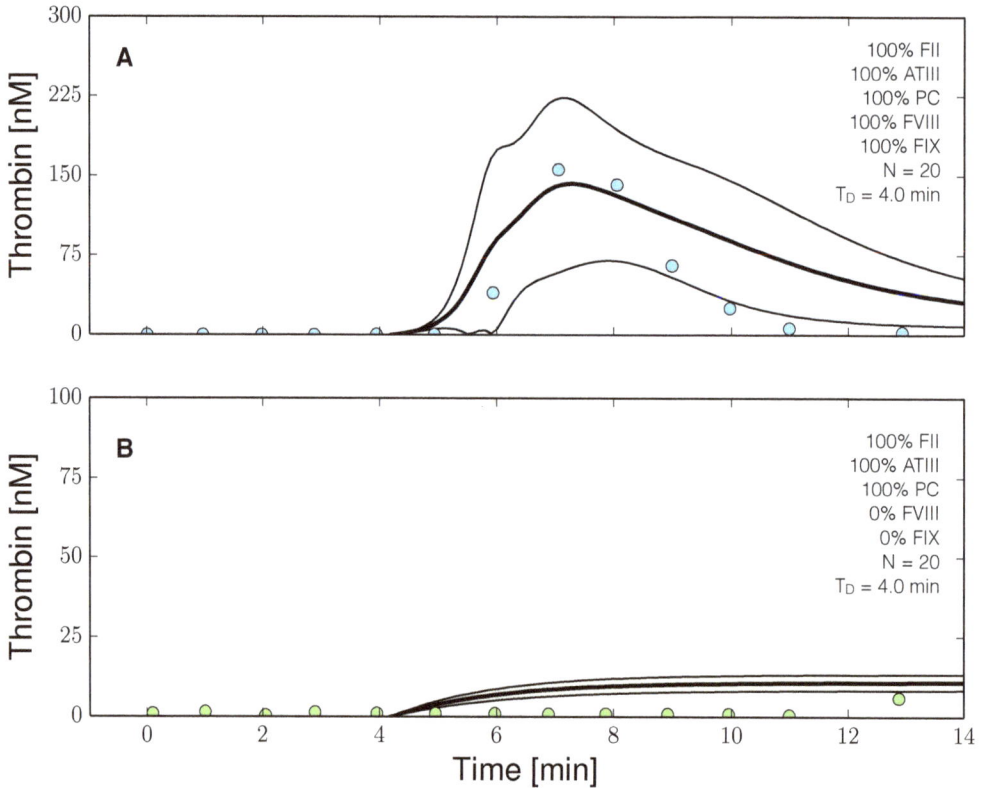

Figure 7. Reduced order coagulation model predictions of rFVIIa administration. (**A**) Simulations of thrombin formation in the presence of ATIII and the protein C pathway were conducted for a range of trigger values (1x–200x nominal) in the absence of fVIII and fIX; (**B**) Comparison of thrombin generation for normal *versus* hemophilia for 10x nominal trigger. Solid lines denote the simulated mean value of the thrombin profile for $N = 20$ independent particles. The peak thrombin time for normal coagulation (t^*) is less than rFVIIa induced coagulation in hemophilia (t^{**}), while the peak thrombin value was greater in normal coagulation. The shaded region denotes the 99% confidence estimate of the mean thrombin value (uncertainty in the model simulation). All factors and control proteins were at their physiological concentration unless others noted.

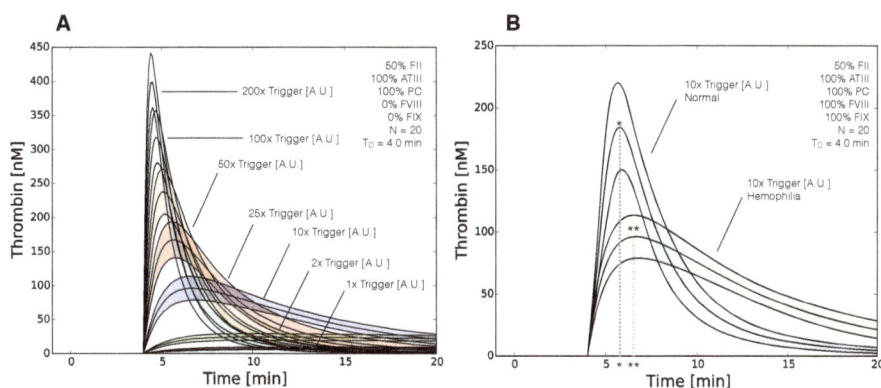

The model ensemble predicted a direct correlation between thrombin generation and rFVIIa addition in hemophilia (Figure 7). In the current model, we cannot distinguish between different initiation sources, e.g., TF/FVIIa *versus* rFVIIa, as we have only a single lumped initiation source (trigger). Thus, we simulated the addition of rFVIIa in hemophilia by removing fVIII and fIX from the model, and modulating the initial level of trigger. Simulations with a baseline level of trigger were consistent with the previous hemophilia simulations, where the only thrombin produced was from initiation (Figure 7A, 1× trigger). However, as we increased the trigger strength, the thrombin profile began to approximate normal coagulation, showing a pronounced peak albeit with a slower peak time (Figure 7B, $t^{**} > t^*$). Further increases in trigger strength resulted in decreased thrombin peak time and increased maximum thrombin values (Figure 7A, 50× trigger). Thus, for large trigger values (200× trigger), the hemophilic thrombin profile approximated normal coagulation, where peak thrombin was achieved shortly after administration and 95% of the thrombin was gone by 20 min after initiation. We performed flux analysis to understand how the reduced order coagulation model balanced initiation, amplification and termination of thrombin generation for normal and hemophilic coagulation. Analysis of the reaction flux through the reduced order network for thrombin generation in normal, hemophilia and rFVIIa-treated hemophilia identified three distinct operational modes (Figure 8). We calculated the flux through four lumped reactions, initiation, amplification, thrombin-induced APC generation and total thrombin inhibition (including both APC and ATIII action). Directly after the addition of a trigger (e.g., TF/FVIIa or rFVIIa), the lumped initiation flux was the largest for all three cases. However, within a few minutes enough thrombin was generated by the initiation mechanism to induce the amplification stage. During amplification, thrombin catalyzes its own formation and inhibition by generating activated protein C (APC), a potent inhibitor of the coagulation cascade. For normal coagulation, amplification and thrombin inhibition were the dominate reactions by 6 min after initiation (Figure 8, left). After 10 min, the dominate reaction had shifted to thrombin inhibition (both ATIII and APC action). In hemophilia (missing both fVIII and fIX), the amplification reaction did not occur, and thrombin was produced only by initiation (Figure 8, center). Initiation was quickly inhibited by APC, and the thrombin level stabilized (eventually decaying at longer times because of ATIII activity). Lastly, when $50\times$ trigger was used to induce thrombin formation in hemophilia (absence of fVIII/fIX), initiation mechanisms dominated for up to 6 min following initiation (Figure 8, right). Similar to hemophilia alone, no amplification occurred in the $50\times$ trigger+hemophilia case, and the rate of thrombin generation was extinguished by the combined action of ATIII and APC. Taken together, the hybrid modeling approach captured the transition between the modes of thrombin generation, as well as the role that inhibitors play in attenuating the thrombin generation rate. Thus, the transfer function approach encoded the inhibitory logic of this cascade in the absence of specific mechanism.

Figure 8. Reaction flux distribution as a function of time for thrombin generation under normal (left), hemophilia (center) and rFVIIa treated hemophilia (right). Reaction flux was calculated for each particle at $T = 0, 4, 6, 8, 10, 12, 14$ min after the initiation of coagulation. Reaction fluxes were calculated for each particle in the parameter ensemble ($N = 20$). Blue colors denote low flux values while red colors denote high flux values.

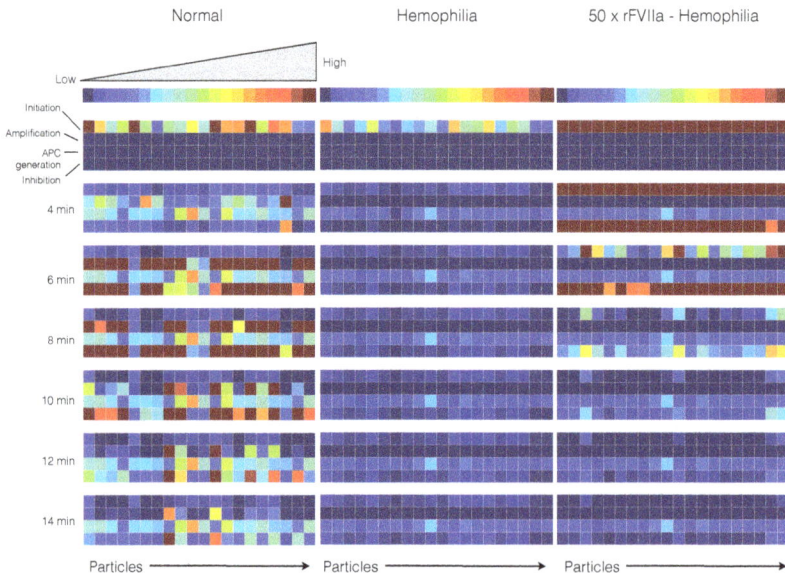

Figure 9. Global sensitivity analysis of the reduced order coagulation model with respect to the model parameters. (**A**) Sensitivity analysis of the thrombin peak time for different prothrombin levels (150%, 100% and 50% of the physiological value) as a function of activated protein C; (**B**) Sensitivity analysis of the thrombin exposure for different prothrombin levels (150%, 100% and 50% of the physiological value) as a function of activated protein C. Points denote the mean total sensitivity value, while the area around each point denotes the uncertainty in the sensitivity value. The gray dashed line denotes the 45° degree diagonal, if sensitivity values are equal for different conditions they will lie on the diagonal. Sensitivity values significantly above or below the diagonal indicate differentially important model parameters. The radius of the shaded region around each total sensitivity value was the maximum uncertainty in that value estimated by the Sobol method.

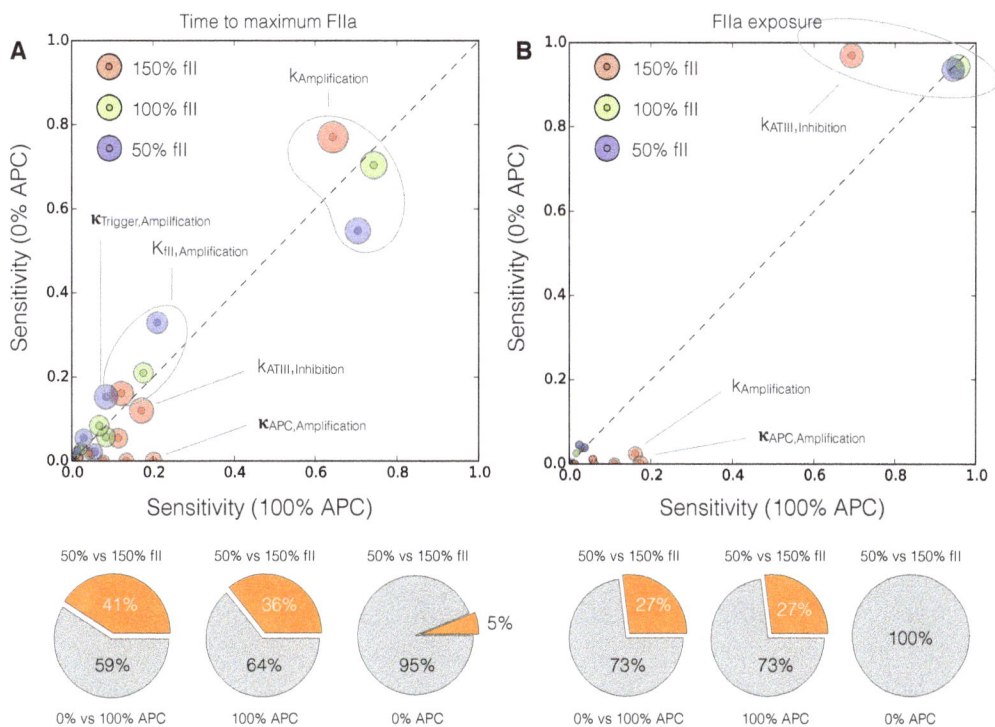

2.4. Global Sensitivity Analysis of the Reduced Order Coagulation Model

We conducted a global sensitivity analysis to estimate which parameters controlled the performance of the reduced order model. We calculated the sensitivity of the time to maximum thrombin (peak time) and the thrombin exposure (area under the thrombin curve) for different levels of prothrombin, and protein C (Figure 9). Globally, 41% of the parameters shifted in importance between the (fII,PC) = (50%, 0%) and (150%, 100%) cases for the peak thrombin time

(Figure 9A). The majority of these shifts involved the interaction between increased prothrombin and the protein C pathway, while only 5% were directly associated with increased prothrombin alone. The rate constant for thrombin amplification was the most important parameter controlling the peak thrombin time. While this parameter was differentially important for different prothrombin levels, and in the presence or absence of the activated protein C pathway, it was consistently the most sensitive parameter in the model. The saturation constant governing thrombin amplification was the second most important parameter, followed by the initiation control gain parameter. Other important parameters influencing the thrombin peak time included the control gain for activated protein C formation, and the rate constant controlling ATIII inhibition of thrombin activity. On the other hand, only 27% of the model parameters were differentially sensitive between the (fII,PC) = (50%, 0%) and (150%, 100%) cases for thrombin exposure (Figure 9B). Of these parameters, all of the shifts were associated with the interplay between thrombin formation and the protein C pathway. The rate constant controlling ATIII inhibition was the most important parameter controlling the thrombin exposure. While this parameter was less important in the presence of protein C for 150% prothrombin levels, it was significantly above all other parameters. Similar to the peak time, for 150% prothrombin, the control gain for activated protein C formation was differentially important along with the rate constant controlling amplification. However, the amplification parameter was much less important for thrombin exposure *vs.* peak time.

3. Discussion

In this study, we developed a reduced order model of the human coagulation cascade. We modeled coagulation because it is well studied, has a complex architecture, and has an abundance of experimental data available for model identification and validation. However, coagulation was just a proof of concept test of our approach. The proposed hybrid framework could also be used to dynamically model other biochemical networks, including intracellular metabolic networks, gene expression programs or potentially even cell free metabolic systems. The model consisted of five differential equations augmented with several logical rules describing regulatory connections between model components and unmodeled interactions in the network. We estimated model parameters from *in vitro* extrinsic coagulation data sets, in the presence of ATIII, with and without the protein C pathway. To estimate parameters, the residual between model simulations and experimental measurements was minimized using particle swarm optimization (PSO). However, not all of the model parameters were uniquely identifiable, given the training data. Instead, we estimated an ensemble of likely parameter sets (N = 20) from eight *in vitro* time-series coagulation data sets with and without the protein C pathway. Ensemble approaches have been used previously for other signal transduction models [43–47], and for metabolic models [48] to estimate the impact of poorly constrained parameter values or poorly understood network structure on simulation performance. Thus, ensemble approaches are common in the dynamic modeling community. However, a unique feature of the current study is the direct connection between our particle swarm approach, and the parameter ensemble; each particle in our swarm uniquely corresponded to a parameter set in our ensemble. Thus, by constraining particles to operate in different parameter regions, giving each particle a different parameter combination to explore, or perhaps even suppling a different model formulation to each particle we can effectively traverse through complex parameter and model spaces. We validated the ensemble using thrombin data sets taken from multiple laboratories for a variety of experimental conditions not used during training. The ensemble predicted thrombin trajectories for conditions not used for model training, including thrombin generation for normal and hemophilic coagulation in the presence of platelets (a significant unmodeled component). We then used flux analysis to understand how the network operated in a variety of conditions, and global sensitivity analysis to identify which parameters controlled the performance of the network. Flux analysis showed the logical rules formulation encoded the transitions between initiation, amplification and termination of thrombin generation. Sensitivity analysis suggested that the amplification rate constant was more important to the time to peak thrombin, while the ATIII inhibition constant controlled thrombin exposure. Taken together, the proposed hybrid framework produced a surprisingly predictive model, suggesting this approach could be used to effectively model other biochemical networks important to human health.

Malfunctions in coagulation can have potentially fatal consequences. Aggressive clotting involved with Coronary Artery Diseases (CADs), collectively accounts for 38% of all deaths in North America [49]. Coagulation management during surgery can also be challenging, particularly with the increase in clinical use of antithrombotic drugs [50]. Insufficient coagulation due to genetic disorders such as hemophilia can also result in recurrent bleeding. The coagulation factors VIII (fVIII) and IX (fIX) are deficient in Hemophilia A and B, respectively [51–53]. People with mild hemophilia have 5%–40% of the normal clotting factor levels while severe hemophiliacs have <1% [53]. Hemophilia can be controlled with regular infusions of the deficient clotting factors. However, clotting factor replacement sometimes leads to the formation of fVIII and fIX inhibitors *in vivo* [54]. Alternatively, recombinant factor VIIa (rFVIIa) has been used to treat bleeding disorders [55,56] including hemophilia with and without factor VIII/IX inhibitors [57]. However, rFVIIa requires frequent administration (every 2–3 h), and many questions remain about its mechanism of action, its effective dosage [54], and its overall utility for the treatment trauma-associated hemorrhage [58]. In this study, we did not model rFVIIa-induced coagulation directly. Rather, we modeled a general trigger which initiated the extrinsic coagulation cascade. Since we identified the model using TF/FVIIa, inherent to our rFVIIa simulations (and the rate constant governing initiation) was the presence of TF. However, even with this complication, the model generated potentially useful insight into the rFVIIa mechanism of action, and its possible shortcomings especially for the treatment of hemophilia. The addition of rFVIIa directly activated thrombin through the initiation pathway. However, no amplification of the thrombin signal occurred without fVIII or fIX. Thus, the peak thrombin signal was lower than normal coagulation, the peak thrombin time was longer, and thrombin generation was eventually inhibited by the combined action of ATIII and the protein C pathway. However, as the dose of rFVIIa increased, the peak thrombin time decreased (eventually saturating around 200×nominal trigger), and the peak thrombin value increased such that the thrombin profile resembled normal coagulation. Butenas *et al.* performed an extensive *in vitro* study of rFVIIa-induced thrombin generation under normal and hemophilic conditions [59]. They found qualitatively similar trends, namely rFVIIa restored normal coagulation (even in the absence of TF) for large enough rFVIIa doses, although rFVIIa-induced coagulation in hemophilia (even for large rFVIIa doses) lagged the normal profile. These results suggest that rFVIIa administration alone might not be able to initiate normal coagulation in recurrent bleeding, unless the dosage is well above a critical threshold. However, defining this threshold, which is likely patient specific, is difficult as there is tremendous patient to patient variability even with a normal coagulation phenotype [60].

The hybrid model simulations of thrombin formation were qualitatively similar to other published coagulation models. Many mathematical models of coagulation dynamics have been built upon the Hockin-Mann model [26]. For example, Brummel-Ziedins and co-workers incorporated the Protein C (PC) pathway into the original Hockin-Mann network to investigate thrombin generation in cases of familial PC deficiency [61]. Simulations using this model showed that PC mutations caused elevated thrombin levels without changing the initiation time or the slope of thrombin generation. This trend was qualitatively captured by our reduced order model. For example, we showed decreased peak thrombin concentration in the presence of PC pathway and similar thrombin generation slopes, although the initiation time was different as these were different experimental trials. Danforth *et al.* simulated normal thrombin generation using 5 pM tissue factor with all other factors at their mean physiological level [60]. The initiation time in this simulation was approximately 4.4 min. When predicting the normal thrombin generation curve using 0.2 nM of TF/FVIIa, we showed an initiation time between 4–5 min in a platelet based system, and an initiation time of 3–4 min in a PCPS system with 5 pM TF/FVIIa, although these times were largely dictated by the time delay parameter TD. Thus, while were not able to explicitly predict initiation time, we did bracket the initiation time predicted by the Danforth study. Mitrophanov *et al.* showed, in a study exploring the mode of action of rfVIIa [62], that increasing amounts of rFVIIa accelerated the maximal slope of thrombin generation, and both the peak thrombin and initiation times. While we failed to capture the effect of rFVIIa on initiation time, we correctly predicted that changes in the rFVIIa concentration affected the maximal thrombin slope and the propagation phase. Lastly, detailed mechanistic coagulation models, for example the model by Luan and co-workers [24] or the recent model by Mitrophanov *et al.* [30], often contain hundreds of proteins and interactions. Thus, it is unlikely that the reduced order hybrid model will replicate all the rich behavior of these other models. However, for qualitatively different cases such as normal *versus* hemophilia, the hybrid approach gave similar results. For example, akin to the hybrid model, Luan *et al.* modeled the normal and hemophilia data from Allen [42]. However, unlike the previous Luan *et al.* study, we used this data for validation rather than training. Hybrid model simulations of the Allen *et al.* data set were surprisingly consistent with the Luan *et al.* model. For example, the amplification of a normal thrombin signal were comparable, and in hemophilia both correctly predicted decreased thrombin amplification. Taken together, the hybrid model performance was similar to other full scale mechanistic models in the literature, although we consistently failed to predict initiation times across data sets.

The performance of the proof of principle coagulation model was impressive given its limited size. However, there are several issues that could be further explored. First, the prediction of initiation time should be investigated. We were able to estimate initiation time within a data set, but unable to predict initiation time *across* independent data sets. This suggested that we should update the initiation module to distinguish between different triggers, e.g., TF/FVIIa *versus* rFVIIa alone, and to include key biological milestones such as FXa activation (a prerequisite to thrombin formation). Next, there are several additional biological modules that could be added to the core model presented here. First, we could include thrombin-induced platelet activation and the role of activated platelets in amplification. We captured thrombin generation data in the presence of platelets, however, the initial shape of the activation curve and the time-scale of activation was not always consistent with the data. Platelets are activated by thrombin through the cleavage of the extracellular domain of protease-activated receptors (PARs) on the platelet surface. Once activated, platelets play an important role in amplification, and are key mediators of the positive feedback driving amplification. Thus, this biology is a potentially important component of an expanded model. We should also add the intrinsic pathway to the model. The intrinsic pathway is triggered by contact activation of the plasma protease factor XI (fXI) by negatively charged surfaces and by thrombin and upstream factors such as activated plasma protease factor XII (FXIIa) [63,64]. Activated platelets may also release polyphosphate which directly activates fXII [65]. Arguably a minor player in acute bleeding, contact activation could also be important in other wound healing contexts. Finally, to make the model more clinically relevant, we should include the biochemical processes responsible for clot formation and clot dissolution (fibrinolysis). Clot formation is driven by thrombin activity, while fibrinolysis is driven by plasmin activity, a key enzyme that cleaves fibrin (one of the main materials in a clot). Similar to coagulation, fibrinolysis is managed by several activating and inhibitory factors which control the balance between clot formation and dissolution. Tissue plasminogen activator (t-PA) and urokinase activate plasmin, along with contact pathway factors such as fXIa. On the other hand, thrombin activatable fibrinolysis inhibitor (TAFI) inhibits the degradation of fibrin by plasmin. Also, similar to coagulation, there is considerable fibrinolysis and contact pathway data sets that can be used to train the model. Lastly, the choice of max/min integration rules or the particular form of the transfer functions could be generalized to include other rule types and functions. Theoretically, an integration rule is a function whose domain is a set of transfer function inputs, and whose range is $v \in [0, 1]$. Thus, integration rules other than max/min could be used, such as the mean or the product, assuming the range of the transfer functions is always $f \in [0, 1]$. Alternative integration rules such as the mean might have different properties which could influence model identification or performance. For example, a mean integration rule would be differentiable, which allows derivative-based optimization approaches to be used. The particular form of the transfer function could also be explored. We choose a Hill-like function because of its prominence in the systems and synthetic biology community. However, the only mathematical requirement for a transfer function is that it map a non-negative continuous or categorical variable into the range $f \in [0, 1]$. Thus, many types of transfer functions are possible.

4. Materials and Methods

4.1. Formulation and Solution of the Model Equations

We used ordinary differential equations (ODEs) to model the time evolution of proteins (x_i) in our reduced order coagulation model:

$$\frac{dx_i}{dt} = \sum_{j=1}^{\mathcal{R}} \sigma_{ij} r_j \left(\mathbf{x}, \epsilon, \mathbf{k} \right) \qquad i = 1, 2, \ldots, \mathcal{M} \tag{1}$$

where $\mathcal{R}$ denotes the number of reactions, $\mathcal{M}$ denotes the number of protein species in the model. The quantity $r_j \left(\mathbf{x}, \epsilon, \mathbf{k} \right)$ denotes the rate of reaction j. Typically, reaction j is a non-linear function of biochemical species abundance, as well as unknown kinetic parameters $\mathbf{k}$ ($\mathcal{K} \times 1$). The quantity σ_{ij} denotes the stoichiometric coefficient for species i in reaction j. If $\sigma_{ij} > 0$, species i is produced by reaction j. Conversely, if $\sigma_{ij} < 0$, species i is consumed by reaction j, while $\sigma_{ij} = 0$ indicates species i is not connected with reaction j. The system material balances were subject to the initial conditions $\mathbf{x} \left(t_o \right) = \mathbf{x}_o$, which were specified by the experimental setup.

Each reaction rate was written as the product of two terms, a kinetic term ($\bar{r}_j$) and a control term (v_j) that could depend upon many regulatory transfer functions:

$$r_j \left(\mathbf{x}, \epsilon, \mathbf{k} \right) = \bar{r}_j v_j \tag{2}$$

We used multiple saturation kinetics to model the reaction term $\bar{r}_j$:

$$\bar{r}_j = k_j^{max} \epsilon_i \left(\prod_{s \in m_j^-} \frac{x_s}{K_{js} + x_s} \right) \tag{3}$$

where k_j^{max} denotes the maximum rate for reaction j, ϵ_i denotes the enzyme abundance which catalyzes reaction j, and K_{js} denotes the saturation constant for species s in reaction j. The product in Equation (3) was carried out over the set of *reactants* for reaction j (denoted as m_j^-). The control term v_j depended upon the combination of factors which influenced the activity of enzyme i. For each enzyme, we used a rule-based approach to select from competing control factors (Figure 2). If an enzyme was activated by m metabolites, we modeled this activation as:

$$v_j = \max \left(f_{1j} \left(\mathcal{Z} \right), \ldots, f_{mj} \left(\mathcal{Z} \right) \right) \tag{4}$$

where $0 \leq f_{ij} \left(\mathcal{Z} \right) \leq 1$ was a regulatory transfer function that calculated the influence of metabolite i on the activity of enzyme j. Conversely, if enzyme activity was inhibited by m metabolites, we modeled this inhibition as:

$$v_j = 1 - \max \left(f_{1j} \left(\mathcal{Z} \right), \ldots, f_{mj} \left(\mathcal{Z} \right) \right) \tag{5}$$

Lastly, if an enzyme had both m activating and n inhibitory factors, we modeled the control term as:

$$v_j = \min \left(u_j, d_j \right) \tag{6}$$

where:

$$u_j = \max_{j^+} \left(f_{1j}\left(\mathcal{Z}\right), \ldots, f_{mj}\left(\mathcal{Z}\right) \right) \tag{7}$$

$$d_j = 1 - \max_{j^-} \left(f_{1j}\left(\mathcal{Z}\right), \ldots, f_{nj}\left(\mathcal{Z}\right) \right) \tag{8}$$

The quantities j^+ and j^- denoted the sets of activating and inhibitory factors for enzyme j. If a process has no modifying factors, we set $v_j = 1$. There are many possible functional forms for $0 \le f_{ij}\left(\mathcal{Z}\right) \le 1$. However, in this study, each individual regulatory transfer function took the form:

$$f_i\left(\mathcal{Z}_j, k_{ij}\right) = \frac{k_{ij}^\eta \mathcal{Z}_j^\eta}{1 + k_{ij}^\eta \mathcal{Z}_j^\eta} \tag{9}$$

where $\mathcal{Z}_j$ denotes the abundance of the j factor (e.g., metabolite abundance), and k_{ij} and η are control parameters. k_{ij} was the species gain parameter, while η was a cooperativity parameter (similar to a Hill coefficient). Applying the general framework to the reduced coagulation network resulted in five ordinary differential equations:

$$\frac{dx_1}{dt} = -\left(r_{init}v_{init} + r_{amp}v_{amp}\right) \tag{10}$$

$$\frac{dx_2}{dt} = r_{amp}v_{amp} + r_{init}v_{init} - r_{inh,ATIII}v_{inh,ATIII} \tag{11}$$

$$\frac{dx_3}{dt} = -r_{apc}v_{apc} \tag{12}$$

$$\frac{dx_4}{dt} = r_{apc}v_{apc} \tag{13}$$

$$\frac{dx_5}{dt} = -r_{inh,ATIII}v_{inh,ATIII} \tag{14}$$

where $\mathbf{x} = \left(fII, FIIa, PC, APC, ATIII\right)^T$. The terms $r_* v_*$ in the balance equations denote corrected kinetic expressions for initiation, amplification and inhibition processes. The rate of initiation $\bar{r}_{init}$ was modeled as:

$$\bar{r}_{init} = k_{init}\left(trigger\right)\frac{x_1}{K_{init,fII} + x_1} \tag{15}$$

where k_{init}, $K_{init,fII}$ are the rate and saturation constants governing initiation, respectively. The rate of initiation was modified by v_{init}, the control parameter governing initiation. Initiation was sensitive to the level of trigger (activator) and TFPI (inhibitor):

$$v_{init} = \min\left(f_{init}^-\left(TFPI\right), f_{init}^+\left(trigger\right)\right) \tag{16}$$

where the transfer functions f took the form of Equation (9). The rate of thrombin amplification was given by:

$$\bar{r}_{amp} = k_{amp}\left(x_2\right)\frac{x_1}{K_{amp,fII} + x_1} \tag{17}$$

where k_{amp}, $K_{amp,fII}$ denote the rate and saturation constants governing amplification, respectively. The amplification control term, which modified amplification rate, was modeled as a combination of multiple inhibition terms and one activation term:

$$v_{amp} = \min\left(f_{amp}^-\left(TFPI\right), f_{amp}^-\left(x_4\right), f_{amp}^+\left(\mathcal{Z}_{amp}\right)\right) \tag{18}$$

where $\mathcal{Z}_{amp} = fV \times fX \times fVIII \times fIX$. Although $f_{amp}^+ (\mathcal{Z}_{amp})$ is an activating term, we included it in the min integration rule; the factors in $\mathcal{Z}_{amp}$ were essential for amplification (if any of these factors was missing the amplification reaction would not occur). Thus, the factors in $\mathcal{Z}_{amp}$ were required components, a classification that we implemented by the min selection rule. The rate activated protein C formation was given by:

$$\bar{r}_{apc} = k_{APC,formation} (TM) \frac{x_3}{K_{formation,PC} + x_3} \tag{19}$$

where $k_{APC,formation}$ and $K_{formation,PC}$ denote the rate and saturation constants governing activated protein C formation, respectively and TM denotes the thrombomodulin abundance. We modeled the control term which governed APC formation as a single thrombin-dependent activation term:

$$v_{apc} = \max \left(f_{apc}^+ (x_2) \right) \tag{20}$$

Lastly, we included direct irreversible inhibition of FIIa by ATIII:

$$\bar{r}_{inh,ATIII} = k_{ATIII,inhibition} \left(x_5 x_2^{\gamma} \right) \tag{21}$$

where γ was estimated to be $\gamma = 1.26$. For ATIII inhibition of FIIa, the control variables $v_{inh,ATIII}$ was taken to be unity. The model equations were encoded using the Python programming language and solved using the ODEINT routine of the SciPy module [66]. The model files can be downloaded from http://www.varnerlab.org.

4.2. Estimation of Model Parameters From Experimental Data

Model parameters were estimated by minimizing the difference between simulations and experimental thrombin measurements (squared residual):

$$\min_{\mathbf{k}} \sum_{\tau=1}^{T} \sum_{j=1}^{S} \left(\frac{\hat{x}_j (\tau) - x_j (\tau, \mathbf{k})}{\omega_j (\tau)} \right)^2 \tag{22}$$

where $\hat{x}_j (\tau)$ denotes the measured value of species j at time τ, $x_j (\tau, \mathbf{k})$ denotes the simulated value for species j at time τ, and $\omega_j (\tau)$ denotes the experimental measurement variance for species j at time τ. The outer summation is with respect to time, while the inner summation is with respect to state. We minimized the model residual using Particle swarm optimization (PSO) [67]. PSO uses a *swarming* metaheuristic to explore parameter spaces. A strength of PSO is its ability to find the global minimum, even in the presence of potentially many local minima, by communicating the local error landscape experienced by each particle collectively to the swarm. Thus, PSO acts both as a local and a global search algorithm. For each iteration, particles in the swarm compute their local error by evaluating the model equations using their specific parameter vector realization. From each of these local points, a globally best error is identified. Both the local and global error are then used to update the parameter estimates of each particle using the rules:

$$\Delta_i = \theta_1 \Delta_i + \theta_2 \mathbf{r}_1 (\mathcal{L}_i - \mathbf{k}_i) + \theta_3 \mathbf{r}_2 (\mathcal{G} - \mathbf{k}_i) \tag{23}$$

$$\mathbf{k}_i = \mathbf{k}_i + \Delta_i \tag{24}$$

where $(\theta_1, \theta_2, \theta_3)$ are adjustable parameters, $\mathcal{L}_i$ denotes the local best solution found by particle i, and $\mathcal{G}$ denotes the best solution found over the entire population of particles. The quantities r_1 and r_2 denote uniform random vectors with the same dimension as the number of unknown model parameters $(\mathcal{K} \times 1)$. In thus study, we used $(\theta_1, \theta_2, \theta_3) = (1.0, 0.05564, 0.02886)$. The quality of parameter estimates was measured using goodness of fit (model residual). The particle swarm optimization routine was implemented in the Python programming language. All plots were made using the Matplotlib module of Python [68].

4.3. Global Sensitivity Analysis of Model Performance

We conducted a global sensitivity analysis, using the variance-based method of Sobol, to estimate which parameters controlled the performance of the reduced order model [69]. We computed the total sensitivity index of each parameter relative to two performance objectives, the peak thrombin time and the area under the thrombin curve (thrombin exposure). We established the sampling bounds for each parameter from the minimum and maximum value of that parameter in the parameter set ensemble. We used the sampling method of Saltelli *et al.* [70] to compute a family of $N(2d + 2)$ parameter sets which obeyed our parameter ranges, where N was the number of trials, and d was the number of parameters in the model. In our case, $N = 10{,}000$ and $d = 22$, so the total sensitivity indices were computed from 460,000 model evaluations. The variance-based sensitivity analysis was conducted using the SALib module encoded in the Python programming language [71].

Acknowledgments

This study was supported by an award from the Army Research Office (ARO #59155-LS).

Author Contributions

A.S. and J.V. conceived of and developed the coagulation model. J.V. supervised the project, and prepared the manuscript.

Conflicts of Interest

The authors declare no conflict of interest.

References

1. Kholodenko, B.; Yaffe, M.B.; Kolch, W. Computational approaches for analyzing information flow in biological networks. *Sci. Signal* **2012**, *5*, re1.
2. Gadkar, K.G.; Varner, J.; Doyle, F.J. Model identification of signal transduction networks from data using a state regulator problem. *Syst. Biol.* **2005**, *2*, 17–30.
3. Bailey, J.E. Complex biology with no parameters. *Nat. Biotechnol.* **2001**, *19*, 503–504.

4. Machta, B.B.; Chachra, R.; Transtrum, M.K.; Sethna, J.P. Parameter space compression underlies emergent theories and predictive models. *Science* **2013**, *342*, 604–607.

5. Tasseff, R.; Nayak, S.; Song, S.O.; Yen, A.; Varner, J.D. Modeling and analysis of retinoic acid induced differentiation of uncommitted precursor cells. *Integr. Biol.* **2011**, *3*, 578–591.

6. Morris, M.K.; Saez-Rodriguez, J.; Sorger, P.K.; Lauffenburger, D.A. Logic-based models for the analysis of cell signaling networks. *Biochemistry* **2010**, *49*, 3216–3224.

7. Saez-Rodriguez, J.; Alexopoulos, L.G.; Zhang, M.; Morris, M.K.; Lauffenburger, D.A.; Sorger, P.K. Comparing signaling networks between normal and transformed hepatocytes using discrete logical models. *Cancer Res.* **2011**, *71*, 5400–5411.

8. Morris, M.K.; Saez-Rodriguez, J.; Clarke, D.C.; Sorger, P.K.; Lauffenburger, D.A. Training signaling pathway maps to biochemical data with constrained fuzzy logic: Quantitative analysis of liver cell responses to inflammatory stimuli. *PLoS Comput. Biol.* **2011**, *7*, e1001099.

9. Morris, M.K.; Shriver, Z.; Sasisekharan, R.; Lauffenburger, D.A. Querying quantitative logic models (Q2LM) to study intracellular signaling networks and cell-cytokine interactions. *Biotechnol. J.* **2012**, *7*, 374–386.

10. Butenas, S.; Mann, K.G. Blood coagulation. *Biochemistry* **2002**, *67*, 3–12.

11. Schenone, M.; Furie, B.C.; Furie, B. The blood coagulation cascade. *Curr. Opin. Hematol.* **2004**, *11*, 272–277.

12. Adams, R.L.C.; Bird, R.J. Review article: Coagulation cascade and therapeutics update: Relevance to nephrology. Part 1: Overview of coagulation, thrombophilias and history of anticoagulants. *Nephrology* **2009**, *14*, 462–470.

13. *Hemostasis and Thrombosis: Basic Principles and Clinical Practice*; Goldhaber, S.Z., Colman, R.W., Clowes, A.W., Eds.; Lippincott Williams and Wilkins: Philadelphia, PA, USA, 2006.

14. Brummel, K.E.; Paradis, S.G.; Butenas, S.; Mann, K.G. Thrombin functions during tissue factor-induced blood coagulation. *Blood* **2002**, *100*, 148–152.

15. Mann, K.; Nesheim, M.; Church, W.; Haley, P.; Krishnaswamy, S. Surface-dependent reactions of Vitamin K-dependent enzyme complexes. *Blood* **1990**, *76*, 1–16.

16. Roberts, H.; Monroe, D.; Oliver, J.; Chang, J.; Hoffman, M. Newer Concepts of Blood Coagulation. *Haemophilia* **1998**, *4*, 331–334.

17. Mann, K. Biochemistry and Physiology of Blood Coagulation. *Thromb. Haemost.* **1999**, *82*, 165–174.

18. Khanin, M.A.; Semenov, V.V. A mathematical model of the kinetics of blood coagulation. *J. Theor. Biol.* **1989**, *136*, 127–134.

19. Willems, G.M.; Lindhout, T.; Hermens, W.T.; Hemker, H.C. Simulation model for thrombin generation in plasma. *Haemostasis* **1991**, *21*, 197–207.

20. Baldwin, S.A.; Basmadjian, D. A mathematical model of thrombin production in blood coagulation, Part I: The sparsely covered membrane case. *Ann. Biomed. Eng.* **1994**, *22*, 357–370.

21. Leipold, R.J.; Bozarth, T.A.; Racanelli, A.L.; Dicker, I.B. Mathematical model of serine protease inhibition in the tissue factor pathway to thrombin. *J. Biol. Chem.* **1995**, *270*, 25383–25387.

22. Kuharsky, A.L.; Fogelson, A.L. Surface-mediated control of blood coagulation: The role of binding site densities and platelet deposition. *Biophys. J.* **2001**, *80*, 1050–1074.

23. Luan, D.; Zai, M.; Varner, J.D. Computationally derived points of fragility of a human cascade are consistent with current therapeutic strategies. *PLoS Comput. Biol.* **2007**, *3*, e142.

24. Luan, D.; Szlam, F.; Tanaka, K.A.; Barie, P.S.; Varner, J.D. Ensembles of uncertain mathematical models can identify network response to therapeutic interventions. *Mol. Biosyst.* **2010**, *6*, 2272–2286.

25. Jones, K.C.; Mann, K.G. A model for the tissue factor pathway to thrombin. II. A mathematical simulation. *J. Biol. Chem.* **1994**, *269*, 23367–23373.

26. Hockin, M.F.; Jones, K.C.; Everse, S.J.; Mann, K.G. A model for the stoichiometric regulation of blood coagulation. *J. Biol. Chem.* **2002**, *277*, 18322–18333.

27. Butenas, S.; Orfeo, T.; Gissel, M.T.; Brummel, K.E.; Mann, K.G. The significance of circulating factor IXa in blood. *J. Biol. Chem.* **2004**, *279*, 22875–22882.

28. Lo, K.; Denney, W.S.; Diamond, S.L. Stochastic modeling of blood coagulation initiation. *Pathophysiol. Haemost Thromb.* **2005**, *34*, 80–90.

29. Chatterjee, M.S.; Denney, W.S.; Jing, H.; Diamond, S.L. Systems biology of coagulation initiation: Kinetics of thrombin generation in resting and activated human blood. *PLoS Comput. Biol.* **2010**, *6*, doi:10.1371/journal.pcbi.1000950.

30. Mitrophanov, A.Y.; Wolberg, A.S.; Reifman, J. Kinetic model facilitates analysis of fibrin generation and its modulation by clotting factors: Implications for hemostasis-enhancing therapies. *Mol. Biosyst.* **2014**, *10*, 2347–2357.

31. Stalker, T.J.; Traxler, E.A.; Wu, J.; Wannemacher, K.M.; Cermignano, S.L.; Voronov, R.; Diamond, S.L.; Brass, L.F. Hierarchical organization in the hemostatic response and its relationship to the platelet-signaling network. *Blood* **2013**, *121*, 1875–1885.

32. Leiderman, K.; Fogelson, A. An overview of mathematical modeling of thrombus formation under flow. *Thromb. Res.* **2014**, *133*, S12–S14.

33. Bannish, B.E.; Keener, J.P.; Fogelson, A.L. Modelling fibrinolysis: A 3D stochastic multiscale model. *Math. Med. Biol.* **2014**, *31*, 17–44.

34. Voronov, R.S.; Stalker, T.J.; Brass, L.F.; Diamond, S.L. Simulation of intrathrombus fluid and solute transport using *in vivo* clot structures with single platelet resolution. *Ann. Biomed. Eng.* **2013**, *41*, 1297–1307.

35. Papadopoulos, K.; Gavaises, M.; Atkin, C. A simplified mathematical model for thrombin generation. *Med. Eng.Phys.* **2014**, *36*, 196–204.

36. Runyon, M.K.; Johnson-Kerner, B.L.; Ismagilov, R.F. Minimal functional model of hemostasis in a biomimetic microfluidic system. *Angew. Chem. Int. Ed. Engl.* **2004**, *43*, 1531–1536.

37. Kastrup, C.J.; Runyon, M.K.; Shen, F.; Ismagilov, R.F. Modular chemical mechanism predicts spatiotemporal dynamics of initiation in the complex network of hemostasis. *Proc. Natl. Acad. Sci. USA* **2006**, *103*, 15747–15752.

38. Runyon, M.K.; Johnson-Kerner, B.L.; Kastrup, C.J.; van Ha, T.G.; Ismagilov, R.F. Propagation of blood clotting in the complex biochemical network of hemostasis is described by a simple mechanism. *J. Am. Chem. Soc.* **2007**, *129*, 7014–7015.

39. Runyon, M.K.; Kastrup, C.J.; Johnson-Kerner, B.L.; Ha, T.G.V.; Ismagilov, R.F. Effects of shear rate on propagation of blood clotting determined using microfluidics and numerical simulations. *J. Am. Chem. Soc.* **2008**, *130*, 3458–3464.

40. Butenas, S.; van't Veer, C.; Mann, K.G. Normal thrombin generation. *Blood* **1999**, *94*, 2169–2178.

41. Van't Veer, C.; Golden, N.J.; Kalafatis, M.; Mann, K.G. Inhibitory Mechanism of the Protein C Pathway on Tissue Factor-induced Thrombin Generation. *J. Biol. Chem.* **1997**, *272*, 7963–7994.

42. Allen, G.A.; Hoffman, M.; Roberts, H.R.; Monroe, D.M. Manipulation of prothrombin concentration improves response to high-dose factor VIIa in a cell-based model of haemophilia. *Br. J. Haematol.* **2006**, *134*, 314–319.

43. Kuepfer, L.; Peter, M.; Sauer, U.; Stelling, J. Ensemble modeling for analysis of cell signaling dynamics. *Nat. Biotechnol.* **2007**, *25*, 1001–1006.

44. Song, S.O.; Varner, J. Modeling and analysis of the molecular basis of pain in sensory neurons. *PLoS One* **2009**, *4*, e6758.

45. Song, S.O.; Chakrabarti, A.; Varner, J.D. Ensembles of signal transduction models using Pareto Optimal Ensemble Techniques (POETs). *Biotechnol. J.* **2010**, *5*, 768–780.

46. Tasseff, R.; Nayak, S.; Salim, S.; Kaushik, P.; Rizvi, N.; Varner, J.D. Analysis of the molecular networks in androgen dependent and independent prostate cancer revealed fragile and robust subsystems. *PLoS One* **2010**, *5*, e8864.

47. Lequieu, J.; Chakrabarti, A.; Nayak, S.; Varner, J.D. Computational modeling and analysis of insulin induced eukaryotic translation initiation. *PLoS Comput. Biol.* **2011**, *7*, e1002263.

48. Tran, L.M.; Rizk, M.L.; Liao, J.C. Ensemble modeling of metabolic networks. *Biophys. J.* **2008**, *95*, 5606–5617.

49. Hansson, G.K. Inflammation, Atherosclerosis and Coronary Artery Disease. *N. Engl. J. Med.* **2005**, *352*, 1685–1695.

50. Tanaka, K.A.; Key, N.S.; Levy, J.H. Blood coagulation: Hemostasis and thrombin regulation. *Anesth. Anal.* **2009**, *108*, 1433–1446.

51. Tuddenham, E.; Cooper, D. *The Molecular Genetics of Haemostasis and Its Inherited Disorders*; Oxford monographs in medical genetics, Oxford University Press: New York, NY, USA, 1994; Volume 25.

52. Mannucci, M.P.; Tuddenham, E.G.D. The Hemophilias–From royal genes to gene therapy. *N. Engl. J. Med.* **2001**, *344*, 1773–1780.

53. Mitchell, J.; Phillott, A. Haemophilia and inhibitors 1: diagnosis and treatment. *Nurs. Times* **2008**, *104*, 26–27.

54. Tomokiyo, K.; Nakatomi, Y.; Araki, T.; Teshima, K.; Nakano, H.; Nakagaki, T.; Miyamoto, S.; Funatsu, A.; Iwanaga, S. A novel therapeutic approach combining human plasma-derived factors VIIa and X for haemophiliacs with inhibitors: Evidence of a higher thrombin generation rate *in vitro* and more sustained haemostatic activity *in vivo* than obtained with factor VIIa alone. *Vox Sanguinis* **2003**, *85*, 290–299.

55. Hedner, U. Factor VIIa and its potential therapeutic use in bleeding-associated pathologies. *Thromb. Haemost* **2008**, *100*, 557–562.

56. Talbot, M.; Tien, H.C. The use of recombinant factor VIIa in trauma patients. *J. Am. Acad. Orthop. Surg.* **2009**, *17*, 477–481.

57. Shapiro, A.D. Single-dose recombinant activated factor VII for the treatment of joint bleeds in hemophilia patients with inhibitors. *Clin. Adv. Hematol. Oncol.* **2008**, *6*, 579–586.

58. Duchesne, J.C.; Mathew, K.A.; Marr, A.B.; Pinsky, M.R.; Barbeau, J.M.; McSwain, N.E. Current evidence based guidelines for factor VIIa use in trauma: The good, the bad, and the ugly. *Am. Surg.* **2008**, *74*, 1159–1165.

59. Butenas, S.; Brummel, K.E.; Branda, R.F.; Paradis, S.G.; Mann, K.G. Mechanism of factor VIIa-dependent coagulation in hemophilia blood. *Blood* **2002**, *99*, 923–930.

60. Danforth, C.M.; Orfeo, T.; Everse, S.J.; Mann, K.G.; Brummel-Ziedins, K.E. Defining the boundaries of normal thrombin generation: Investigations into hemostasis. *PLoS One* **2012**, *7*, e30385.

61. Brummel-Ziedins, K.E.; Orfeo, T.; Callas, P.W.; Gissel, M.; Mann, K.G.; Bovill, E.G. The prothrombotic phenotypes in familial protein C deficiency are differentiated by computational modeling of thrombin generation. *PLoS One* **2012**, *7*, e44378.

62. Mitrophanov, A.Y.; Reifman, J. Kinetic modeling sheds light on the mode of action of recombinant factor VIIa on thrombin generation. *Thromb. Res.* **2011**, *128*, 381–390.

63. Naito, K.; Fujikawa, K. Activation of Human Blood Coagulation Factor XI Independent of Factor XII. *J. Biol. Chem.* **1991**, *266*, 7353–7358.

64. Gailani, D.; Broze, G.J., Jr. Factor XI activation in a revised model of blood coagulation. *Science* **1991**, *253*, 909–912.

65. Smith, S.A.; Mutch, N.J.; Baskar, D.; Rohloff, P.; Docampo, R.; Morrissey, J.H. Polyphosphate modulates blood coagulation and fibrinolysis. *Proc. Natl. Acad. Sci. USA* **2006**, *103*, 903–908.

66. Jones, E.; Oliphant, T.; Peterson, P. SciPy: Open source scientific tools for Python. Available online: http://www.scipy.org/ (accessed on 13 March 2015).

67. Kennedy, J.; Eberhart, R. Particle swarm optimization. In Proceedings of the International Conference on Neural Networks, Perth, WA, Australia, 27 November–1 December 1995; pp. 1942–1948.

68. Hunter, J.D. Matplotlib: A 2D Graphics Environment. *Comput. Sci. Eng.* **2007**, *9*, 90–95.

69. Sobol, I. Global sensitivity indices for nonlinear mathematical models and their Monte Carlo estimates. *Math. Comput. Simul.* **2001**, *55*, 271–280.

70. Saltelli, A.; Annoni, P.; Azzini, I.; Campolongo, F.; Ratto, M.; Tarantola, S. Variance based sensitivity analysis of model output. Design and estimator for the total sensitivity index. *Comput. Phys. Commun.* **2010**, *181*, 259–270.

71. Herman, J.D. Available online: https://github.com/jdherman/SALib (accessed on 13 March 2015).

A Quantitative Systems Pharmacology Perspective on Cancer Immunology

Christina Byrne-Hoffman and David J. Klinke II

Abstract: The return on investment within the pharmaceutical industry has exhibited an exponential decline over the last several decades. Contemporary analysis suggests that the rate-limiting step associated with the drug discovery and development process is our limited understanding of the disease pathophysiology in humans that is targeted by a drug. Similar to other industries, mechanistic modeling and simulation has been proposed as an enabling quantitative tool to help address this problem. Moreover, immunotherapies are transforming the clinical treatment of cure cancer and are becoming a major segment of the pharmaceutical research and development pipeline. As the clinical benefit of these immunotherapies seems to be limited to subset of the patient population, identifying the specific defect in the complex network of interactions associated with host immunity to a malignancy is a major challenge for expanding the clinical benefit. Understanding the interaction between malignant and immune cells is inherently a systems problem, where an engineering perspective may be helpful. The objective of this manuscript is to summarize this quantitative systems perspective, particularly with respect to developing immunotherapies for the treatment of cancer.

Reprinted from *Processes*. Cite as: Byrne-Hoffman, C.; Klinke, D.J., II. A Quantitative Systems Pharmacology Perspective on Cancer Immunology. *Processes* **2015**, *3*, 235-256.

1. Introduction

Motivated by a desire to improve human health, the pharmaceutical industry leverages biological discoveries to develop drugs that aim to restore health at significant financial investment. Overall the costs associated with pharmaceutical research and development, as represented by the US share of the market, has been increasing exponentially. The current estimate to bring a new medical entity to market requires upwards of approximately a $1–2.5 billion investment in research and development [1,2]. To recoup these financial investments, pharmaceutical companies are provided with protection from competition for a limited time by patenting their inventions. However, the estimated return on investment by the pharmaceutical industry has been experiencing an exponential decline in the last several decades. This trend, sometimes referred to as the innovation gap, presents a challenge for economic sustainability of the current model for innovation within the pharmaceutical industry.

The cost of pharmaceutical development escalates as drugs progress further from bench to market. In particular, Phase II clinical trials have become a key pinch point in the research and development pipeline, as it combines both significant risk and cost. This phase is the first time the efficacy of the drug is tested in real patients within the target disease and therefore has the highest probability of failure. It is also one of the phases with the highest out-of-pocket cost for the developer and other stakeholders [3]. Clinical trials are predicated largely on positive pre-clinical studies using animal models of disease. Unfortunately, animal modeling, which presents lower financial and human health hazards, may not always recreate the specific molecular and cellular networks associated with the pathophysiology and adverse reactions in human subjects [4]. Therefore, to lessen the costs of development and risk to humans in clinical trials, it is important to use appropriate models of the disease to maximize efficacy, safety, and benefit to patients.

In the last 50 years, computer-aided modeling and simulation has transformed a variety of industries, including financial portfolio management and the aerospace industry. Modeling and simulation in the financial sector has enabled real-time evaluation of economic performance measures using a mathematical model of the particular business sector to predict future performance and to optimize financial return [5]. In the aerospace industry, modeling and simulation is used to design new airframes, which eliminates the need for multiple physical prototypes constructed at intermediate points during design and reduces the time from concept to production [6]. In both cases, mathematical modeling and simulation provide a quantitative framework to capture our conceptual understanding of the modeled process and interpret heterogeneous data acquired from the process. These two examples also represent extremes of our conceptual understanding. Financial markets are complex systems that are influenced by a variety of observed and unobserved factors. Assuming that the underlying structure of the market is not changing, future behavior can be predicted using empirical mathematical models that are constructed using historical data. At the opposite end of the spectrum, computer-aided design of airframes captures physical principles such as the conservation of mass, which implies two physical objects cannot occupy the same space, and the governing physics associated with the performance objectives of the airframe. Similar to these industries, mathematical modeling and simulation has been proposed as approach to improve our understanding of the biological mechanisms targeted by a particular therapy [7]. This could help predict the outcomes of human clinical trials and, thereby, help bridge the innovation gap between cell and animal models and human pathophysiology, while also providing a cost savings in development, as computationally "expensive" modeling is inherently more cost effective than additional physical and biological models of disease [8]. In a 2011 National Institutes of Health White Paper, recommendations for quantitative systems pharmacology using quantitative experimental studies and model-based computational analyses that also incorporate clinical "omics data" were given in hopes of addressing clinical Phase II study failures in drug development and physiological, chemical, and biological disconnects in preclinical research [9].

Given the oncology slice of pharmaceutical research and development and the recent shift towards immunotherapies for cancer, the objective of this review is to summarize how modeling and simulation has aided in the understanding of biological changes associated with oncogenesis as it relates to immunity. In subsequent paragraphs, we provide a brief overview of the cancer and immune systems, tumor somatic and clonal evolution properties, and how cancer cell heterogeneity complicates therapeutic aims. We will also discuss recent immunotherapeutic advancements and the computational models used to describe the interactions between cancer and the immune system.

2. Emerging View of Cancer as a System

Oncogenesis is attributed to the accumulation of genetic mutations that lead to uncontrolled cell growth and proliferation. These mutations alter function of the modified gene through overexpression of the corresponding protein or rearrangement of a gene to create an entirely new protein that has dysregulated activity [10]. Mutations in specific genes that can, in isolation, transform a normal cell into a malignant cell are called oncogenes. Cancer drug development over the past several decades has been focused on targeting these oncogene mutations by inhibiting the function of corresponding proteins using small molecule drugs [11]. Researchers have scrutinized the altered signaling pathways in malignant cells in hopes of finding the key protein conserved in oncogenesis and metastasis but that plays minimal role in normal cells [12]. However, these drugs are rarely as efficacious in the clinic, where de novo and emergent drug resistance is common [13].

The small molecule inhibitor segment of the pharmaceutical industry is also associated with a view of cancer as a disease driven by malignant alterations that are intrinsic to or driven by the cancer cell [14,15]. This view can be represented by the six hallmarks of cancer discussed by Hanahan and Weinberg 2000 [16]. The six hallmarks summarize how genetic alterations change how a malignant cell senses and responds to extracellular signals in ways detrimental to the host. Assuming cancer is driven by the autonomous actions of malignant cells, the *in vitro* study of a cell line can be an appropriate model for identifying new therapeutic leads. This idea underpins using a collection of cell lines as a way to screen drugs that inhibit cell proliferation of exhibit cytotoxic activity in a high-throughput manner, such as the NCI-60 [17–20].

Previously to recent breakthroughs in immunotherapy, small molecule inhibitors were standard of care for many non-resectable metastatic diseases. B-Raf, a Raf kinase member of the MAP Kinase/ERK signaling pathway and involved in cell growth and proliferation, is a commonly mutated gene in many human cancers, such as metastatic melanoma. Vemurafenib and Dabrafenib are two FDA-approved B-Raf inhibitors used in the clinic that targets cancer with the B-RAF V600E (valine at amino acid position 600 to glutamic acid) mutation. However, cancer cells without the V600E B-Raf mutation may proliferate more in response to the vemurafenib drug [21]. Additionally, most metastatic melanoma patients become chemoresistant to both of these B-Raf inhibitors within 6 to 7 months of treatment. Therefore, combination therapies, such as vemurafenib with MEK-inhibitors like FDA-approved trametinib, are preferred to overcome the resistance mechanism in advanced melanoma, and may extend progression-free survival in patients by an average of about 3 months [22]. However, resistance will eventually reoccur and the patient will fatally relapse.

More recently, cancer research has expanded to include factors external to the malignant cell that contribute to oncogenesis. In 2011, Hanahan and Weinberg updated the hallmarks to include four new "emerging hallmarks and enabling characteristics", which focus on changes associated with the malignancy that alter interactions among cells of the host [23]. The immune system was identified as having an influential role on tumor progression, and changes in metabolism and inflammation in the tumor microenvironment and throughout the body are known to have an effect on clinical outcomes, as well. This shift in perspective represents a malignancy as part of an integrated but dysfunctional system, rather than as an isolated mass of malignant clones. By incorporating the emerging hallmarks into the collective understanding of carcinogenesis, the ability of a malignant cell to manipulate its local environment and the immune system that it interacts with are being recognized as integral to tumor development and support the description of cancer as an evolutionary process (Figure 1). The ability of a malignant cell to maintain the tumor microenvironment hinges on dysfunctional intercellular communication [24,25].

A tumor is comprised of a variety of cell types, including a heterogeneous collection of malignant clones, various stromal cells that provide nutrients and facilitate remodeling of the extracellular matrix, and immune cells. Heterogeneity among malignant clones can exist within various morphologies or cellular phenotypes, producing cells originating from a similar origin, but may yet exhibit various structural, gene expression, signaling network, proliferative, metabolic, and metastatic differences (Figure 1). The existence of heterogeneity is a key element of evolutionary process. Somatic evolution is thought to change the dynamics of tissues [26,27], while evolutionary processes also maintains their own dynamics [28], both of which can influence development of heterogeneity within a particular cellular population. Resolving the issues surrounding multiple dynamics of biological systems and the influence of the immune and other cellular systems on tumors is technically difficult to replicate, particularly in *in vitro* models of cancer. The dynamics of cell communication can also influence the hallmarks of cancer of a potential malignancy and emerging hallmarks can alter the fitness, adaptive, or phenotypic landscapes [29–31]. Timing of response to signaling, especially that of the immune system, can mean the difference between malignancy proliferation and tumor eradication [32].

As part of the evolutionary process, recognizing the heterogeneity of the environment is crucial to understanding how alterations in cell signaling and immune response may promote tumors. Cancer cell heterogeneity within some tumors is linked to epigenetic differences in tumor cell genomes, perhaps from variances in cancer stem cells, while other examples of divergence from a single phenotype can be accounted for through clonal evolution, or even a combination of the two models of tumor propagation [33,34]. Therefore, we see that clonal evolution and heterogeneity are directly proportional to one another and will determine impact of immunological eradication or pharmaceutical treatment of the tumor. The remainder of this section will discuss models of somatic and clonal evolution and influence of heterogeneity of the tumor microenvironment on metastatic progression from a systemic perspective of cancer.

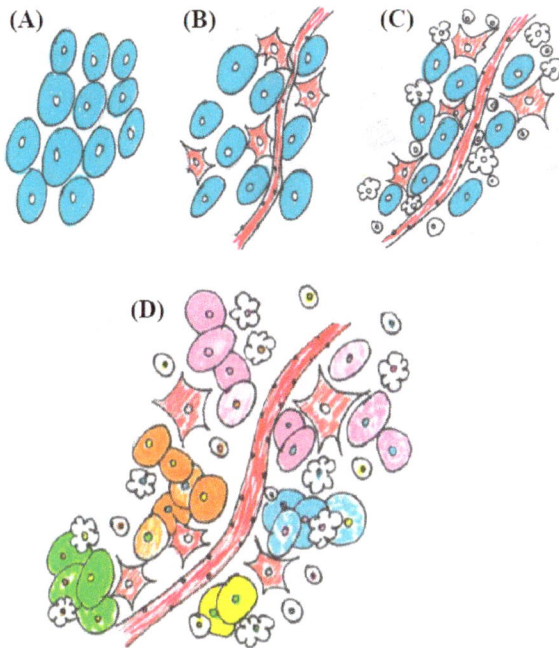

Figure 1. An illustration of the conceptual progression of the tumor microenvironment from a reductionist view to cancer as a dynamic system. (**A**) A highly reductionist view. Tumors viewed as a homogenous population malignant clones; (**B**) Cancer cells integrated into the circulatory system, as represented by including blood vessels and fibroblasts, as seen in Hanahan and Weinberg 2000; (**C**) The tumor microenvironment becomes more complex. In addition to blood vessels and fibroblasts, immune cell types of various kinds are introduced into the system, as seen in Hanahan and Weinberg 2011; (**D**) The emerging view of a malignancy as a heterogeneous and dynamic system. Not only are blood vessels, fibroblasts, and immune cells in the tumor microenvironment, but there are distinct clonal populations of cancer cells intermingled with a variety of different immune cells and other cell types.

2.1. Somatic Evolution

Somatic evolution, as distinct from classical Darwinian evolution, is a driving force for carcinogenesis [35]. The majority of somatic mutations within the genome are neutral, but accumulations of even non-deleterious mutations can alter the fitness landscape of an organism via genetic drift [36,37]. Changes in the fitness landscape include alterations in cellular communication [38], such as upregulation or downregulation of receptors, intermediate players, or downstream products, alterations in pathway activity, or mutations in pathway protein partners. This can promote selection of advantageous clones [39]. More importantly, these malignant clones can change the fitness landscape in such a way that the local microenvironments become a favorable niche for malignant cells. A particular case of somatic evolution is clonal evolution, which considers how the environment influences the selection of pre-existing clones within a heterogeneous population and neglects the influence of mutagens to further diversify the clonal population. Somatic and clonal evolution may provide many benefits to a cell. Changes in the environment of the cell may force it down a particular path, providing a fitness advantage for growth and proliferation. While most of these particular mutations may provide benefit for survival of the host, these alterations can promote further genomic instability, leading to tumorigenesis.

Reductionist theories and experimentation usually examines one input and one output of a particular system, and therefore may be inadequate in describing cell processes with appropriate models, especially in relation to somatic and clonal evolution within a population. A cell is ever-sensing of its microenvironment, ready to respond to relevant extracellular signals present in its vicinity, whether the cues are coming from other cell types, other tissues, or from itself. These cues can also have a profound impact on the evolution of a cellular population, depending on duration, strength, and ability of the cell machinery to respond to the signal. This communication may come in the form of a soluble ligand or direct cell contact and provide positive or negative feedback to the cell.

As a heterogeneous environment, a variety of signals are occurring and being processed by the target cell at any time and are difficult to observe with traditional models. Various factors, driven by cellular communication, can affect the fitness of a particular population of tumor cells, such as availability of nutritional or metabolic resources, presence and strength of immune system influences, and ability of the cell to interact with the extracellular matrix. To combat this hurdle, development of a hybrid-discrete continuum (HDC) model to examine tumor morphology and metastatic potential was developed by Anderson *et al.* [40], which leverages the impact of multiple factors of the tumor microenvironment on somatic or clonal evolution. This mathematical model allows for "random" influence of particular set of variables, while retaining deterministic properties of known inputs with known outputs.

Related to factors surrounding use of the HDC computational model is the "clonal dominance theory" developed by Kerbel and colleagues. While clonal dominance was easily observed *in vivo* in mammary tumors, replication under ideal growth conditions was difficult. The group examined the growth of two subpopulations, non-metastatic SP1 and metastatic variant C1, of cells under both ideal and non-optimal growth conditions. Under optimal conditions, both subpopulations of cells grew at the same rate, while under the "stressed" conditions, emergence of the "dominant" metastatic subpopulation C1 became evident. In addition, transforming growth factor-beta (TGFβ) and extracellular matrix-driven cell-to-cell communication between the two subpopulations was indicated as being required for assistance of the dominant metastatic subpopulation to maximum capacity [41]. This study highlighted the importance of culture conditions and physiological dimensions, such as serum concentration and spheroid growth, for *in vitro* work to best replicate *in vivo* tumors, given the importance of cell-to-cell communication in tumor development and maintenance. However, researchers often overlook the effect of these conditions on clonal evolution, and therefore, how these conditions may affect whether a certain model is appropriate for judging efficacy of a treatment in *in vitro* or *in vivo* experiments.

Other features that impact clonal dominance in the evolution of cancer include consideration of compartment size. All tissues can be viewed as "compartments". Clonal expansion readily occurs within these compartments and can be restricted by neighboring clones [42]. As shown by Michor *et al.* in a mathematical modeling study, clonal expansion within a small compartment is driven by genetic drift and is quickly dominated by a certain genetically unstable subpopulation. However, the expansion is readily limited by boundaries of the compartment and is therefore likely to be contained. Conversely, large compartments will contain a variety of subpopulations, some tumorigenic; however, the effect of clonal expansion is diluted unless both alleles of a particular gene are affected, or selective advantage favors the tumorigenic subpopulation [43]. Once a tumorigenic subpopulation has taken over the compartment, success of the neoplasm and resulting metastasis is a more probable scenario. This highlights the importance of cell-communication through cell-to-cell contact and local regulation of homeostasis in limiting genomic instability. Experimental cell models may reflect these effects, and therefore cellular confluency *in vitro*, or tumor transplant location *in vivo* may affect treatment outcomes.

Driver mutations and passenger mutations are another area of investigation in regards to computational models of the evolution of cancer. Driver mutations result in functional or morphological differences of cells, while passenger mutations are a neutral consequence of the evolutionary process [44]. In a meta-analysis study of lung and ovarian cancer, Youn and Simon found that metastasis as a result of cancer evolution was related to the age of the tumor and number of cell generations. The difference in metastatic potential as a function of generation or age of the tumor was related to the ability of the parental lineage to self-renew. For example, more generations were required for ovarian cancer to metastasize *versus* the lung cancer cell, which self-renews more gradually. The study was conducted by using passenger mutations, which do not confer a selective growth advantage in clonal expansion of the tumor, to estimate when the driver mutation, developed in the early stages of tumorigenesis and linked to causation of the tumor, had occurred and therefore, the approximate generational age of the tumor [45]. This study highlights the tumor progression differences in cell line or cancer types are largely attributed to clonal differences and how this may have an effect on treatment outcomes, not only in cell and animal model experiments, but also in the clinic.

2.2. Heterogeneity of the Tumor Microenvironment

We have examined the effects of clonal expansion in tumorigenesis and how compartment size and tissue dynamics may affect this process. Clonal expansion implies a somewhat homogenous subpopulation of cells dominates a certain microenvironment, compartment, or tissue. However, the heterogeneity of tumor microenvironments and tissues is well described [46]. Not only does the tumor stroma contain different cell types, such as immune cells, and structural support cells, but tumor cells themselves are part of a heterogeneous population. Different subpopulations of various clones exist, each expressing a variety of biomarkers and each clone with their own signaling dynamics, within the total tumor cell population.

Analysis of National Cancer Institute (NCI) and National Human Genome Research Institute (NHGRI) TCGA (the Cancer Genome Atlas) Project has revealed that sequencing of human tumors has provided a wealth of genomic information. The goal of this project was to determine common genetic alterations in various cancer types and subtypes, and use this information to improve clinical outcomes and tailor therapies to individuals. However, the impact of this data on improving clinical outcomes or identifying successful drug targets within corresponding patient groups remains limited. While the TCGA project was conceptually born out of the oncogene perspective, recognizing that tumors contain a variety of heterogeneous tumor clonal populations and prevalence within these clonal subsets of varying abundance could have a profound impact on the information gathered and possibly skew data interpretation [12]. Molecular-targeted therapies focusing on one set of genetic alterations would have little effect on a subpopulation expressing an entirely different set of alterations, as seen within the heterogeneous tumor. This may be one reason combination therapies have been so effective in treating certain types of cancer, as seen with the synergistic effect of p53 vaccines and chemotherapy in an evolutionary double bind study conducted by Anderson and colleagues [47].

As discussed previously, clonal evolution plays a critical role in the development of a tumor, but tumor heterogeneity may be a critical aspect for tumor persistence. Different tumor clonal populations may allow for evasion of the immune system. If tumor antigens produced vary from cell to cell within the tumor, a targeted immune system response will become diluted, allowing immunoescape of the tumor. This type of clonal selection provides for a phenomenon described as immunoselective pressure. By allowing for targeted killing of a specific tumor cell population expressing a given antigen, other tumor cell populations may be able to fill the tumor niche, especially cell populations that do not express an immunogenic antigen. Tumor cells sensitive to the inflammatory response, generalized apoptosis signals, or other factors in the microenvironment, may also be eliminated, while more resistant cells are allowed to proliferate. Additionally, cytokines may provide selective pressure to develop mutations that result in clones that are able to overcome these immune signals targeting the cells for destruction [48]. This phenomenon is called immunoediting. A deadly combination of elimination of immune-sensitive tumor cells, equilibrium of tumor cells that have survived elimination, and escape of resistant clones limits tumor eradication by the immune system [49].

Clonal heterogeneity can also provide survival advantages depending on the location of the clonal subset within the tumor microenvironment. Mutations that promote angiogenesis, for example, would provide a survival advantage in a tumor region not within immediate contact of the vasculature and perfusion of blood and nutrient resources. Alternatively, mutations that upregulate multidrug resistance transporters would provide a survival advantage in areas of the tumor that is most susceptible to contact with chemotherapeutic agents, such as in the periphery of the tumor that is most accessible to drug delivery. To explore this relationship between tumor heterogeneity and persistence, a computational study by Michor and colleagues examined the effect of immune system response and chemotherapeutic interventions in tumor escape [50]. They developed a mathematical model to describe original tumor cell and variant tumor cell fitness and number, mutation rate of the cancer cells, competition between variants, tumor cell elimination rate, interactions between tumor cells and cytotoxic T lymphocytes (CTLs), and the proliferation and decay of CTLs. They found that the more variants that exist within a tumor, the more likely the tumor will catastrophically or partially escape from immune system or chemotherapy attack. However, a certain level of homeostasis in variation must be maintained to support a functional genome for survival of the tumor.

In summary, a tumor is not only heterogeneous in its various clonal populations and expression of various tumor antigens, but is also heterogeneous in terms of weaknesses. By directing therapy at several weaknesses at once, instead of one at a time, or by targeting the interplay between tumor and other systems it interacts with, this lessens the probability of a certain clonal population gaining proliferative momentum in the tumor niche. Tumor heterogeneity is extremely important for the success of a tumor cell in the microenvironment and relies on the dynamics between the immune system, rates of genetic alterations, and sensitivity of variants to chemotherapeutic or immune system attack.

3. The Re-Emergence of Cancer Immunotherapy

The immune system has long been suspected as influential on cancer development. In the early 1890s, Dr. William Coley noticed a reduction in tumor size of patients with bacterial infections. He began injecting live bacteria into the tumors, and later developed a safer concoction, termed "Coley's toxins", with mixed results [51]. Over time, the field shifted away from cancer immunology research, as cytokines and immune cell types became well defined, yet cancer treatment with interleukins and other cytokines yielded variable results. Interleukin-2 showed efficacy in treating metastatic melanoma. However, its adverse side effects, inefficacy in large tumors, and high rates of toxicity such as allergic reactions and seizures discouraged widespread use in patients [52–55]. Interleukin-12 was also investigated as a cancer drug for its robust anti-angiogenic and immune cell-promoting activity. Although it had shown promise in preclinical trials, a patient death in a study designed to test safety diminished enthusiasm to pursue systemic delivery of IL-12 as a potential therapeutic [56,57].

Most of the failures of immune cell-promoting cytokines to eradicate tumors in patients could be attributed to inappropriate animal models and the lack of target specificity of these immunotherapeutic agents, as seen with the inability of cytokines to possess the sensitivity and specificity to function as appropriate biomarkers [58]. In an effort to gain specificity, antibodies, such as Rituximab, were developed and approved for the treatment of cancer [59]. This development of antibodies to target cancer has continued to this day, with the break-through of ipilimumab in metastatic melanoma [60]. Through this process of repeated success and failures, the view of cancer immunology has changed, and once considered a flame extinguished, has reignited from the embers.

3.1. The Cancer Immunology System and Systems Therapeutics

The view of cancer and the immune system as an integrated system is an important perspective to develop additional targets for cancer immunotherapy. Being cognizant of the interplay of systems is the first step to developing an appropriate model to test hypotheses. As described by Chen and colleagues, the "cancer-immunity cycle" involves an intricate process of cancer cell antigen recognition by the immune system, immune cell trafficking and tumor infiltration, and a localized immune response to eradicate the tumor [61]. At each step of the process, a variety of cytokines may be expressed to promote or inhibit the immune response. Checkpoints in this process exist to support a balance of effective activity to maintain tissue homeostasis and discourage fluctuations that could lead to the extremes of immunosenescence or autoimmunity. The following paragraphs, we will discuss oncolytic viruses, adoptive cell transfer, chimeric antigen receptors, and immune checkpoint modulators, in context of their role in the cancer-immunity cycle. Each is able to alter feedback mechanisms to promote the cancer-immunity cycle via various methods. The advantages and disadvantages of each of these therapies will also be discussed.

3.1.1. Oncolytic Viruses

Oncolytic viruses are used to target and kill cancer cells. There are two methods used to target the oncolytic viruses to kill only cancer cells, and they can be used in combination to promote efficacy of tumor-targeted killing. Transductional targeting of oncolytic viruses entails modifying the viral coat proteins of the virus to target malignant cells preferentially by inhibiting the entry of the virus into non-cancerous cells. Non-transductional targeting genetically alters the virus so that it may only replicate in the targeted cancer cell, whereby tumor-specific transcriptional promoters are used [62]. However, the host immune response to oncolytic viruses may vary between individuals, and mechanisms of resistance to and effective delivery of oncolytic viruses are poorly understood [63]. The variation among patients in immune response to oncolytic viruses provides a significant clinical barrier to their use and efficacy. However, oncolytic viruses may be used as an experimental tool to provide insight to mechanisms of cancer evolution and immune system response in virally-mediated cancers.

3.1.2. Adoptive Cell Transfer and Chimeric Antigen Receptors

A number of approaches have been proposed to jumpstart the cancer-immunity cycle and maintain its efficacy. Adoptive Cell Transfer (ACT) expands T cell immunity to a particular cancer antigen in the patient through isolation of T lymphocytes, population expansion, and reinfusion. Cells can be genetically modified to recognize certain antigens, to infiltrate tumors more readily, or to respond more robustly to cytokine cues [64]. One example of genetic modifications that can be integrated into T lymphocytes in ACT involves chimeric antigen receptors (CARs). These receptors are engineered to recognize a specific antigen, which allows this treatment method to become tailored to an individual type of cancer and the antigens it expresses [65]. Use of CARs in conjunction with ACT also overcomes a potential barrier with ACT and other immune modulation therapies, which assumes that a patient's T lymphocytes or that of a donor will recognize the patient's cancer cells and target them for killing. As shown by Chacon *et al.* [64], using tumor-infiltrating CD8+ lymphocytes in ACT can also have an impact on the rest of the tumor microenvironment, expanding a dynamically regulated and more competent T cell population for killing of the tumor.

ACT and CARs can be modeled mathematically in two ways: (1) through mathematical modeling of the system effects and response of directed ACT and CAR activity; and (2) by modeling molecular level CAR interactions. When designing CARs, it is important to remember the effect of this type of therapy on the host's system. The plasmids used to express classical CARs use a signaling fragment, an extracellular spacer, a co-stimulating domain, and an antibody to direct a specific response against the tumor. However, using single-chain fragment variable (scFv) antibodies can cause activation of the immune system in an undesirable manner. Single variable heavy chain domains (VHH) have been shown to avoid immunogenicity. However, changing any component of the CAR can lead to poor interaction of the receptor with its target. To improve CAR targeting, molecular modeling can be used to predict interactions between targets and the CAR with various components, such as a VHH directed against a particular target [66]. At the system level, a recent study by James and colleagues modeled the target lysis achieved by alterations in chimeric T cell receptor (cTCR) expression density, target antigen density, and activation of the cTCR. They found that approximately 20,000 cTCRs per cell was an ideal density of expression of the surface of the T cell. Anything above this did not increase target lysis or increase sensitivity to the target antigen, and anything below this expression impaired target lysis activity, perhaps by causing antigen-induced T cell death [67].

3.1.3. Immune Checkpoint Modulators

Based on remarkable clinical success, the current immunotherapies that hold the most promise are immune checkpoint modulators. Immune checkpoints are considered natural negative feedback mechanisms that limit an adaptive immune response to minimize the risk of autoimmunity. Investigation of immune checkpoint modulators has been on the rise in academia, biotechnology group, and pharmaceutical company research [68]. It is also known that in late stage cancers, the immune system appears to be turned "off". Part of this "immune switching" phenomenon can be credited to the tumor actively evading the immune response through its clonal evolution of certain cellular subpopulations [69]. As an example, metastatic melanoma is arguably one of the hardest cancers to treat and has very few treatment options in the clinic. As an immunogenic cancer, it is known the immune system plays a role in progression of the disease [70]. Therefore, the immune system was targeted as a potential solution to molecular inhibitor resistance in metastatic melanoma and other cancers. The particular proteins currently being examined for targeting include CTLA-4, cytotoxic T-lymphocyte antigen-4, and PD-1, programed death receptor-1, two regulatory molecules on the surface of T cells [71].

A neutralizing antibody against CTLA-4, called ipilimumab, improved overall and progression-free survival in metastatic melanoma [72] and was the first FDA-approved therapy to target cancer through an immune checkpoint blockade. While clinical benefit was observed in only a subset of patients, ipilimumab demonstrates an important proof-of-principle especially in metastatic melanoma. Anti-CTLA4 immunotherapy, such as found with ipilimumab, is able to target senescence of CD8+ cytotoxic and CD4+ T cell populations. It does this by blocking the effect of CTLA4, a T cell inhibitory molecule similar in function to costimulatory protein CD28, on the surface of the T lymphocyte that limits clonal expansion [73]. As this therapy globally increases T cell numbers, serious adverse side effects, such as autoimmunity, are directly linked to this function, by causing over-activation of the immune system. Ipilimumab is currently being investigated for potential treatment of other cancer types, while other immunotherapies targeting T cell response are also being tested for efficacy against melanoma and other cancer types to improve efficacy and safety [68].

Similar to the actions of CTLA-4, PD-1 is also an inhibitory protein expressed on the surface of chronically activated T cells, which show decreased TCR-mediated proliferative and cytokine-releasing ability. Its expression is increased, compared to peripheral blood cells, on tumor-infiltrating T-lymphocytes [74,75]. PD-1, along with its ligand that is also upregulated in the tumor microenvironment, PD-L1, is currently being investigated as a potential immunotherapy target in clinical trials. An anti-PD-L1 antibody was administered to patients with melanoma, colorectal, renal-cell, ovarian, pancreatic, gastric, breast, or non-small-cell lung cancers. Patients were treated for an average of 12 weeks and evaluated. A durable tumor regression response was recorded at 6%–17%, while 12%–41% of patients experienced prolonged disease stabilization over a period of 24 weeks [76]. Drugs targeting PD-1 or PD-L1 seem to have a slightly better safety profile compared with anti-CTLA4 and may prove to be more effective as combination therapies [32,77].

3.1.4. Summarizing the Rise of Immunotherapies

Overall, immune system modulators, ACT therapies, and oncolytic viruses all show promise clinically, but their considerable development and clinical costs, safety profiles, and limited efficacy in certain patient populations are a substantial obstacle to a broad clinical impact on patient response and management. However, these technologies and molecular targets may not only be useful as therapeutics to treat patients in the clinic, but can be used as preclinical tools to investigate and better define the interplay between cancer and immunology systems for further pharmaceutical development. To ensure these tools are used in a way that minimizes financial and patient risk in pharmaceutical development, it is necessary to use these tools in appropriate preclinical models for investigation and success of future cancer immunotherapies. In the next section, the use of appropriate system models that should be used with these potential preclinical and clinical tools for cancer treatment will be discussed.

3.2. Modeling the System for Pharmaceutical Aims

Knowing the particular causal suppression mechanisms at work in a cancer from observations of biological state is one of the most pervasive problems in the analysis of physiological systems. In engineering, this problem is called an identification problem, where causal relationships between system elements are inferred from a set of input cues and output responses [78]. In context of cancer, an input cue may be antibodies against tumor-specific epitopes and an output response may be tumor regression. Many approaches exist for the identification of simple-input-simple-output (SISO) systems—where a change in input causes a unique change in output.

As a consequence of reductionist methods, there is a wealth of experimental data that characterize how isolated elements of physiological systems respond to inputs. However, approaches for identifying causal relationships among elements of more complex integrated closed-loop systems, like the immune system, are less well developed. Typically, a closed-loop system is defined as a multi-element system where the output (*i.e.*, response) of one element provides the input (*i.e.*, biochemical cue) to another element. A schematic diagram of a closed-loop system comprised of two cell types is shown in Figure 2. Closed-loop systems are particularly challenging as it is impossible to identify the relationships among cells of a system based upon overall input (e.g., tumor vaccines) and output (e.g., tumor regression) measurements. One of the reasons for this is that changes in the internal state of the system may alter the response of the system to a defined input, such that there is not a direct causal relationship between overall system input and output.

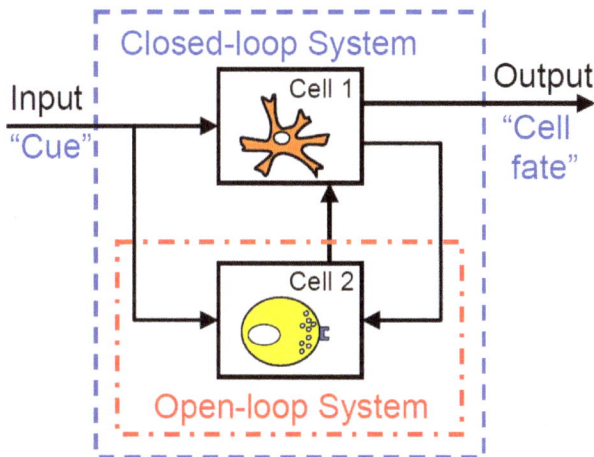

Figure 2. Studying open-loop cellular systems, as indicated by the red open-dash box, involves directly examining at a target cell or a cell population, without regard to the other cells, inputs, or outputs that may be affecting the behavior of that particular cell. Closed-loop systems examine multiple components of the overall system, including cues going into the system, interactions within the cellular environment, and outputs resulting from the multiple dynamics of cellular signaling and communication.

Historically, the causal mechanisms underlying the behavior of closed-loop systems in physiology have been identified via ingenious methods for isolating elements within the integrated system (*i.e.*, "opening the loop"). A classic example of this is the discovery of insulin and its role in connecting food intake to substrate metabolism. As insulin is only produced by the endocrine pancreas, measuring plasma insulin provides a direct measure of the organ-level communication between food intake and substrate metabolism in the peripheral tissues. The pancreas can then be approximated as a SISO system where the glucose concentration in the portal vein is the input and insulin release into the plasma is the output, as depicted in the Minimal Model for the regulation of blood glucose [79]. Measuring changes in insulin in the blood in response to changes in plasma glucose provide the basis for partitioning alterations in system response (e.g., diabetes) into deficiencies in insulin production (*i.e.*, type 1 diabetes) and insulin action (*i.e.*, type 2 diabetes). Treatment for diabetes is tailored to the deficiency in component function that exists in the patient. In diabetes, "opening the loop" means identifying organ-level cross-talk using blood measurements. In contrast, the cell-level cross-talk between tumor and immune cells occurs locally within the tumor microenvironment and may not reach a titer sufficient enough to detect using blood measurements.

3.2.1. Methods to Quantify and Characterize Cross-Talk: Developing Appropriate Models

Because biological systems are dynamic and require inputs from a variety of sources to maintain homeostasis, pathway cross-talk is essential for regulation of equilibrium. Biologists have often depended on methods to quantify this cross-talk that introduce scientific bias. Immunohistology, for example, requires the researcher to know what they are probing for. The same concern exists in Western blotting, classical reverse transcriptase-polymerase chain reaction (RT-PCR), ELISAs, and flow cytometry, which require labeling or probing for a specific molecular component. Because levels of gene expression and proteins are dynamic in that they change with respect to both time and response to a stimulus, it is difficult to say with certainty that a particular stimulus is direct cause of protein expression or whether cross-talk between molecular pathways is involved and causing an upregulation or downregulation of a particular gene and its translation into protein gene product. This requires the researcher to have extensive knowledge of "known" pathways and relies on previous data to support the hypothesis and conclusions.

Discovery-based research, using high-throughput methods to monitor expression level changes in a wide variety of targets, was one way to combat this "known pathway" requirement in cancer biomarker research. Discovery-based research, because it looks at a variety of targets, which may or may not be known to have pathway association, may be thought of as a way to combat methodological bias. However, in his Nature opinion paper regarding biomarker research, Ransohoff argued target-driven discovery-based research methods still portray a bias component, as "cancer group" samples may be handled in a particular way or undergo particular procedures during collection and storage, whereas "normal tissue" would not. In sensitive assays, which themselves may have these molecular pathway biases, this can make data interpretation, statistical analyses, and conclusions difficult and a technical bias may be inadvertently contributed to the protocol. Ransohoff also discusses "fitting" of the model during biomarker research and how this can become a bias in data analysis and conclusions [80]. We expect certain associations of molecules or patterns in signaling, and the researcher then tries to fit these patterns to a certain disease. However, as protein and mRNA expression levels are highly variable and individual samples or subjects can have divergent baselines, fitting the data to a certain model of gene or protein expression may not be a dependable method of determining disease status conclusions.

One of the great challenges of biochemical research is that the biological activity of many gene products are unknown, and therefore, it is unclear as to how they may influence cellular response mechanisms. Proteomics workflow methods may remove some of the bias of probing for known protein products. Studies conducted by our laboratory on the secretomes of a "normal" breast epithelial line compared to the breast cancer lines BT474 and SKBR3 provided a wealth of information on protein products that may be selectively secreted by breast cancers. We performed this workflow by isolating and enriching the secretomes from the cell lines, running the samples on a 2DE-gel and using MALDI-TOF analysis to identify proteins secreted [81]. To gain insight into common mechanisms for altering local intercellular communication in breast cancer, we used computational tools to identify common pathway alterations an as a way to counter uncertainty due to the inherent variability among samples and limits to sensitivity of this experimental approach.

An alternative approach to identify alterations in intracellular communication is to develop phenotypic screening assays. While the contemporary focus has largely been on target-based drug discovery, phenotypic screening produced greater than 60% of first-in-class small molecule drugs approved by the FDA between 1999 and 2008 [82]. Phenotypic screening assays are one method of discover-based research that does not rely on target-based screening, as do many discovery-based research methods, and is making a return to the field of pharmaceutical research [83]. Phenotypic screening assays rely on observing the effects of a particular stimulus on a given outcome, rather than the players that may cause that outcome. It allows for some uncertainty on what particular pathways are involved to effect disease status and can give insight to unknown molecular interactions and molecules acting via previously undiscovered mechanisms. To illustrate the approach, we developed a phenotypic screening assay to identify biochemical cues responsible for local immunosuppression *in vitro*, validated these mechanisms in human tumor biopsy specimens, and correlated these putative immunosuppressive mechanisms with clinical outcomes [84,85]. The phenotypic screening assay also incorporated a mathematical model that provides a quantitative prediction of a T cell response to Interleukin-12 in terms of cytokine production and cell fate. As multiple factors were observed in the phenotypic screening assay, the mathematical model provided a quantitative context to determine whether the observed factors were sufficient to explain the observed changes in T cell response or whether there were additional behaviors that were unexplained.

3.2.2. Computer Simulation Provides a Translational Bridge Across Model Systems

Even after defining signaling pathway components, it is still difficult for scientists to predict the overall behavior of the system. Animal models are most commonly used to predict the human response to a particular therapy and to replicate human systems. However, success of therapeutic interventions seems to be more easily attained in animals and brings to question the translational viability of animal models of disease [86]. As discussed previously, methodological bias may play a role in these discrepancies. Moreover, humans are known to have fundamental differences in biology, regardless of whether molecular components are conserved, and animal models are usually made of clones of a particular population. Therefore, this does not account for the heterogeneity of the human population, which will attribute different clinical responses. Additionally, it is unclear how conserved cellular signaling networks across different model systems. The heterogeneity of tumor cell populations and individual metabolic and disease state differences may further cloud this issue, even when "normal" human cellular networks are well-defined.

Novel drug targets are often difficult to predict, due to the interplay of multiple genes and systems and the differences between human and animal models. To overcome these barriers to disease modeling, computer simulation using previous molecular interaction and human clinical data may be able to elucidate these signaling network differences [9,87]. Quantitative systems pharmacology is now being used to analyze drug interactions and provide insight to adverse effects, and has been suggested as a method of selecting better drug candidates for development [9]. Two recent studies highlight the different functions computer modeling may perform in aiding in drug development.

In a study examining prostate cancer malignancies, human and mouse model data was analyzed to determine master regulators of prostate cancer using the Algorithm for the Reconstruction of Accurate Cellular Networks (ARACNe), which uses microarray data to predict direct molecular interactions [88,89]. A Master Regulator Inference algorithm (MARINa) was then used to determine gene regulators of prostate cancer. Co-expression of FOXM1 and CENPF was determined to promote prostate cancer malignancy by synergistically acting via the PI3K and MAPK pathways. This suggests a complicated feedback system that may be therapeutically difficult to target with current treatments [88]. Using this analysis could serve as a way to determine gene interactions in tumor development and to help to better classify tumors based on their phenotypes and genotypes.

Knowing gene regulators of malignancy could also be helpful in finding new targets for therapy and determining adverse effects before they are observed in the clinic. In a study regarding drug-induced peripheral neuropathy, data from DrugBank and Therapeutic Target Database was used to create pharmacological networks of peripheral neuropathy-inducing drugs, their known targets, and the diseases for which they are used to treat. The Database for Annotation, Visualization, and Integrated Discovery (DAVID) was then used to identify connections between drug targets and cellular pathways. Regulation of two genes, MYC and PAF15, were found to be correlated with a higher incidence of drug-induced peripheral neuropathy and suggested drugs that may produce this adverse effect, as well as targets for future studies to alleviate neuropathy symptoms in the clinic [90].

4. Conclusions

One of the most costly decisions a pharmaceutical company can make in regards to a cancer immunotherapy is selecting the patient population to conduct the clinical trial. Success of the drug is often determined by which patient populations the study is conducted in. If an inappropriate cancer phenotype is selected for the study, adverse events or inefficacy can stop the study and discourage the drug from continuing in the development pipeline. Troubleshooting and optimizing the drug could prove more costly than developing an entirely different drug. Most cell lines do not contain heterogeneous populations of cancer cells, let alone T cells and other stromal cells, which can lead to misleading results in preclinical trials for determining optimal cancer phenotypes.

Given validation of these computational methods, and by combining methods to investigate both drug response and molecular signaling, we can begin to properly examine cellular networks to infer on whether our biological experimental methods correctly forecast patient outcomes and portray an accurate representation of biological activity of a certain therapy or whether that therapy is appropriate for use in a certain patient population. Using computational tools to model relevant signaling networks and cancer immunity may also remove some of the bias associated with sample processing and inappropriate biological method selection. To improve the translational value of these approaches, they should account for clonal evolution and heterogeneity of the system. Combining system-targeting drugs with system modeling and phenotypic assays is in line with NIH recommendations for more quantitative and systems pharmacology research, and could prove to be our best resource in engaging host immunity in the fight against cancer. Given our currently known limitations and the recent developments in cancer immunotherapy research, the pharmaceutical industry can reinvigorate their model for innovation with the help of quantitative and systems pharmacology techniques and focusing on a systems perspective with respect to cancer.

Acknowledgments

This work was supported by grants from the National Science Foundation (NSF) CAREER 1053490 and the National Cancer Institute (NCI) R15CA123123. The content is solely the responsibility of the authors and does not necessarily represent the official views of the NSF, the NCI, or the National Institutes of Health.

Author Contributions

C.B.H. and D.J.K. wrote, reviewed, and edited the manuscript.

Conflicts of Interest

The authors declare no conflict of interest.

References

1. Pharmaceutical Research and Manufacturers of America (PhRMA). *2013 Biopharmaceutical Research Industry Profile*; PhRMA: Washington, DC, USA, 2013.
2. Mullin, R. Tufts study finds big rise in cost of drug development. *Chemical & Engineering News*, 20 November 2014.
3. Chen, C.; Beckman, R.A. Maximizing return on socioeconomic investment in phase ii proof-of-concept trials. *Clin. Cancer Res.* **2014**, *20*, 1730–1734.
4. Wartha, K.; Herting, F.; Hasmann, M. Fit-for purpose use of mouse models to improve predictivity of cancer therapeutics evaluation. *Pharmacol. Ther.* **2014**, *142*, 351–361.
5. Dailami, M.; Lipkovich, I.; Dyck, V.J. *Infrisk: A Computer Simulation Approach to Risk Management in Infrastructure Project Finance Transactions*; Economic Development Institute of the World Bank: Washington, DC, USA, 1999.
6. Mavris, D.; Bandte, O.; DeLaurentis, D.A. Robust design simulation: A probabilistic approach to multidisciplinary design. *J. Aircr.* **1999**, *36*, 298–307.
7. Lauffenburger, D.; Giacomini, K. Systems biology and systems pharmacology. *Bridge Converg. Eng. Life Sci.* **2013**, *43*, 26–33.
8. Ananthakrishnan, R.; Gona, P. Pharmacological modeling and biostatistical analysis of a new drug. *Open Access J. Clin. Trials* **2010**, *2*, 59–82.

9. Sorger, P. *Quantitative and Systems Pharmacology in the Post-Genomic Era: New Approaches to Discovering Drugs and Understanding Therapeutic Mechanisms*; National Institutes of Health: Bethesda, MD, USA, 2011.

10. Radulescu, R.T. Oncoprotein metastasis: An expanded topography. *Romanian J. Morphol. Embryol.* **2013**, *54*, 237–239.

11. Arnedos, M.; Soria, J.C.; Andre, F.; Tursz, T. Personalized treatments of cancer patients: A reality in daily practice, a costly dream or a shared vision of the future from the oncology community? *Cancer Treat. Rev.* **2014**, *40*, 1192–1198.

12. Li, S.C.; Tachiki, L.M.; Kabeer, M.H.; Dethlefs, B.A.; Anthony, M.J.; Loudon, W.G. Cancer genomic research at the crossroads: Realizing the changing genetic landscape as intratumoral spatial and temporal heterogeneity becomes a confounding factor. *Cancer Cell Int.* **2014**, *14*, 115.

13. Ramos, P.; Bentires-Alj, M. Mechanism-based cancer therapy: Resistance to therapy, therapy for resistance. *Oncogene* **2014**, doi:10.1038/onc.2014.314.

14. Weinstein, I.B.; Joe, A.K. Mechanisms of disease: Oncogene addiction—A rationale for molecular targeting in cancer therapy. *Nat. Clin. Pract. Oncol.* **2006**, *3*, 448–457.

15. Ellis, L.M.; Hicklin, D.J. Resistance to targeted therapies: Refining anticancer therapy in the era of molecular oncology. *Clin. Cancer Res.* **2009**, *15*, 7471–7478.

16. Hanahan, D.; Weinberg, R.A. The hallmarks of cancer. *Cell* **2000**, *100*, 57–70.

17. Shoemaker, R.H. The nci60 human tumour cell line anticancer drug screen. *Nat. Rev. Cancer* **2006**, *6*, 813–823.

18. Monks, A.; Scudiero, D.; Skehan, P.; Shoemaker, R.; Paull, K.; Vistica, D.; Hose, C.; Langley, J.; Cronise, P.; Vaigro-Wolff, A.; *et al.* Feasibility of a high-flux anticancer drug screen using a diverse panel of cultured human tumor cell lines. *J. Natl. Cancer Inst.* **1991**, *83*, 757–766.

19. Paull, K.D.; Shoemaker, R.H.; Hodes, L.; Monks, A.; Scudiero, D.A.; Rubinstein, L.; Plowman, J.; Boyd, M.R. Display and analysis of patterns of differential activity of drugs against human tumor cell lines: Development of mean graph and compare algorithm. *J. Natl. Cancer Inst.* **1989**, *81*, 1088–1092.

20. Wong, C.C.; Cheng, K.W.; Rigas, B. Preclinical predictors of anticancer drug efficacy: Critical assessment with emphasis on whether nanomolar potency should be required of candidate agents. *J. Pharmacol. Exp. Ther.* **2012**, *341*, 572–578.

21. Chapman, P.B.; Hauschild, A.; Robert, C.; Haanen, J.B.; Ascierto, P.; Larkin, J.; Dummer, R.; Garbe, C.; Testori, A.; Maio, M.; *et al.* Improved survival with vemurafenib in melanoma with braf v600e mutation. *N. Engl. J. Med.* **2011**, *364*, 2507–2516.

22. Flaherty, K.T.; Infante, J.R.; Daud, A.; Gonzalez, R.; Kefford, R.F.; Sosman, J.; Hamid, O.; Schuchter, L.; Cebon, J.; Ibrahim, N.; *et al.* Combined braf and mek inhibition in melanoma with braf v600 mutations. *N. Engl. J. Med.* **2012**, *367*, 1694–1703.

23. Hanahan, D.; Weinberg, R.A. Hallmarks of cancer: The next generation. *Cell* **2011**, *144*, 646–674.

24. Polyak, K.; Haviv, I.; Campbell, I.G. Co-evolution of tumor cells and their microenvironment. *Trends Genet.* **2009**, *25*, 30–38.

25. Tlsty, T.D.; Coussens, L.M. Tumor stroma and regulation of cancer development. *Annu. Rev. Pathol.* **2006**, *1*, 119–150.

26. Gatenby, R.A.; Vincent, T.L. An evolutionary model of carcinogenesis. *Cancer Res.* **2003**, *63*, 6212–6220.

27. Michor, F.; Iwasa, Y.; Nowak, M.A. Dynamics of cancer progression. *Nat. Rev. Cancer* **2004**, *4*, 197–205.

28. Nowak, M.A.; Michor, F.; Iwasa, Y. The linear process of somatic evolution. *Proc. Natl. Acad. Sci. USA* **2003**, *100*, 14966–14969.

29. Klinke, D.J., II. An evolutionary perspective on anti-tumor immunity. *Front. Oncol.* **2012**, *2*, 202.

30. Mahadevan, N.R.; Zanetti, M. Tumor stress inside out: Cell-extrinsic effects of the unfolded protein response in tumor cells modulate the immunological landscape of the tumor microenvironment. *J. Immunol.* **2011**, *187*, 4403–4409.

31. Schreiber, R.D.; Old, L.J.; Smyth, M.J. Cancer immunoediting: Integrating immunity's roles in cancer suppression and promotion. *Science* **2011**, *331*, 1565–1570.

32. Rech, A.J.; Vonderheide, R.H. Dynamic interplay of oncogenes and t cells induces pd-l1 in the tumor microenvironment. *Cancer Discov.* **2013**, *3*, 1330–1332.

33. LaBarge, M.A. The difficulty of targeting cancer stem cell niches. *Clin. Cancer Res.* **2010**, *16*, 3121–3129.

34. Shackleton, M.; Quintana, E.; Fearon, E.R.; Morrison, S.J. Heterogeneity in cancer: Cancer stem cells *versus* clonal evolution. *Cell* **2009**, *138*, 822–829.

35. Meacham, C.E.; Morrison, S.J. Tumour heterogeneity and cancer cell plasticity. *Nature* **2013**, *501*, 328–337.

36. Merlo, L.M.; Pepper, J.W.; Reid, B.J.; Maley, C.C. Cancer as an evolutionary and ecological process. *Nat. Rev. Cancer* **2006**, *6*, 924–935.

37. Knudson, A.G. Two genetic hits (more or less) to cancer. *Nat. Rev. Cancer* **2001**, *1*, 157–162.

38. Gatenby, R.A.; Gillies, R.J. A microenvironmental model of carcinogenesis. *Nat. Rev. Cancer* **2008**, *8*, 56–61.

39. Casas-Selves, M.; Degregori, J. How cancer shapes evolution, and how evolution shapes cancer. *Evolution (N.Y.)* **2011**, *4*, 624–634.

40. Anderson, A.R.; Weaver, A.M.; Cummings, P.T.; Quaranta, V. Tumor morphology and phenotypic evolution driven by selective pressure from the microenvironment. *Cell* **2006**, *127*, 905–915.

41. Rak, J.W.; Kerbel, R.S. Growth advantage ("clonal dominance") of metastatically competent tumor cell variants expressed under selective two- or three-dimensional tissue culture conditions. *In Vitro Cell Dev. Biol. Anim.* **1993**, *29A*, 742–748.

42. Mintz, B. Gene control of mammalian pigmentary differentiation. I. Clonal origin of melanocytes. *Proc. Natl. Acad. Sci. USA* **1967**, *58*, 344–351.

43. Michor, F.; Frank, S.A.; May, R.M.; Iwasa, Y.; Nowak, M.A. Somatic selection for and against cancer. *J. Theor. Biol.* **2003**, *225*, 377–382.

44. Gatenby, R.A.; Cunningham, J.J.; Brown, J.S. Evolutionary triage governs fitness in driver and passenger mutations and suggests targeting never mutations. *Nat. Commun.* **2014**, *5*, 5499.

45. Youn, A.; Simon, R. Using passenger mutations to estimate the timing of driver mutations and identify mutator alterations. *BMC Bioinform.* **2013**, *14*, 363.

46. Peterson, E.A.; Chavan, S.S.; Bauer, M.A.; Heuck, C.J.; Johann, D.J. Revealing the inherent heterogeneity of human malignancies by variant consensus strategies coupled with cancer clonal analysis. *BMC Bioinform.* **2014**, *15*, S9.

47. Basanta, D.; Gatenby, R.A.; Anderson, A.R. Exploiting evolution to treat drug resistance: Combination therapy and the double bind. *Mol. Pharm.* **2012**, *9*, 914–921.

48. Mumm, J.B.; Oft, M. Cytokine-based transformation of immune surveillance into tumor-promoting inflammation. *Oncogene* **2008**, *27*, 5913–5919.

49. Dunn, G.P.; Old, L.J.; Schreiber, R.D. The three es of cancer immunoediting. *Annu. Rev. Immunol.* **2004**, *22*, 329–360.

50. Iwami, S.; Haeno, H.; Michor, F. A race between tumor immunoescape and genome maintenance selects for optimum levels of (epi)genetic instability. *PLoS Comput. Biol.* **2012**, *8*, e1002370.

51. Zacharski, L.R.; Sukhatme, V.P. Coley's toxin revisited: Immunotherapy or plasminogen activator therapy of cancer? *J. Thromb. Haemost.* **2005**, *3*, 424–427.

52. Heywood, G.R.; Rosenberg, S.A.; Weber, J.S. Hypersensitivity reactions to chemotherapy agents in patients receiving chemoimmunotherapy with high-dose interleukin 2. *J. Natl. Cancer Inst.* **1995**, *87*, 915–922.

53. Karp, B.I.; Yang, J.C.; Khorsand, M.; Wood, R.; Merigan, T.C. Multiple cerebral lesions complicating therapy with interleukin-2. *Neurology* **1996**, *47*, 417–424.

54. Alexandrescu, D.T.; Maddukuri, P.; Wiernik, P.H.; Dutcher, J.P. Thrombotic thrombocytopenic purpura/hemolytic uremic syndrome associated with high-dose interleukin-2 for the treatment of metastatic melanoma. *J. Immunother.* **2005**, *28*, 144–147.

55. Moreno-Ramirez, D.; Ales-Martinez, M.; Ferrandiz, L. Fast-growing in-transit melanoma metastasis after intratumoral interleukin-2. *Cancer Immunol. Immunother.* **2014**, *63*, doi:10.1007/s00262-014-1583-2.

56. Zagozdzon, R.; Golab, J. Immunomodulation by anticancer chemotherapy: More is not always better (review). *Int. J. Oncol.* **2001**, *18*, 417–424.

57. Zagozdzon, R.; Golab, J.; Mucha, K.; Foroncewicz, B.; Jakobisiak, M. Potentiation of antitumor effects of il-12 in combination with paclitaxel in murine melanoma model *in vivo*. *Int. J. Mol. Med.* **1999**, *4*, 645–648.

58. Schetter, A.J.; Heegaard, N.H.; Harris, C.C. Inflammation and cancer: Interweaving microRNA, free radical, cytokine and p53 pathways. *Carcinogenesis* **2010**, *31*, 37–49.

59. Grillo-Lopez, A.J.; White, C.A.; Dallaire, B.K.; Varns, C.L.; Shen, C.D.; Wei, A.; Leonard, J.E.; McClure, A.; Weaver, R.; Cairelli, S.; *et al.* Rituximab: The first monoclonal antibody approved for the treatment of lymphoma. *Curr. Pharm. Biotechnol.* **2000**, *1*, 1–9.

60. Wolchok, J.D.; Hodi, F.S.; Weber, J.S.; Allison, J.P.; Urba, W.J.; Robert, C.; O'Day, S.J.; Hoos, A.; Humphrey, R.; Berman, D.M.; *et al.* Development of ipilimumab: A novel immunotherapeutic approach for the treatment of advanced melanoma. *Ann. N.Y. Acad. Sci.* **2013**, *1291*, 1–13.

61. Chen, D.S.; Mellman, I. Oncology meets immunology: The cancer-immunity cycle. *Immunity* **2013**, *39*, 1–10.

62. Verheije, M.H.; Rottier, P.J. Retargeting of viruses to generate oncolytic agents. *Adv. Virol.* **2012**, *2012*, 798526.

63. Wong, H.H.; Lemoine, N.R.; Wang, Y. Oncolytic viruses for cancer therapy: Overcoming the obstacles. *Viruses* **2010**, *2*, 78–106.

64. Chacon, J.; Sarnaik, A.; Chen, J.; Creasy, C.; Kale, C.; Robinson, J.; Weber, J.; Hwu, P.; Pilon-Thomas, S.; Radvanyi, L.G. Manipulating the tumor microenvironment *ex vivo* for enhanced expansion of tumor-infiltrating lymphocytes for adoptive cell therapy. *Clin. Cancer Res.* **2014**, *21*, 611–621.

65. Kalos, M.; June, C.H. Adoptive T cell transfer for cancer immunotherapy in the era of synthetic biology. *Immunity* **2013**, *39*, 49–60.

66. Pirooznia, N.; Hasannia, S.; Taghdir, M.; Rahbarizadeh, F.; Eskandani, M. The construction of chimeric t-cell receptor with spacer base of modeling study of vhh and muc1 interaction. *J. Biomed. Biotechnol.* **2011**, *2011*, 578128.

67. James, S.E.; Greenberg, P.D.; Jensen, M.C.; Lin, Y.; Wang, J.; Budde, L.E.; Till, B.G.; Raubitschek, A.A.; Forman, S.J.; Press, O.W. Mathematical modeling of chimeric tcr triggering predicts the magnitude of target lysis and its impairment by tcr downmodulation. *J. Immunol.* **2010**, *184*, 4284–4294.

68. Pardoll, D.M. The blockade of immune checkpoints in cancer immunotherapy. *Nat. Rev. Cancer* **2012**, *12*, 252–264.

69. Pawelec, G.; Derhovanessian, E.; Larbi, A. Immunosenescence and cancer. *Crit. Rev. Oncol. Hematol.* **2010**, *75*, 165–172.

70. Bombelli, F.B.; Webster, C.A.; Moncrieff, M.; Sherwood, V. The scope of nanoparticle therapies for future metastatic melanoma treatment. *Lancet Oncol.* **2014**, *15*, e22–e32.

71. Ott, P.A.; Hodi, F.S.; Robert, C. Ctla-4 and pd-1/pd-l1 blockade: New immunotherapeutic modalities with durable clinical benefit in melanoma patients. *Clin. Cancer Res.* **2013**, *19*, 5300–5309.

72. Hodi, F.S.; O'Day, S.J.; McDermott, D.F.; Weber, R.W.; Sosman, J.A.; Haanen, J.B.; Gonzalez, R.; Robert, C.; Schadendorf, D.; Hassel, J.C.; *et al.* Improved survival with ipilimumab in patients with metastatic melanoma. *N. Engl. J. Med.* **2010**, *363*, 711–723.

73. Riley, J.L. Combination checkpoint blockade—Taking melanoma immunotherapy to the next level. *N. Engl. J. Med.* **2013**, *369*, 187–189.

74. Mockler, M.B.; Conroy, M.J.; Lysaght, J. Targeting T cell immunometabolism for cancer immunotherapy; understanding the impact of the tumor microenvironment. *Front. Oncol.* **2014**, *4*, 107.

75. Ahmadzadeh, M.; Johnson, L.A.; Heemskerk, B.; Wunderlich, J.R.; Dudley, M.E.; White, D.E.; Rosenberg, S.A. Tumor antigen-specific cd8 t cells infiltrating the tumor express high levels of PD-1 and are functionally impaired. *Blood* **2009**, *114*, 1537–1544.

76. Brahmer, J.R.; Tykodi, S.S.; Chow, L.Q.; Hwu, W.J.; Topalian, S.L.; Hwu, P.; Drake, C.G.; Camacho, L.H.; Kauh, J.; Odunsi, K.; *et al.* Safety and activity of anti-pd-l1 antibody in patients with advanced cancer. *N. Engl. J. Med.* **2012**, *366*, 2455–2465.

77. Chen, D.S.; Irving, B.A.; Hodi, F.S. Molecular pathways: Next-generation immunotherapy—Inhibiting programmed death-ligand 1 and programmed death-1. *Clin. Cancer Res.* **2012**, *18*, 6580–6587.

78. Khoo, M.C.K. Identification of physiological control systems. In *Physiological Control Systems*; John Wiley & Sons, Inc.: New York, NY, USA, 1999; pp. 159–202.

79. Bergman, R.N. Toward physiological understanding of glucose tolerance: Minimal-model approach. *Diabetes* **1989**, *38*, 1512–1527.

80. Ransohoff, D.F. Bias as a threat to the validity of cancer molecular-marker research. *Nat. Rev. Cancer* **2005**, *5*, 142–149.

81. Klinke, D.J., 2nd; Kulkarni, Y.M.; Wu, Y.; Byrne-Hoffman, C. Inferring alterations in cell-to-cell communication in her2+ breast cancer using secretome profiling of three cell models. *Biotechnol. Bioeng.* **2014**, *111*, 1853–1863.

82. Swinney, D.C.; Anthony, J. How were new medicines discovered? *Nat. Rev .Drug Discov.* **2011**, *10*, 507–519.

83. Kotz, J. Phenotypic screening, take two. *SciBX* **2012**, *5*, doi:10.1038/scibx.2012.380.

84. Kulkarni, Y.M.; Chambers, E.; McGray, A.J.; Ware, J.S.; Bramson, J.L.; Klinke, D.J., II. A quantitative systems approach to identify paracrine mechanisms that locally suppress immune response to interleukin-12 in the b16 melanoma model. *Integr. Biol. (Camb.)* **2012**, *4*, 925–936.

85. Klinke, D.J., II. Induction of wnt-inducible signaling protein-1 correlates with invasive breast cancer oncogenesis and reduced type 1 cell-mediated cytotoxic immunity: A retrospective study. *PLoS Comput. Biol.* **2014**, *10*, e1003409.

86. Wen, F.T.; Thisted, R.A.; Rowley, D.A.; Schreiber, H. A systematic analysis of experimental immunotherapies on tumors differing in size and duration of growth. *Oncoimmunology* **2012**, *1*, 172–178.

87. Coumans, J.V.; Gau, D.; Poljak, A.; Wasinger, V.; Roy, P.; Moens, P.D. Profilin-1 overexpression in mda-mb-231 breast cancer cells is associated with alterations in proteomics biomarkers of cell proliferation, survival, and motility as revealed by global proteomics analyses. *Omics* **2014**, *18*, 778–791.

88. Aytes, A.; Mitrofanova, A.; Lefebvre, C.; Alvarez, M.J.; Castillo-Martin, M.; Zheng, T.; Eastham, J.A.; Gopalan, A.; Pienta, K.J.; Shen, M.M.; *et al.* Cross-species regulatory network analysis identifies a synergistic interaction between foxm1 and cenpf that drives prostate cancer malignancy. *Cancer Cell* **2014**, *25*, 638–651.

89. Margolin, A.A.; Nemenman, I.; Basso, K.; Wiggins, C.; Stolovitzky, G.; Dalla Favera, R.; Califano, A. Aracne: An algorithm for the reconstruction of gene regulatory networks in a mammalian cellular context. *BMC Bioinform.* **2006**, *7*, S7.

90. Hur, J.; Guo, A.Y.; Loh, W.Y.; Feldman, E.L.; Bai, J.P. Integrated systems pharmacology analysis of clinical drug-induced peripheral neuropathy. *CPT Pharmacomet. Syst. Pharmacol.* **2014**, *3*, e114.

Network Analysis Identifies Crosstalk Interactions Governing TGF-β Signaling Dynamics during Endoderm Differentiation of Human Embryonic Stem Cells

Shibin Mathew, Sankaramanivel Sundararaj and Ipsita Banerjee

Abstract: The fate choice of human embryonic stem cells (hESCs) is controlled by complex signaling milieu synthesized by diverse chemical factors in the growth media. Prevalence of crosstalks and interactions between parallel pathways renders any analysis probing the process of fate transition of hESCs elusive. This work presents an important step in the evaluation of network level interactions between signaling molecules controlling endoderm lineage specification from hESCs using a statistical network identification algorithm. Network analysis was performed on detailed signaling dynamics of key molecules from TGF-β/SMAD, PI3K/AKT and MAPK/ERK pathways under two common endoderm induction conditions. The results show the existence of significant crosstalk interactions during endoderm signaling and they identify differences in network connectivity between the induction conditions in the early and late phases of signaling dynamics. Predicted networks elucidate the significant effect of modulation of AKT mediated crosstalk leading to the success of PI3K inhibition in inducing efficient endoderm from hESCs in combination with TGF-β/SMAD signaling.

Reprinted from *Processes*. Cite as: Mathew, S.; Sundararaj, S.; Banerjee, I. Network Analysis Identifies Crosstalk Interactions Governing TGF-β Signaling Dynamics during Endoderm Differentiation of Human Embryonic Stem Cells. *Processes* **2015**, *3*, 286-308.

1. Introduction

Human embryonic stem cells (hESCs) are a promising raw material for regenerative medicine applications because of their potential for directed differentiation into clinically relevant cell types. In differentiating hESCs to lineages of pancreas, liver, *etc.*, an important first step is the induction of Definitive Endoderm (called endoderm henceforth) [1]. The quality of later stage maturation is dictated by the initial pathway of endoderm differentiation [2]. Extensive research over a decade have given rise to alternate protocols for endoderm induction of hESCs that employ unique combinations of growth factors and small molecules [3–7]. These protocols aim to recapitulate signaling dynamics mimicking *in vivo* development of endoderm. Among these signaling mediators, activation of TGF-β/SMAD2,3 pathway by ligand molecules like Activin A (called Activin henceforth) has been identified as necessary to induce endoderm differentiation of hESCs [8]. Additionally, secondary modulation of parallel pathways have been shown to enhance endoderm specific differentiation; these include inhibition of PI3K pathway [4], or activation of FGF + BMP4 pathways [3], or activation of WNT pathway [6]. Among these commonly used supplementary constituents, inhibition of PI3K pathway has been consistently reported to enhance endoderm differentiation of hESCs in conjunction with Activin induction [9]. The resulting endoderm cells show good potential to transform into pancreatic β-like cells with continued differentiation [2,10].

The process of differentiation is induced by activation of specific signaling molecules by growth factors and small molecules. Activin in culture media binds to type 1 and type 2 receptors on the cell surface to form active heteromeric complexes [11]. These receptors then regulate downstream signals using the same mechanism as TGF-β receptor complexes. Briefly, the ligand receptor complex is responsible for C-terminal phosphorylation of regulatory SMAD molecules, namely SMAD2 and SMAD3 [12]. Active regulatory SMADs form homomeric complexes and they also complex with co-regulatory molecules like SMAD4 to form heteromeric complexes. These complexes shuttle to the nucleus to orchestrate a host of gene transcriptional events that maintain homeostasis and activate developmental programs [13]. Previous studies have shown that SMAD molecules play a major role in fate choice of hESCs [14–16]. However, the context of survival pathways like PI3K/AKT and mitogen activated pathways like MAPK/ERK ultimately decides whether active SMAD complexes support self-renewal or differentiation of hESCs. This is because of critical crosstalk interactions between TGF-β/SMAD and these other pathways participating in development [14,17].

Multiple experimental reports have shown the existence of significant crosstalk interfering with the TGF-β pathway during endoderm differentiation [17,18]. The efficiency of endoderm differentiation is consequently diminished without appropriate removal of interaction with parallel pathways. In spite of several experimental reports, there has not been a thorough mathematical and network level analysis of the existing interactions, which is the focus of the current paper. The primary goal of this work is to quantitatively evaluate the existence of network interactions and the direction of interaction from signaling dynamics of hESCs differentiating to endoderm. Due to the high variability associated with hESC systems, it is also necessary to infer robust connections from noisy data. Bayesian models provide a natural framework to investigate the causal dependence between nodes in a network and derive probabilistic relationships that most likely explain experimental observations [19]. These models have proven successful in network reconstruction from noisy signal transduction data [20,21]. Among the different Bayesian models, non-stationary Dynamic Bayesian Networks (DBNs) provide the best representation of the adaptive nature of signal transduction networks [22].

In the present work, interactions between the signaling molecules belonging to the TGF-β/SMAD, PI3K/AKT and MAPK/ERK pathways controlling endoderm differentiation of hESCs were identified. The nature of interactions present in the signaling network and the sequence of signal propagation events are cumulatively captured in the dynamics of key molecules in a signaling pathway [23]. As a first step, a multiplex measurement platform was used to measure detailed dynamics of multiple signaling molecules of the TGFβ pathway along with key crosstalk molecules. The measurements were made under Activin induction condition along with a perturbed case where PI3K pathway was simultaneously inhibited. DBN analysis conducted on the entire time series of key signaling molecules from these pathways elucidated the presence of multiple crosstalk interactions regulating the endoderm induction conditions. The results show that the receptor levels play an important role in controlling majority of the intracellular signaling molecules in the early and late phases of the signaling dynamics. Further, molecule from PI3K/AKT pathway (p-AKT) shows significant crosstalk interactions in the high PI3K condition, which is removed in the low PI3K condition. The early signaling dynamics contained enough information to recapitulate the key crosstalk interactions. Overall, the analysis provided explanations for the network level differences between the Activin alone and Activin + PI3K inhibition condition used for endoderm induction. The DBNs showed the key crosstalk interactions removed under Activin + PI3K inhibition providing an explanation for its good endoderm differentiation potential at the signaling level.

2. Methods

2.1. Cell Culture and Treatment

2.1.1. Human ESC Maintenance

H1 hESCs were placed on hESC certified Matrigel (BD Biosciences, Billerica, MA, USA)-coated tissue culture plate for 5–7 days in mTESR1 (Stemcell Technologies, Vancouver, BC, Canada) at 37 °C and 5% CO_2 before passaging. Cells were examined under the microscope every day and colonies with observable differentiation were picked and removed before the media changes. The maintenance protocol was adopted from our previous studies [2,9,10].

2.1.2. Experimental Induction of Endoderm from hESCs

Human ESCs were allowed to grow to 60%–70% confluency before experiments were started. Once confluency was reached, endoderm differentiation was induced by adding 100 ng/mL Activin A (R & D Systems, Minneapolis, MN, USA) in the presence or absence of 1 μM Wortmannin (PI3K inhibitor; Sigma-Aldrich, St. Louis, MO, USA) for 24 h (or otherwise indicated). The differentiation media were made using DMEM/F12 (Life Technologies, Grand Island, NE, USA), supplemented with 0.2% bovine serum albumin (BSA; Sigma-Aldrich, St. Louis, MO, USA) and 1xB27 (Life Technologies, Grand Island, NE, USA). The induction protocol for endoderm was adopted from our previous study [2,9].

2.1.3. Measuring Experimental Dynamics of Signaling Molecules

Intracellular expression of signaling proteins were measured by MagPix analysis using the TGFβ Signaling Pathway Magnetic Bead 6-Plex Cell Signaling Multiplex Assay (EMD Millipore, Catalog no.: 48-614MAG) according to manufacturer's instructions. The detailed protocol for MagPix is described in our previous study [10]. Mean fluorescence intensity (MFI) was measured using the xMAP (Luminex, Madison, WI, USA) instrument. Measurements were obtained for 6 analytes, namely total TGFβ receptor 2 (t-TGFβRII), total SMAD4 (t-SMAD4), phosphorylated SMAD2 (p-SMAD2 Ser465/Ser467), p-SMAD3 (Ser423/Ser425), p-AKT (Ser473) and p-ERK (Thr185/Tyr187). The time points selected for analysis were: 0, 0.5, 1, 1.5, 2, 3, 6, 12, 18 and 24 h (10 time points, each from a different well of tissue culture plate). Three repeats were conducted per experimental condition and quantitative analysis was performed on each repeat separately.

Total protein content of the sample was measured using BCA total protein kit (Thermo Scientific, Grand Island, NE, USA), according to manufacturer's instructions.

2.2. Identification of Network Interactions from Experimental Time Series Signaling Data

A DBN analysis was performed to identify the network structure that can explain the dynamics of signaling molecules during endoderm differentiation. Bayesian networks are probabilistic graphical models that relate nodes via directed edges, with the direction showing the causal relationship between the nodes [19]. These relationships are stronger as compared to correlative methods. Graphical models have nodes that represent entities that can interact (here molecules) and edges show how the nodes influence each other. The node where the edge originates is commonly called a parent node and the node where the edge ends is called a child node. Each node in the network is described by conditional probabilities as tables or functions. In continuous space, the relationship is represented by conditional probability distributions, and Gaussian distributions are commonly used to model the relationships [19,24]. Bayesian networks however cannot represent cyclic loops like feedbacks that are common in signal transduction networks. The problem of cyclic loops can be overcome by use of a generalization of Bayesian networks via DBNs [25].

2.2.1. Details of DBN Algorithm

DBNs relate variables between adjacent time points such that a child node at a given time point is related to the parent nodes at a previous time point, thereby expanding the network in time. Based on the system and the dynamics, the relationship can go back one or several time steps. A common approach to construct DBN is by using score equivalence criterion [24]. Here, a scoring metric (for example, maximum likelihood (ML) estimate in combination with regularization strategies) is used to evaluate how well a graph reconstructs the experimental data. Although DBNs provide good representation of biological networks, they are computationally expensive. Recently Grzegorczyk *et al.* developed a computationally efficient algorithm to identify non-stationary DBNs [25]. Specifically, in non-stationary DBNs, the network structure is kept constant between different time points, but the model parameters are allowed to vary between different time segments. The method has been successful in discovering biologically relevant interactions from diverse biological data sets including times series of gene expression and Milliplex protein concentrations across species [25–31]. The model systems are diverse, including circadian rhythms in *A. thaliana*, morphogenesis in *D. melanogaster*, synthetic metabolic networks in *S. cerevisiae*, serum inflammatory cytokine mediators in pediatric acute liver injury *etc.* [27,32]. Full details of the algorithm are presented in the manuscript and supplementary material of Grzegorczyk *et al.* [25]. A brief discussion of the algorithm based on the original manuscript is presented below.

Consider a set of N interacting nodes of a signaling network represented by $X_1, X_2, ..., X_n$ and a directed graph structure G. An edge pointing from X_i to X_j in a DBN with time lag equal to one time step shows that the realization of X_j at time step t is dependent on the realization of its parent X_i at time step $t-1$. It is commonly assumed that a time lag equal to one time step is sufficient to represent the relationship, indicating that the data have to be sampled at the right time intervals for the dynamics to be represented correctly. The parent node set, π_j, of a node X_j is the set of all nodes from which an edge points to X_j in G. Grzegorczyk *et al.* proposed a non-stationary generalization of the Bayesian Gaussian with score equivalence model (called BGe), and it is a node-specific mixture of BGe models [25]. A linear Gaussian distribution is chosen for the local conditional distributions. The non-stationary DBN is based on the following Markov chain expansion:

$$P\left(D|G,\underline{V},\underline{K},\theta\right) = \prod_{n=1}^{N}\prod_{t=2}^{m}\prod_{k=1}^{\kappa_n}\psi\left(D_n^{\pi_n}\left[t,\underline{\theta}_n^k\right]\right)^{\delta_{V_n(t),k}} \tag{1}$$

$$\psi\left(D_n^{\pi_n}\left[t,\underline{\theta}_n^k\right]\right)^{\delta_{V_n(t),k}} = P\left(X_n(t) = D_{n,t}|\pi_n(t-1) = D_{\pi_n,t-1},\underline{\theta}_n^k\right) \tag{2}$$

where, D is the time course data, $\delta_{V_n(t),k}$ is the Kronecker delta, $\underline{V}$ is a matrix of latent variables that indicate which BGe mixture component generates a data point, $\underline{K} = (\kappa_1,...,\kappa_n)$ is a vector of mixture components, m is the total number of time points. Vectors and matrices are denoted by single underbars in the symbols of all the equations of this manuscript. Each column of matrix $\underline{V}$ is the vector $\underline{V}_n$, which divides the time series for a node into different time segments. The endpoints of these time segments are called as change-points. Each time segment between change-points is a different BGe model with parameters θ_n^k, which includes the mean and covariance matrix of the conditional dependences for the mixture component. The allocation scheme in Equation (1) provides representation of a nonlinear regulatory process by a piecewise linear process. From Equation (1), the marginal likelihood conditional on the latent variables is given by:

$$P\left(D|G,\underline{V},\underline{K}\right) = \int P\left(D|G,\underline{V},\underline{K},\theta\right)P(\theta)d\theta = \prod_{n=1}^{N}\psi^*\left(D_n^{\pi_n}\left[\kappa_n,\underline{V}_n\right]\right) \tag{3}$$

$$\psi^*\left(D_n^{\pi_n}\left[\kappa_n,\underline{V}_n\right]\right) = \prod_{k=1}^{\kappa_n}\psi\left(D_n^{\pi_n}\left[k,\underline{V}_n\right]\right) \tag{4}$$

$$\psi\left(D_n^{\pi_n}\left[k,\underline{V}_n\right]\right) = \int\prod_{t=2}^{m}P\left(X_n(t) = D_{n,t}|\pi_n(t-1) = D_{(\pi_n,t-1)},\theta_n^k\right)P\left(\theta_n^k|\pi_n\right)d\underline{\theta} \tag{5}$$

Equation (4) is the local change-point BGe score (called as cpBGe) for node n. In this work, a Gibbs MCMC sampling scheme was followed to sample from the local posterior distributions. Although, the location of change-points is inferred, the actual values of the parameters are not directly obtained since they are integrated out as seen from Equation (3). In this manuscript, correlation analysis was used in selected time segments to evaluate the nature of influence. In the algorithm, the change-points were sampled from a point process prior using dynamic programming and the graphs were sampled by sampling parent node set (restricted to 3 parents per node) from a Boltzmann posterior distribution using the cpBGe score. Additional details of the sampling procedure is given in [25]. The codes provided online by Grzegorczyk *et al.* were used in this work [25]. The sampling parameters were kept at nominal values suggested by Grzegorczyk *et al.* All simulations were performed in MATLAB® (Natick, MA, USA) on Linux 64-bit platform and single core of INTEL® (Santa Clara, CA, USA) Core™ 2 Quad CPU (Q8400 @ 2.66 GHz).

2.2.2. Constructing the DBNs

The DBN analysis was performed on each of the two experimental conditions separately to identify the network interactions that exist in each condition. The data were preprocessed as described in the results section prior to the DBN analysis (see Section 3.2). To construct the DBN, the marginal edge probability was monitored for each Gibbs sampling step. The marginal edge probability for a given edge denotes the fraction of the graphs in which that edge was present. Each Gibbs sampling step represents an instance of the network that can best explain the experimental time series. In the early phases of the simulation, the network is not yet stabilized and hence, the likelihood scores and the marginal edge probabilities fluctuate. The marginal edge probabilities of the final network were calculated after a burn-in phase when the distributions have stabilized. The DBN algorithm was applied to each repeat of the three available for each condition. The marginal edge probability scores from networks obtained for the three repeats were averaged to obtain a consensus network for a given condition or time zone. Finally, only those edges that were present in more than 50% of the sampled graphs were kept in the consensus DBN, a criterion used in the study by Azhar *et al.* [27]. Any value less than 50% indicates that the number of samples in which the associated edge was absent is more than the number of samples in which it is present. Specific details of the application of DBN for the dataset evaluated in this paper are explained in Section 3.2.

3. Results and Discussion

3.1. Dynamics of Signaling Molecules during Endoderm Induction

Figure 1 shows the dynamics of six signaling molecules after Activin A addition in the presence (shown by blue dashed line) and absence (shown by red continuous line) of PI3K inhibitor, called as low and high PI3K conditions respectively. The original data was normalized by time 0 values to obtain the fold change. The mean levels and standard deviation from 3 experimental repeats are plotted here. The time points selected for the study include: 0, 0.5, 1, 1.5, 2, 3, 6, 12, 18 and 24 h. The high PI3K condition represents the differentiation protocol where only the TGF-β/SMAD2,3 pathway is externally activated while the PI3K/AKT pathway is left unperturbed. In this condition, p-AKT levels are maintained near the basal levels, only slightly lower (Figure 1A). For the purpose of this manuscript, the basal levels are defined as the protein levels at time 0. It is seen that the mean levels of p-AKT fluctuate in the early time points (<6 h). Levels of t-TGFβRII (Figure 1B) also remain close to basal levels under high PI3K signaling. For p-SMAD2 (Figure 1C), an overshoot behavior is seen with levels reaching the maximum within 2–3 h and settling at intermediate levels by 6 h. For p-SMAD3 (Figure 1D), the dynamics shows a different behavior than p-SMAD2 even though both are activated by the same ligand-receptor complex. In general, the dynamics shows a continuous increase instead of the overshoot behavior seen for p-SMAD2. t-SMAD4 (Figure 1E) is maintained near the basal levels for this condition. p-ERK shows a minimal and delayed increase (Figure 1F) under high PI3K.

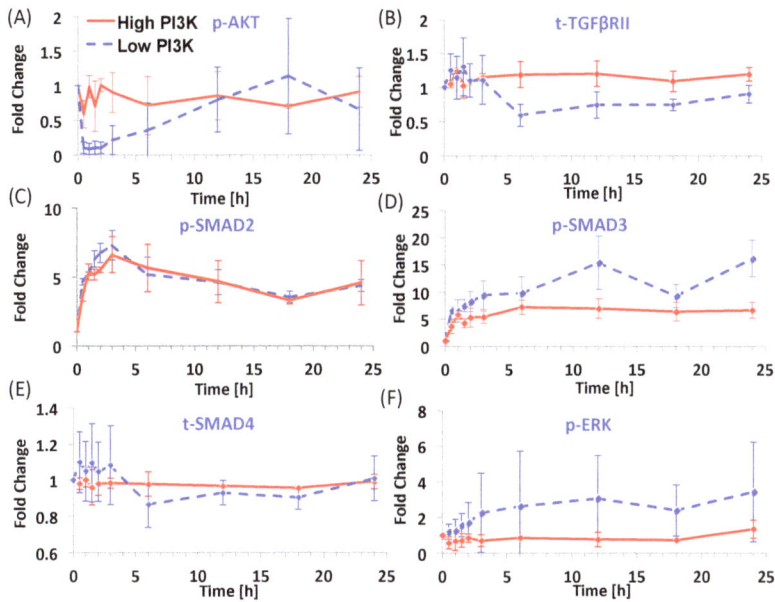

Figure 1. Dynamics of key molecules from the TGF-β/SMAD, PI3K/AKT and MAPK/ERK pathways for two endoderm induction conditions. (**A–F**) p-AKT, t-TGFβRII, p-SMAD2, p-SMAD3, t-SMAD4 and p-ERK dynamics under high and low PI3K conditions. H1 hESCs were treated with 100 ng/mL Activin A in the presence or absence of 1 μM Wortmannin (PI3K inhibitor) for 24 h. The protein levels were quantified using Multiplex MagPix Assay. The mean and standard deviation (number of repeats = 3) in protein levels are represented here as fold change over time 0 levels.

The low PI3K condition represents a modulation over the high PI3K condition with the PI3K/AKT pathway externally inhibited in addition to activation of TGF-β/SMAD2,3 pathway. In this condition, we see a considerable decrease in p-AKT levels since it is a downstream effector of PI3K signal (Figure 1A). However, interestingly this decrease is short-lived. Even after continued inhibition of PI3K, the levels of p-AKT start increasing from 3 h with the levels reaching near basal levels by 12 h. The levels of t-TGFβRII in this condition are lower than high PI3K condition at time points from 6 h (Figure 1B). The dynamics of p-SMAD2 is similar to the high PI3K condition (Figure 1C) with slightly higher fold-change at early time points. The fold-change in p-SMAD3 is higher compared to high PI3K signaling and it also shows substantial increase at later time points. t-SMAD4 (Figure 1E) shows fluctuations at early time points and a slight reduction at later time points (from 6 h). p-ERK (Figure 1F) shows a slow rise as compared to p-SMAD2,3 and the increase is substantial as compared to high PI3K signaling. Thus, overall, the low PI3K condition results in higher fold-changes in levels of phosphorylated SMAD3 and ERK than high PI3K condition. Further, low PI3K results in higher expression of two important endoderm genes, *SOX17* and *CER* (Figure S1, also see [9] for more marker comparisons).

The dynamics shown in Figure 1 is the first detailed study of signaling dynamics obtained for hESCs under endoderm induction conditions. Two unique features are observed for hESCs, namely the rise in p-AKT levels under continued PI3K inhibition and the divergent dynamics of p-SMAD2 and p-SMAD3 under high Activin levels. Further, as is typical for hESC system, there is high degree of variability in the levels of most molecules and the degree of variability is different at different time points. The variability is higher for low PI3K condition, a possible effect resulting from high degree of cell death observed in this condition since PI3K is an important cell survival pathway. The differences in the levels and dynamics of molecules between high and low PI3K conditions indicate existence of crosstalk interactions between the TGF-β/SMAD2,3, PI3K/AKT and MAPK/ERK pathways. Previous reports from the Dalton group has indicated interactions between these pathways using static end-point analysis [17]. Here, we use a computational framework to identify all possible interactions from the information contained in the signaling dynamics.

3.2. Predictions of Network Interactions by DBN Analysis on Entire Time Series

We employed DBN algorithm developed by Grzegorczyk et al. for network identification [25]. As detailed in Section 2.2.1, the algorithm infers a causal relationship (hence directed graph) between the nodes in the network across any two adjacent time points from a given time series data. To apply the algorithm to the high and low PI3K data, the data were preprocessed by normalizing the raw MFI values of each protein by its maximum MFI value for the given time series. The data consists of 10 time points, namely 0, 0.5, 1, 1.5, 2, 3, 6, 12, 18 and 24 h for 6 nodes in the network and 3 experimental repeats per condition. The normalized data are presented in the supplementary figures, Figures S2 and S3 for high and low PI3K conditions respectively. A common concern with biological datasets is the inherent variability arising from batch-to-batch and well-to-well variability. This is further enhanced in hESC systems, used in the current work, due to inherent variations in differentiation, which cannot be conveniently controlled in the current experimental setting. However, even though the individual repeats elicited high variability in measured MFI values, many features of the overall protein dynamics was largely conserved (see Figures S2 and S3). Hence we performed the DBN on normalized data for each individual repeats separately. Figure 2 presents the directed graphs (or digraphs, used interchangeably in this manuscript) obtained from the DBN analysis.

3.2.1. DBN Analysis and Consensus Graph

Figure 2A,C shows the consensus digraphs for the high and low PI3K data. To construct the graph, DBN is first performed for each of the three repeats separately. This gives rise to three DBNs for each condition. The consensus graph in Figure 2 is constructed by averaging the edge probability across the repeats and keeping the edges that occur $\geq$ 50% of the times. The convergence diagnostics for the DBNs for each repeat is presented in Figure S4. The log likelihood score stabilized very early in the sampling runs for both conditions (see Figure S4A,B). For the current data, it was found that 250 Gibbs sampling steps were sufficient to converge to the marginal edge posterior distribution (see Figure S4C,D). This was confirmed over independent sampling runs, due to stochastic nature of the algorithm. Then, 500 sampling steps were performed to obtain enough samples in the converged region to calculate the marginal edge probabilities. At the end of 500 Gibbs sampling steps, the final marginal edge probabilities were calculated using the latter half of the 500 samples (the early half belongs to the burn-in phase of the simulation). The mean marginal edge probabilities from the three samples are presented in Figure 2B,D. Any edge, which was present in less than 50% of the samples, was removed from the consensus graph. Note that the DBN for each sample represents the network that can explain the entire time series of that sample, with only network parameters allowed to vary between time segments.

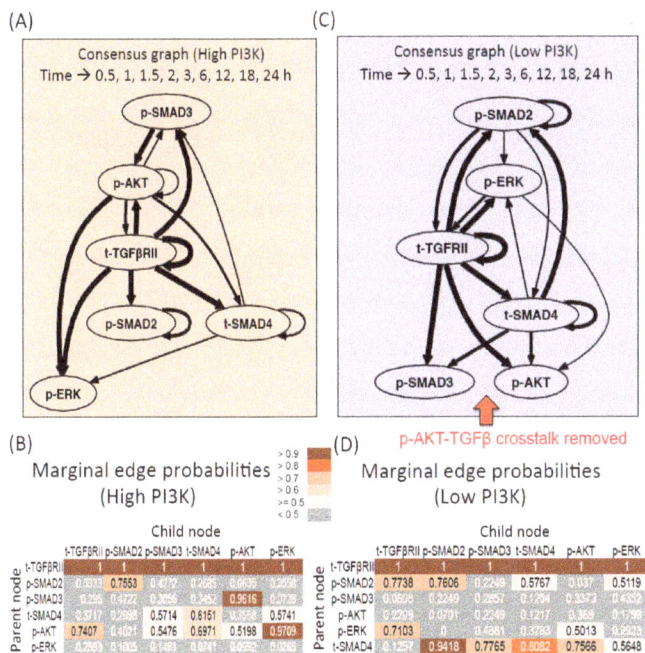

Figure 2. Dynamic Bayesian Networks inferred for endoderm induction conditions (**A**) Consensus graph for high PI3K data. The thickness of the edges reflects the value of edge probabilities (≥0.5); (**B**) Marginal edge probability table for high PI3K data. The parent node is the node whose value at time step $(t-1)$ affects the value of child node at time step t; (**C**) Consensus graph for low PI3K data; (**D**) Marginal edge probability table for low PI3K data.

3.2.2. High PI3K Condition

The consensus graph shows the average interactions that are present for the given experimental condition across the three samples. As seen from Figure 2A, the dynamics of the receptor influences all the other molecules in the network, both molecules of the TGF-β pathway (p-SMAD2,3, SMAD4) and molecules of parallel pathways (p-AKT and p-ERK). The receptor is also self-regulated. The mean marginal edge probabilities in Figure 2B show that these edges are present in 100% of the sampled graphs (t-TGFβRII as the parent node and all the other molecules including the receptor being the child node). Next common edges include p-ERK regulation by p-AKT and p-AKT regulation by p-SMAD3 present in 97% and 96% of the graphs respectively. Remaining possible interactions include: Regulation of the receptor levels by p-AKT (74%), t-SMAD4 by p-AKT and t-SMAD4 (60%–70%), p-SMAD3 by p-AKT (55%), p-AKT self-regulation (52%), p-SMAD3 and p-ERK by t-SMAD4 (57%). The graphs for the individual repeats are presented in Figure S5.

3.2.3. Low PI3K Condition

For the low PI3K condition, the edges originating from the receptor are similar to the high PI3K case and are also reflected in 100% of the graphs (Figure 2C,D). Next highly represented edges include p-SMAD2 regulation by t-SMAD4 (94%) and t-SMAD4 self-regulation (81%). Remaining possible interactions include: t-SMAD4 as a parent node for p-SMAD3 (77%), p-AKT (76%), p-ERK (56%), p-ERK as the parent node for t-TGFβRII (71%) and p-AKT (50%), p-SMAD2 as a parent node for t-TGFβRII (77%), p-SMAD2 (76%), t-SMAD4 (58%) and p-ERK (51%). The graphs for the individual repeats are presented in Figure S6.

3.2.4. Comparison between Digraphs of High and low PI3K Conditions

Influence of Total Receptor Levels

The DBN analysis identified several similarities and differences in the interactions present in the two conditions. Firstly, the dynamics of the total receptor levels affect the downstream molecules in both the conditions. This influence of total receptor levels is reflected in all the individual samples across both the conditions (Figures S5 and S6). This indicates that the changes in the receptor levels are important in influencing the downstream molecules during endoderm induction.

Interactions between Intracellular Molecules

Among the TGF-β pathway molecules, p-SMAD2 has increased regulatory interactions in the low PI3K condition, especially influencing the receptor levels. Further, p-SMAD2 shows influence on p-SMAD3 and p-ERK in sample 1 of high PI3K (Figure S5). p-SMAD3 shows interactions with p-AKT in the high PI3K condition. This interaction is removed in the low PI3K condition. The low PI3K condition also shows increased role for t-SMAD4 in influencing the p-SMAD2 and p-SMAD3 dynamics. p-AKT shows striking differences in the connections between the two conditions. For example, p-AKT interacts with and regulates majority of the nodes in the high PI3K condition. However, in the low PI3K condition, p-AKT does not regulate other nodes, but instead acts as a child node for all of its interactions. This is an interesting prediction, because the levels of p-AKT increased back in spite of continued inhibition in the low PI3K condition. The current analysis indicates that a short-term decrease in p-AKT levels is sufficient to remove the influence of p-AKT on TGF-β pathway molecules. Next important difference is in the regulatory role of p-ERK. p-ERK is not regulating any of the other nodes in the high PI3K condition. This is also reflected in each of the repeats in high PI3K condition (Figure S5). Interestingly in the low PI3K condition, p-ERK takes an important role in regulating the receptors. p-ERK shows increased regulatory role on p-SMAD3 and t-SMAD4 in one of the samples (see sample 3 in Figure S6). This sample also had the highest increase in p-ERK by 24 h among all the samples (data not shown), indicating that this specific sample is crossing the threshold for p-ERK mediated regulation of SMAD molecules.

3.2.5. Change-points Inferred by cpBGe Model

During DBN analysis, the algorithm segments the time series data in a non-supervised, node-specific manner. The ends of these time segments are called as change-points. Mathematically, the parameters of the distribution change at a change-point, but the specific structure of the network is not allowed to vary. Therefore, the nature (strength and/or direction) of the regulatory relation between the nodes for all the time points of a segment remains the same but different from the time points in the preceding and succeeding segments. Currently, the algorithm fits a Gaussian mixture model for each node separately and assigns the time points in the data to a specific mixture component. Two time points belonging to the same mixture component of a node will show the same regulatory relation with its parent nodes at these two time points. The algorithm also calculates the posterior probability of pairs of time points being co-allocated in this way. The co-allocation matrices for the time steps of high and low PI3K condition for each repeat are shown in Figure 3. The axes of each plot represent the time step (whose actual value is given at the bottom of the figure). The black/white shading of the plot shows the posterior probability of two time points being assigned to the same mixture component of the cpBGe model. Black region shows 0 probability while white region shows a probability of 1. This plot can be made for each node in the network. For the current data, it was observed that all the nodes showed identical change-points, indicating that these nodes are regulated together. The co-allocation plot in Figure 3 is representative of all the nodes in the network.

Based on Figure 3, many of the adjacent time points are well correlated as the high probability regions fall along the diagonals of the plot. This is an important observation given that each time point is obtained from different tissue culture well in the same experiment. More importantly, here we concentrate on those time segments that maintain the same regulatory relation for at least three consecutive time points. The high PI3K condition shows four main segments for each repeat (Figure 3A). Out of these, (0.5, 1, 1.5 h) and (6, 12, 18 h) are the segments containing three time points in at least two of the repeats. In each of these time segments, the parent node is active for the first two time points and the associated child node is active for the last two time points. Hence, the same regulatory relation between the parent and child node is active for at least two time points. Although the regulatory relation is changing frequently along the time series, the most repeated network edges are not changing along the time series, as they are kept fixed by the algorithm. The low PI3K data shows more variability in the number of change-points. Repeats 1, 2 and 3 show four, five and three time segments respectively. Of these, repeat 2 contains very short time segments. Only one of the time segments, containing (0.5, 1 and 1.5 h) is repeated twice in repeats 1 and 3.

The presence of many change-points in the same condition may be because of high variability in data, which is common for hESC systems. However, the presence of highly likely interactions identified by the DBNs indicates that there is a high degree of correlation between the nodes in spite of the variability in the data. In addition, the presence of correlation between many of the adjacent time points indicate that in spite of the uncontrolled variations, the dynamic information of signal transfer is still maintained.

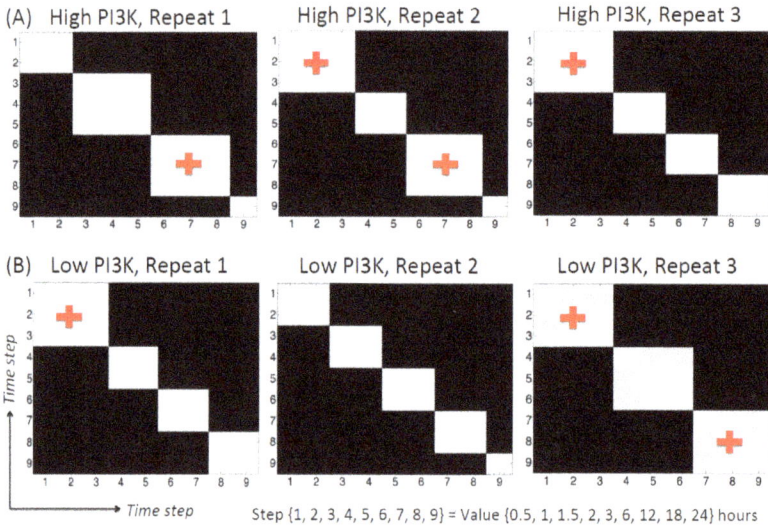

Figure 3. Co-allocation matrices for the high and low PI3K time series. (**A**) High PI3K condition; (**B**) Low PI3K condition. The axes represent time step. The actual time values corresponding to the time step are given below the plots. The black/white shading indicates the posterior probability of two time points being assigned to the same mixture component, ranging from 0 (black) to 1 (white). As seen from the figure, there are several time segments inferred from the data, 4 for the high PI3K condition and 3–5 for the low PI3K condition. All nodes show identical change-point behavior (data not shown), although this was not pre-fixed in the algorithm. The crosses indicate the time segments selected for network inference in different time zones.

3.3. Changes in Regulatory Structure across Time Zones

Previous section showed that the same regulatory relationship is maintained within some of the early and late time zones. This includes (0.5, 1, 1.5 h) which for both conditions is taken as early. The time points (6, 12, 18 h) are taken as late for high PI3K data and (12, 18, 24 h) for the low PI3K data. Since these early and late time segments have a consistent sampling interval of 0.5 and 6 h respectively, they were selected for further analysis to check if the regulatory interactions existing in the early and late time zones of the dynamics is the same (these segments are marked by crosses in Figure 3). This is necessary to check if the crosstalk interactions exist throughout the 24 h time series, or only in certain time zones. Therefore, DBN analysis was done on each zone separately. It is important to note that each of the resulting networks is particular to the time segment of interest since the algorithm has not seen data from the other zone. Nevertheless, the regulatory structure identified in each segment will confirm if these segments contain similar information as any other portion of the dynamics.

3.3.1. High PI3K Condition

Figure 4A,B presents the consensus graph and marginal edge probabilities respectively for the early time points averaged over repeats 2 and 3. The network is very similar to the network obtained using the entire time series of high PI3K condition (Figure 2A), with some minor differences. The key regulations by the receptor as well as supplementary crosstalk interactions are identified from the early time points. Figure 4C,D presents the consensus graph and marginal edge probabilities respectively for the late time points, averaged over repeats 1 and 2. The network obtained only contains regulation by the receptor and some repeats contain the regulation by p-AKT on the receptor and p-ERK levels.

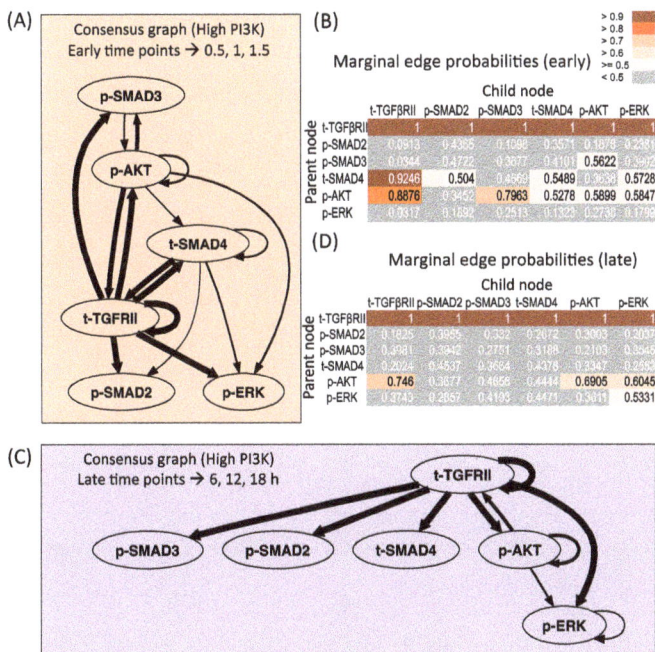

Figure 4. Dynamic Bayesian Network inferred for endoderm induction conditions under different time zones and high PI3K. (**A**) Consensus graph for high PI3K data, early dynamics (t = 0.5, 1, 1.5 h); (**B**) Marginal edge probability table for high PI3K data, early dynamics; (**C**) Consensus graph for high PI3K data, late dynamics (t = 6, 12, 18 h); (**D**) Marginal edge probability table for high PI3K data, late dynamics.

The performance of DBN algorithm is dependent on the sampling resolution of the dataset, and increased time points to a certain extent have been shown to reduce falsely identified connections [33]. Since in our current analysis in Figure 4 we are restricted to low number of time points, we further verified our prediction by artificially increasing the sampling resolution of the dataset by linear interpolation of the original experimental data. It is important to note that the original sampling time points were based on the fact that the TGF-β signaling pathway shows a slow response (with dynamic changes of the order of hours in many cell lines [34]) and at the current sampling resolution, the key dynamic features like slow increase/decrease, first order and overshoot behavior are captured fairly well. Hence, a linear interpolation is a good assumption for the current analysis. The resulting consensus graphs obtained by tripling the data points (Figure S7A) shows that the most repeated connections involving the receptor mediated and p-AKT mediated dependences are re-captured in the resolved data-set. However, some of the less represented connections (mostly those with edge probability less than 0.58) originating from p-SMAD3 and t-SMAD4 are lost and p-SMAD2 regulation is regained. When additional time points were added, no further changes in the graph were observed (data not shown), as also seen by Yu *et al.* where the number of true and false positive connections reached a plateau after a certain point [33]. Similar conclusions are seen for the late time points (Figure S7B). Hence this increases confidence on the current predictions, indicating the robustness of the more repeated connections.

3.3.2. Low PI3K Condition

Figure 5A,B presents the consensus graph and marginal edge probabilities respectively for the early time points, averaged over repeats 1 and 3. The network is very similar to the network obtained using the entire time series of low PI3K condition (Figure 2B), with some minor differences. The key regulations by the receptor as well as supplementary crosstalk interactions are identified from the early time points. Figure 5C,D presents the consensus graph and marginal edge probabilities respectively for the late time points. Here late time points of 12, 18 and 24 h for repeat 3 were chosen based on Figure 3. It is seen that only the receptor-mediated regulation is identified in this region with no additional crosstalk interactions identified. When the sampling resolution is increased to contain triple the current number of time points in the early and late phases each, similar conclusions (see Figure S8) are obtained as the high PI3K case, with most represented connections from Figure 5 retained and some less represented connections (<0.51 in early and <0.63 in the late phase) lost.

3.4. Correlation between Molecule Pairs in the Early and Late Time Zones

The DBNs do not directly infer the strength and direction of regulation (positive or negative). This is because the parameters are integrated out during the calculation of the cpBGe scores (see Equation (3)). Alternately, this can be investigated by using correlation metrics. Since the algorithm here fits a linear Gaussian mixture model at each time segment between the change-points, the nature of interaction between pairs of nodes (parent and child nodes) within a time segment can be measured using linear correlation coefficients. For the purpose of comparison, Pearson correlation coefficient between pairs of molecules was calculated at the early and late time zones for each condition and averaged over the time zones for repeats selected from Figure 3. These coefficients are presented in Figure 6. Note that in Figure 6 the influence of a selected node on all the other nodes in the network (not just the child node) is presented. Based on the identified DBNs, the correlation coefficients are presented in three major groups showing the influence of receptors, p-AKT and p-ERK on the nodes in the network.

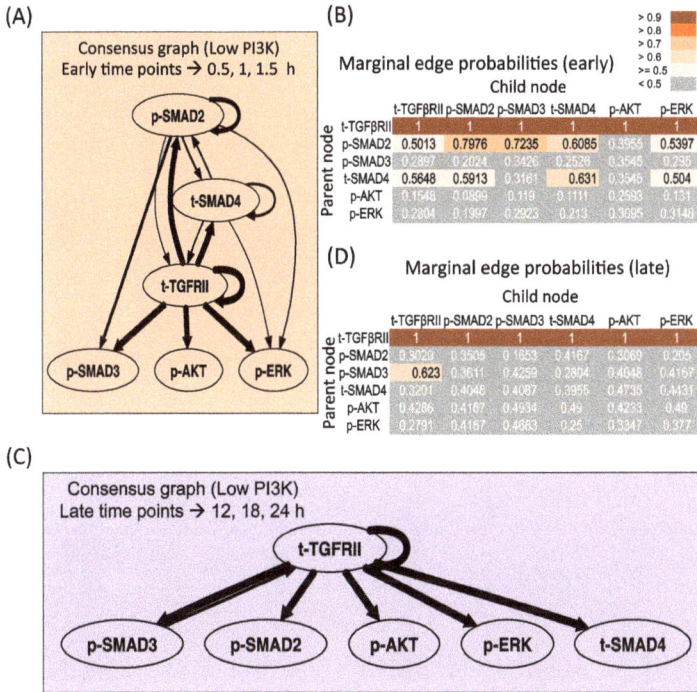

Figure 5. Dynamic Bayesian Network inferred for endoderm induction conditions under different time zones and low PI3K. (**A**) Consensus graph for low PI3K data, early dynamics ($t = 0.5$, 1, 1.5 h); (**B**) Marginal edge probability table for low PI3K data, early dynamics; (**C**) Consensus graph for low PI3K data, late dynamics ($t = 12$, 18, 24 h); (**D**) Marginal edge probability table for low PI3K data, late dynamics.

3.4.1. Influence of Total Receptor Levels

As seen from Figure 6A, most of the molecules show positive and strong correlation with the receptor in the early and late time zones of both conditions. This indicates that the receptor is positively influencing the downstream molecules. It is interesting to note that the correlation is heavily dependent on the interaction with parallel pathways. For example, presence of high PI3K weakens the correlation between the receptors and downstream molecules both in the early and late time points. Suppression of PI3K significantly increases most of the correlation coefficients.

3.4.2. Intracellular Regulation by p-AKT

As shown from Figure 6B, p-AKT shows negative correlation with the p-SMAD2,3 molecules in the high PI3K condition. Among the p-SMADs, the correlation is stronger for p-SMAD3. The correlation is weak in the low PI3K case and from the DBNs, it is seen that edges from p-AKT to p-SMADs are absent in this condition. Therefore our results show that in the low PI3K condition, although the p-AKT levels increase to basal levels at later time points, the influence of p-AKT on p-SMADs is lost. A mostly negative correlation is also seen between the p-AKT and t-SMAD and positive correlation between p-AKT and the receptors in the DBNs. For p-ERK, DBN analysis showed regulation by p-AKT. Correlation coefficients show that the regulation is negative in the early time zone and positive in the late time zone.

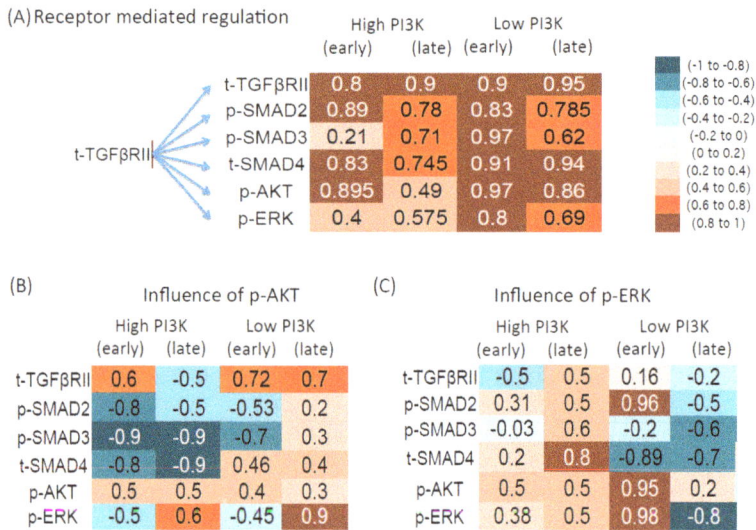

(A) Receptor mediated regulation

	High PI3K (early)	High PI3K (late)	Low PI3K (early)	Low PI3K (late)
t-TGFβRII	0.8	0.9	0.9	0.95
p-SMAD2	0.89	0.78	0.83	0.785
p-SMAD3	0.21	0.71	0.97	0.62
t-SMAD4	0.83	0.745	0.91	0.94
p-AKT	0.895	0.49	0.97	0.86
p-ERK	0.4	0.575	0.8	0.69

Legend: (-1 to -0.8), (-0.8 to -0.6), (-0.6 to -0.4), (-0.4 to -0.2), (-0.2 to 0), (0 to 0.2), (0.2 to 0.4), (0.4 to 0.6), (0.6 to 0.8), (0.8 to 1)

(B) Influence of p-AKT

	High PI3K (early)	High PI3K (late)	Low PI3K (early)	Low PI3K (late)
t-TGFβRII	0.6	-0.5	0.72	0.7
p-SMAD2	-0.8	-0.5	-0.53	0.2
p-SMAD3	-0.9	-0.9	-0.7	0.3
t-SMAD4	-0.8	-0.9	0.46	0.4
p-AKT	0.5	0.5	0.4	0.3
p-ERK	-0.5	0.6	-0.45	0.9

(C) Influence of p-ERK

	High PI3K (early)	High PI3K (late)	Low PI3K (early)	Low PI3K (late)
t-TGFβRII	-0.5	0.5	0.16	-0.2
p-SMAD2	0.31	0.5	0.96	-0.5
p-SMAD3	-0.03	0.6	-0.2	-0.6
t-SMAD4	0.2	0.8	-0.89	-0.7
p-AKT	0.5	0.5	0.95	0.2
p-ERK	0.38	0.5	0.98	-0.8

Figure 6. Correlation tables for high and low PI3K condition. (**A**) Receptor mediated regulation; (**B**) p-AKT mediated regulation; (**C**) p-ERK mediated regulation. The Pearson correlation is calculated between the parent nodes at time step $(t-1)$ and all other nodes at time step t. The early time points 0.5, 1, 1.5 h (both conditions) and the late time points correspond to 6, 12, and 18 for high PI3K and 12, 18, 24 for low PI3K. The average correlation coefficients across 2 repeats (selected in Figure 3) are used for high and low PI3K. However, for the low PI3K late time points, only repeat 3 is used.

3.4.3. Intracellular Regulation by p-ERK

The consensus DBNs from all the conditions showed that p-ERK has a minor role in regulating the levels of most molecules. The influence if it exists is mainly for the low PI3K condition as seen from some samples in this condition. Based on correlation analysis, the correlation coefficients are weak in the high PI3K condition (Figure 6C). But, comparatively stronger correlation coefficients are seen in the low PI3K condition. The type of regulation is however mixed. The correlation with p-SMAD2 is positive in the early time zone and negative in the late time zone. The correlation with p-SMAD3 and t-SMAD4 are negative. In the early low PI3K condition, p-ERK shows a positive correlation with p-AKT.

3.5. Network Regulation during Endoderm Differentiation

This work is the first account in identifying specific signaling interactions governing endoderm differentiation of hESCs using network analysis tools. The DBNs inferred for the high and low PI3K data accomplished two major tasks: (1) They identified molecular interactions within the TGF-β pathway along-with crosstalk interactions with parallel pathways; and (2) They identified distinct pathway regulations during the early and late phases of the signaling dynamics. One key prediction from the entire analysis is the influence of receptor levels on downstream molecules including SMAD, AKT and ERK. In the canonical pathway, TGFβRII is known to activate SMAD molecules after formation of the ligand-receptor complex [35]. TGFβ signaling also participates in several non-canonical signaling leading to activation of PI3K/AKT and MAPK/ERK pathways in many cell lines [35,36]. Our analysis indicates that the levels of the receptors (TGFβRII) are still in the regime where they are limiting and any change in their level is reflected downstream. In addition, it is interesting to note that in each repeat, we repeatedly obtain the receptor-mediated edges, indicating that the connections are maintained in spite of sample variability.

Several important interactions from p-AKT were identified indicating the existence of p-AKT mediated crosstalk in high PI3K condition and its removal under low PI3K. Ours is the first systematic study to identify these crosstalk interactions in differentiating hESCs. The regulation of p-SMAD3 by p-AKT is well known in other cell lines, mainly inhibition of p-SMAD3 phosphorylation by mTORC1 and sequestration of non-phospho SMAD3 by p-AKT [37–41]. The regulation of p-SMAD2 by p-AKT is observed only in one sample of the high PI3K condition (Figure S5). Literature shows that most negative regulation of p-AKT is on p-SMAD3 and not p-SMAD2 [39], however some reports indicate negative regulation of both p-SMAD2 and p-SMAD3 by p-AKT in neuroblastoma and CHO cell lines [42,43]. The removal of crosstalk interactions with p-AKT in the low PI3K condition is interesting although the actual mechanism needs further study. One possibility is that the SMADs have undergone predominant nuclear translocation under inhibition of p-AKT and p-SMAD is no longer accessible to p-AKT [37,38,40]. The regulation of the receptors and t-SMAD4 by p-AKT was also seen but these interactions are not as widely studied as those of p-AKT and p-SMADs.

The DBNs showed regulation of p-ERK by p-AKT in the high PI3K condition. It is well reported that p-ERK is inhibited by p-AKT and many of its downstream effectors (via mTORC1) in multiple cell lines [44]. Previous study has shown the interaction between AKT1 and cRAF in hESCs leading to inhibition of RAF/MEK/ERK signaling [17]. Our experiments show that the levels of p-ERK are higher in low PI3K condition and the influence of p-AKT on p-ERK is also absent from the low PI3K DBNs. This indicates that this interaction negatively influences endoderm induction. However, a positive correlation was seen between p-ERK and p-AKT in the early low PI3K condition. ERK is known to influence the AKT pathway based on the context [44] indicating that this interaction may be seen during endoderm differentiation via additional networks interactions in these pathways.

4. Conclusions

Network inference is an area of active research in systems biology, with a multitude of methods currently being explored, starting from correlative and clustering methods to more involved algorithms utilizing Boolean, Bayesian and Differential Equation frameworks [45–50]. Many attempts to compare different approaches have led to the conclusion that there is no true gold standard method currently available that can identify all true positive connections (those present and inferred) with minimal false positive (not present but inferred) and false negative (present and not inferred) connections [51,52]. During benchmarking tests on *in silico* signal transduction dataset to check the network reconstruction accuracy of the DBN algorithm used in this work, Grzegorczyk *et al.* obtained favorable values for several measures of accuracy for datasets with wide range of signal to noise ratio (SNR) [25,32] and these measures were comparable to other methods for similar node numbers, connectivity and SNR [51]. Due to the success on a variety of datasets, the algorithm by Grzegorczyk *et al.* was used to determine the network connections in this work. Significant care was taken to infer connections from independent repeats of the same experimental condition due to high variability of the system and emphasis was given to most repeated connections across independent repeats. While current analysis indicates robustness of the predictions across independent repeats, verification of such robustness is only possible by running selective perturbation experiments [53,54], which will be considered in our future work.

Precise control of differentiation of hESCs is a difficult problem due to the high variability and multiple signaling mediators associated with this process. While experiments have identified features of signal transduction that orchestrate this process, rational design using network level properties is not well studied. This work is an important contribution in this direction. Using network analysis methods, we uncovered signaling interactions existing amongst the common mediators of endoderm differentiation of hESCs. One of the most common predictions across all conditions and repeats was that the receptor levels are most influential in governing the downstream pathway molecules in most cases. The TGF-βRII levels correlated well with the canonical and non-canonical molecules. The influence of receptor levels on downstream signals provides an avenue to understand the origins of variability that is common in hESC system and it is reasonable to expect that cell-to-cell variability in receptor levels will lead to variability of downstream signals and eventually cell fate specification. In mouse ES cells, it was previously shown that variability in the activity of TGF-β pathway (Activin/Nodal and BMP) controlled the variability in the level of pluripotency marker, *NANOG*, in the self-renewal state and modulation of receptor activity using small molecule inhibitors influenced the heterogeneity of *NANOG* in subpopulations [55].

A significant observation was the strong crosstalk interactions of p-AKT with the mediators of TGF-β pathway during Activin induction condition and complete removal of any of these interactions under PI3K inhibition. The introduction of p-AKT mediated interactions and their removal is captured in the independent repeats of the high and low PI3K conditions. Importantly, p-AKT levels could not be continuously suppressed experimentally by continued inhibition of PI3K pathway, and it regained its basal expression. However in spite of the increase in p-AKT expression levels, the correlation remained weak with most molecules like p-SMAD2,3 and p-ERK, indicating continued removal of crosstalk. This is counter-intuitive and demonstrates the necessity of network level analysis to comprehend experimental data, in particular for such complex and dynamic systems. The influence of p-AKT on p-ERK was seen in the high PI3K condition. Report from Dalton Lab has shown increases in p-ERK under p-AKT inhibition in hESCs leading to activation of WNT signaling supporting endoderm differentiation [17]. This can explain the increase in p-ERK levels under PI3K inhibition in our system. In addition, a weaker regulatory role for p-ERK was predicted in the high PI3K condition, with some enhancement under the low PI3K condition. Possibility of p-ERK mediated interactions under low PI3K signaling is interesting. It is known that p-ERK has additional roles in linker phosphorylation of SMAD molecules which can affect the nucleo-cytoplasmic shuttling and ultimately their dynamics as modeled by Liu *et al.* in this issue of the journal [56]. This could be the reason for seeing increasing p-ERK influence on SMAD molecules under low PI3K condition in some samples. But, since this was observed in only some samples of low PI3K that used the entire time series information, additional investigation needs to be done. Future studies of long-term p-ERK dynamics (>24 h) and perturbation experiments will enable further exploration of this portion of the network. Overall, the identified DBNs demonstrate significant biologically relevant interactions. Such agreement with literature observations along with prediction of additional interactions prove the applicability of quantitative methods for teasing out the network level properties of complex systems like hESCs.

Another important result from DBN analysis was inference of time zones (change-points) where the network parameters changed thereby, indicating an adaptive nature of the network. The nature of the data indicated multiple change-points, with two major change-points in the early and late phases of the dynamics. Application of DBN analysis in the early and late segments showed that the early dynamics is more informative and could adequately identify the network inferred by using the entire time series. This hints at the importance of measuring early dynamics of signal transduction. Further, the late time segments showed strong influence of the receptors levels on other molecules with weaker influence of any crosstalk interaction. It was also seen that the performance of the DBN algorithm in each time zone was better when the number of time points was increased. However, there was no change in the most representative connections, but it enabled loss of some less represented connections.

We recognize that the additional interactions identified here require further testing by perturbation experiments. Further, the interactions identified by DBN need not be direct associations from a biological standpoint, but the resulting effect via intermediated processes. The interacting molecules are solely dependent on the molecules tracked in the experiments. These observations, however, need further investigation in hESCs. An important point to note is that DBN analysis is a useful tool to generate hypothesis based on existing experimental data. The identified networks can inform future experimentation and the network themselves may undergo refinement, a common workflow in systems biology. Hypotheses provided by DBNs can be utilized by detailed modeling approaches like differential equations to investigate the kinetics of the process.

216

Acknowledgments

The authors would like to thank the Stem Cell Core Facility at the University of Pittsburgh for supply of H1 hESCs. This project was supported by the National Institute of Health New Innovator Award (1DP2OD006491 to IB) and National Science Foundation EAGER fund (CBET-1455800 to IB). The authors would like to thank Yoram Vodovotz and Nabil Azhar at the University of Pittsburgh School of Medicine for valuable discussions on the DBN algorithm.

Author Contributions

Conceived and designed the study: S.M., S.S. and I.B.; Conducted wet-lab experiments: S.S.; Performed computational analysis: S.M.; Supplied reagents and materials: I.B.; Wrote the manuscript: S.M. and I.B.

Nomenclature

DBN	Dynamic Bayesian Network
DE	Definitive Endoderm
hESCs	Human Embryonic Stem Cells
p-SMAD2	phosphorylated SMAD2
p-SMAD3	phosphorylated SMAD3
p-AKT	phosphorylated AKT
p-ERK	phosphorylated ERK
TGFβ	Transforming growth factor-beta
t-TGFβRII	total TGFβ receptor 2
t-SMAD4	total SMAD4

SMAD, PI3K, AKT and ERK are common signaling proteins

Conflicts of Interest

The authors declare no conflict of interest.

References

1. Semb, H. Definitive endoderm: A key step in coaxing human embryonic stem cells into transplantable beta-cells. *Biochem. Soc. Trans.* **2008**, *36*, 272–275.
2. Jaramillo, M.; Mathew, S.; Task, K.; Barner, S.; Banerjee, I. Potential for pancreatic maturation of differentiating human embryonic stem cells is sensitive to the specific pathway of definitive endoderm commitment. *PLoS One* **2014**, *9*, e94307.
3. Xu, X.; Browning, V.L.; Odorico, J.S. Activin, bmp and fgf pathways cooperate to promote endoderm and pancreatic lineage cell differentiation from human embryonic stem cells. *Mech. Dev.* **2011**, *128*, 412–427.

4.   McLean, A.B.; D'Amour, K.A.; Jones, K.L.; Krishnamoorthy, M.; Kulik, M.J.; Reynolds, D.M.; Sheppard, A.M.; Liu, H.; Xu, Y.; Baetge, E.E. Activin a efficiently specifies definitive endoderm from human embryonic stem cells only when phosphatidylinositol 3-kinase signaling is suppressed. *Stem Cells* **2007**, *25*, 29–38.

5.   D'Amour, K.A.; Agulnick, A.D.; Eliazer, S.; Kelly, O.G.; Kroon, E.; Baetge, E.E. Efficient differentiation of human embryonic stem cells to definitive endoderm. *Nat. Biotechnol.* **2005**, *23*, 1534–1541.

6.   Nostro, M.C.; Sarangi, F.; Ogawa, S.; Holtzinger, A.; Corneo, B.; Li, X.; Micallef, S.J.; Park, I.-H.; Basford, C.; Wheeler, M.B. Stage-specific signaling through tgfβ family members and wnt regulates patterning and pancreatic specification of human pluripotent stem cells. *Development* **2011**, *138*, 861–871.

7.   Basma, H.; Soto-Gutierrez, A.; Yannam, G.R.; Liu, L.; Ito, R.; Yamamoto, T.; Ellis, E.; Carson, S.D.; Sato, S.; Chen, Y.; *et al.* Differentiation and transplantation of human embryonic stem cell-derived hepatocytes. *Gastroenterology* **2009**, *136*, 990–999.

8.   Sulzbacher, S.; Schroeder, I.S.; Truong, T.T.; Wobus, A.M. Activin a-induced differentiation of embryonic stem cells into endoderm and pancreatic progenitors—The influence of differentiation factors and culture conditions. *Stem Cell Reviews Rep.* **2009**, *5*, 159–173.

9.   Mathew, S.; Jaramillo, M.; Zhang, X.; Zhang, L.A.; Soto-Gutierrez, A.; Banerjee, I. Analysis of alternative signaling pathways of endoderm induction of human embryonic stem cells identifies context specific differences. *BMC Syst. Biol.* **2012**, *6*, 154.

10. Richardson, T.; Kumta, P.N.; Banerjee, I. Alginate encapsulation of human embryonic stem cells to enhance directed differentiation to pancreatic islet-like cells. *Tissue Eng. Part A* **2014**, *20*, 3198–3211.

11. Attisano, L.; Wrana, J.L.; Montalvo, E.; Massague, J. Activation of signalling by the activin receptor complex. *Mol. Cell. Biol.* **1996**, *16*, 1066–1073.

12. Massague, J.; Seoane, J.; Wotton, D. Smad transcription factors. *Genes Dev.* **2005**, *19*, 2783–2810.

13. Wu, M.Y.; Hill, C.S. Tgf-beta superfamily signaling in embryonic development and homeostasis. *Dev. Cell* **2009**, *16*, 329–343.

14. Dalton, S. Signaling networks in human pluripotent stem cells. *Curr. Opin. Cell Biol.* **2013**, *25*, 241–246.

15. Avery, S.; Zafarana, G.; Gokhale, P.J.; Andrews, P.W. The role of smad4 in human embryonic stem cell self-renewal and stem cell fate. *Stem Cells* **2010**, *28*, 863–873.

16. Sakaki-Yumoto, M.; Liu, J.M.; Ramalho-Santos, M.; Yoshida, N.; Derynck, R. Smad2 is essential for maintenance of the human and mouse primed pluripotent stem cell state. *J. Biol. Chem.* **2013**, *288*, 18546–18560.

17. Singh, A.M.; Reynolds, D.; Cliff, T.; Ohtsuka, S.; Mattheyses, A.L.; Sun, Y.; Menendez, L.; Kulik, M.; Dalton, S. Signaling network crosstalk in human pluripotent cells: A smad2/3-regulated switch that controls the balance between self-renewal and differentiation. *Cell Stem Cell* **2012**, *10*, 312–326.

18. Pauklin, S.; Vallier, L. The cell-cycle state of stem cells determines cell fate propensity. *Cell* **2013**, *155*, 135–147.

19. Needham, C.J.; Bradford, J.R.; Bulpitt, A.J.; Westhead, D.R. A primer on learning in bayesian networks for computational biology. *PLoS Comput. Biol.* **2007**, *3*, e129.

20. Woolf, P.J.; Prudhomme, W.; Daheron, L.; Daley, G.Q.; Lauffenburger, D.A. Bayesian analysis of signaling networks governing embryonic stem cell fate decisions. *Bioinformatics* **2005**, *21*, 741–753.

21. Zielinski, R.; Przytycki, P.F.; Zheng, J.; Zhang, D.; Przytycka, T.M.; Capala, J. The crosstalk between egf, igf, and insulin cell signaling pathways—Computational and experimental analysis. *BMC Syst. Biol.* **2009**, *3*, 88.

22. Murphy, K.P. Dynamic Bayesian Networks: Representation, Inference and Learning. Ph.D. Thesis, University of California, Berkeley, CA, USA, 2002.

23. Heinrich, R.; Neel, B.G.; Rapoport, T.A. Mathematical models of protein kinase signal transduction. *Mol. Cell* **2002**, *9*, 957–970.

24. Koller, D.; Friedman, N. *Probabilistic Graphical Models: Principles and Techniques*; MIT Press: Cambridge, MA, USA, 2009.

25. Grzegorczyk, M.; Husmeier, D. Improvements in the reconstruction of time-varying gene regulatory networks: Dynamic programming and regularization by information sharing among genes. *Bioinformatics* **2011**, *27*, 693–699.

26. Grzegorczyk, M.; Husmeier, D. Non-homogeneous dynamic bayesian networks for continuous data. *Mach. Learn.* **2011**, *83*, 355–419.

27. Azhar, N.; Ziraldo, C.; Barclay, D.; Rudnick, D.A.; Squires, R.H.; Vodovotz, Y.; Pediatric Acute Liver Failure Study Group. Analysis of serum inflammatory mediators identifies unique dynamic networks associated with death and spontaneous survival in pediatric acute liver failure. *PLoS One* **2013**, *8*, e78202.

28. Emr, B.; Sadowsky, D.; Azhar, N.; Gatto, L.A.; An, G.; Nieman, G.F.; Vodovotz, Y. Removal of inflammatory ascites is associated with dynamic modification of local and systemic inflammation along with prevention of acute lung injury: *In vivo* and *in silico* studies. *Shock* **2014**, *41*, 317–323.

29. Aerts, J.M.; Haddad, W.M.; An, G.; Vodovotz, Y. From data patterns to mechanistic models in acute critical illness. *J. Crit. Care* **2014**, *29*, 604–610.

30. Dojer, N.; Gambin, A.; Mizera, A.; Wilczynski, B.; Tiuryn, J. Applying dynamic bayesian networks to perturbed gene expression data. *BMC Bioinf.* **2006**, *7*, 249.

31. Chang, R.; Shoemaker, R.; Wang, W. Systematic search for recipes to generate induced pluripotent stem cells. *PLoS Comput. Biol.* **2011**, *7*, e1002300.

32. Grzegorczyk, M.; Husmeier, D. A non-homogeneous dynamic bayesian network with sequentially coupled interaction parameters for applications in systems and synthetic biology. *Stat. Appl. Genet. Mol. Biol.* **2012**, *11*, doi:10.1515/1544-6115.1761.

33. Yu, J.; Smith, V.A.; Wang, P.P.; Hartemink, A.J.; Jarvis, E.D. Advances to bayesian network inference for generating causal networks from observational biological data. *Bioinformatics* **2004**, *20*, 3594–3603.

34. Schmierer, B.; Hill, C.S. Tgfbeta-smad signal transduction: Molecular specificity and functional flexibility. *Nat. Rev. Mol. Cell Biol.* **2007**, *8*, 970–982.

35. Guo, X.; Wang, X.-F. Signaling cross-talk between tgf-β/bmp and other pathways. *Cell Res.* **2008**, *19*, 71–88.

36. Zhang, Y.E. Non-smad pathways in tgf-β signaling. *Cell Res.* **2008**, *19*, 128–139.

37. Conery, A.R.; Cao, Y.; Thompson, E.A.; Townsend, C.M., Jr.; Ko, T.C.; Luo, K. Akt interacts directly with smad3 to regulate the sensitivity to tgf-beta induced apoptosis. *Nat. Cell Biol.* **2004**, *6*, 366–372.

38. Danielpour, D.; Song, K. Cross-talk between igf-i and tgf-β signaling pathways. *Cytokine Growth Factor Rev.* **2006**, *17*, 59–74.

39. Song, K.; Wang, H.; Krebs, T.L.; Danielpour, D. Novel roles of akt and mtor in suppressing tgf-beta/alk5-mediated smad3 activation. *EMBO J.* **2006**, *25*, 58–69.

40. Remy, I.; Montmarquette, A.; Michnick, S.W. Pkb/akt modulates tgf-β signalling through a direct interaction with smad3. *Nat. Cell Biol.* **2004**, *6*, 358–365.

41. Zhang, L.; Zhou, F.; ten Dijke, P. Signaling interplay between transforming growth factor-β receptor and pi3k/akt pathways in cancer. *Trends Biochem. Sci.* **2013**, *38*, 612–620.

42. Qiao, J.; Kang, J.; Ko, T.C.; Evers, B.M.; Chung, D.H. Inhibition of transforming growth factor-beta/smad signaling by phosphatidylinositol 3-kinase pathway. *Cancer Lett.* **2006**, *242*, 207–214.

43. Sun, T.; Ye, F.; Ding, H.; Chen, K.; Jiang, H.; Shen, X. Protein tyrosine phosphatase 1b regulates tgf beta 1-induced smad2 activation through pi3 kinase-dependent pathway. *Cytokine* **2006**, *35*, 88–94.

44. Aksamitiene, E.; Kiyatkin, A.; Kholodenko, B.N. Cross-talk between mitogenic ras/mapk and survival pi3k/akt pathways: A fine balance. *Biochem. Soc. Trans.* **2012**, *40*, 139–146.

45. Villaverde, A.F.; Banga, J.R. Reverse engineering and identification in systems biology: Strategies, perspectives and challenges. *J. R. Soc. Interface* **2014**, *11*, 20130505.

46. Banerjee, I.; Maiti, S.; Parashurama, N.; Yarmush, M. An integer programming formulation to identify the sparse network architecture governing differentiation of embryonic stem cells. *Bioinformatics* **2010**, *26*, 1332–1339.

47. Chemmangattuvalappil, N.; Task, K.; Banerjee, I. An integer optimization algorithm for robust identification of non-linear gene regulatory networks. *BMC Syst. Biol.* **2012**, *6*, 119.

48. Guillen-Gosálbez, G.; Miró, A.; Alves, R.; Sorribas, A.; Jiménez, L. Identification of regulatory structure and kinetic parameters of biochemical networks via mixed-integer dynamic optimization. *BMC Syst. Biol.* **2013**, *7*, 113.

49. Rodriguez-Fernandez, M.; Rehberg, M.; Kremling, A.; Banga, J.R. Simultaneous model discrimination and parameter estimation in dynamic models of cellular systems. *BMC Syst. Biol.* **2013**, *7*, 76.

50. Penfold, C.A.; Wild, D.L. How to infer gene networks from expression profiles, revisited. *Interface Focus* **2011**, *1*, 857–870.

51. Lee, W.P.; Tzou, W.S. Computational methods for discovering gene networks from expression data. *Brief Bioinform.* **2009**, *10*, 408–423.

52. Bansal, M.; Belcastro, V.; Ambesi-Impiombato, A.; Di Bernardo, D. How to infer gene networks from expression profiles. *Mol. Syst. Biol.* **2007**, *3*, doi:10.1038/msb4100120.

53. Tegner, J.; Yeung, M.K.; Hasty, J.; Collins, J.J. Reverse engineering gene networks: Integrating genetic perturbations with dynamical modeling. *Proc. Natl. Acad. Sci. USA* **2003**, *100*, 5944–5949.

54. Molinelli, E.J.; Korkut, A.; Wang, W.; Miller, M.L.; Gauthier, N.P.; Jing, X.; Kaushik, P.; He, Q.; Mills, G.; Solit, D.B.; *et al.* Perturbation biology: Inferring signaling networks in cellular systems. *PLoS Comput. Biol.* **2013**, *9*, e1003290.

55. Galvin-Burgess, K.E.; Travis, E.D.; Pierson, K.E.; Vivian, J.L. Tgf-β-superfamily signaling regulates embryonic stem cell heterogeneity: Self-renewal as a dynamic and regulated equilibrium. *Stem Cells* **2013**, *31*, 48–58.

56. Liu, J.; Dai, W.; Hahn, J. Mathematical modeling and analysis of crosstalk between mapk pathway and smad-dependent tgf-β signal transduction. *Processes* **2014**, *2*, 570–595.

MDPI AG
Klybeckstrasse 64
4057 Basel, Switzerland
Tel. +41 61 683 77 34
Fax +41 61 302 89 18
http://www.mdpi.com/

Processes Editorial Office
E-mail: processes@mdpi.com
http://www.mdpi.com/journal/processes